D0778740

DATE DUE

DC 13'01			
NO 6'07			

DEMCO 38-296

BIP+
9/15/93

THOUGHT EXPERIMENTS

THOUGHT EXPERIMENTS

ROY A. SORENSEN

New York   Oxford
OXFORD UNIVERSITY PRESS
1992

/ersity Press

York   Toronto
tta   Madras   Karachi
re   Hong Kong   Tokyo
laam   Cape Town
Melbourne   Auckland

and associated companies in
Berlin   Ibadan

Copyright © 1992 by Roy A. Sorensen

Published by Oxford University Press, Inc.,
200 Madison Avenue, New York, NY 10016

Oxford is a registered trademark of Oxford University Press

Library of Congress Cataloging-in-Publication Data
Sorensen, Roy A.
Thought experiments / Roy A. Sorensen.
p. cm.   Includes bibliographical references and index.
ISBN 0-19-507422-X
1. Thought and thinking.
2. Logic.   3. Philosophy and science.
I. Title.
B105.T54S67   1992   101—dc20
91-36760

1 3 5 7 9 8 6 4 2

Printed in the United States of America
on acid-free paper

For Julia,
a woman of fine distinctions

ACKNOWLEDGMENTS

This book has indebted me to many people. The first group consists of the individuals who attended colloquia given at the Graduate Center of the City University of New York, Columbia University, Dartmouth College, Rutgers University, the State University of New York at Stony Brook, Virginia Tech, and the 1990 Inter-University Conference for Philosophy of Science in Dubrovnik. Several of my colleagues at New York University have earned my gratitude for their encouragement and advice, among them Raziel Abelson, Frances Kamm, John Richardson, William Ruddick, Peter Unger, and especially John Carroll. I have also acquired far-flung debts to Lars Bergstrom, James Robert Brown, and Martin Bunzl, who scrutinized earlier drafts. Since I used the latter in philosophy of science courses, this book has benefited from student input. Anonymous, detailed referee reports led me to make painful but necessary excisions, precisifications and elaborations. This book has been through enough drafts to win rave reviews from lumberjacks!

A New York University presidential Fellowship allowed me to experience the hospitality of the Virginia Tech philosophy department. Richard Burian guided me through the ins and outs of evolutionary theory and philosophy of biology (though I may still have zigged where I should have zagged); Joseph Pitt was my Galileo hotline; Peter Barker and Roger Ariew filled in assorted gaps in my knowledge of the history of science; and John Christman, Julia Driver, and Jim Klagge kept me in focus about ethical thought experiments.

My last group of creditors are editors. I thank the editor of the *American Scientist* for permission to use a portion of "Thought Experiments" in chapters 3 and 6 and the editors of *Philosophical Studies* for the use of parts of "Moral Dilemmas, Thought Experiments, and Conflict Vagueness" in chapter 7. Finally, at Oxford University Press I thank senior editor Cynthia Read and assistant editor Peter Ohlin for shepherding this book into its final form.

New York   R.A.S.
September 1991

CONTENTS

THOUGHT EXPERIMENTS

Given a sufficient constancy of environment, there is developed a corresponding constancy of thought. By virtue of this constancy our thoughts are spontaneously impelled to *complete* all incompletely observed facts. The impulse in question is not prompted by the individual facts as observed at the time; nor is it intentionally evoked; but we find it operative in ourselves entirely without our personal intervention. . . . By this impulse we have always a *larger* portion of nature in our field of vision than the inexperienced man has, with the single fact alone. For the human being, with his thoughts and impulses, is himself merely a piece of nature, which is added to the single fact. . . . When an electric current flows round a magnetic needle situated in its plane, the north pole of the needle is deflected to my left. I imagine myself as Ampere's swimmer in the current. I enrich the fact (current and needle) which is insufficient in itself to define the direction of my thought, by introducing *myself* into the experiment by an inner reaction.

Ernst Mach

I am a little piece of nature.

Albert Einstein

Introduction

This book presents a general theory of thought experiments: what they are; how they work; their virtues and vices. Since my aim is synoptic, a wide corpus of thought experiments has been incorporated. There is a special abundance of examples from ethics and the metaphysics of personal identity because thought experiments in these areas have recently attracted heavy commentary. But the emphasis is on variety, rather than quantity. Thus, the discussion ranges over thought experiments from many disparate fields, from aesthetics to zoology.

Scientific thought experiments—especially those in physics—are the clear cases, so my primary goal is to establish true and interesting generalizations about them. Success here will radiate to my secondary goal of understanding philosophical thought experiments. The reason for this optimism is subscription to a gradualistic metaphilosophy: philosophy differs from science in degree, not kind. Understand science, understand the parameters to be varied, and you understand philosophy.

My basic means of reaching these two goals is to let the surface grammar of 'thought experiment' be my guide and to pitch this book as part of the growing literature on *experiment*. Philosophers and historians of science have long followed the elder statesmen of science in concentrating on theory; experimentation has been dismissed as a rather straightforward matter of following directions and looking at gauges. Within the last ten years, the "just look and see" picture has been rejected in favor of one that assigns deeper roles for experimenters: creating and stabilizing phenomena; atheoretical exploration; and defining concepts by immersion in laboratory practice. Sympathy with this movement, coupled with the belief that thought experiments are experiments, led me to suspect a corresponding oversimplification of the thought experimenter's role.

The main theme of this book is that thought experiment is experiment (albeit a limiting case of it), so that the lessons learned about experimentation carry over to thought experiment, and vice versa. For the symmetry of 'similar' underwrites a two-way trade: if thought experiments are surprisingly similar to experiments, then experiments are surprisingly similar to thought experiments. In particular, experiments exploit many of the organizational effects associated with the products of armchair inquiry. Study of thought experiments draws attention to these neglected features of ordinary experiments; for when we explain the informativeness of thought experiment, we cannot appeal to the inflow of fresh information. We are forced to look for ways that old information can be rendered more informative. (Consider how nineteenth-century investigations into animal behavior illuminated human psychology just because researchers had to make do with behavior; the distraction of introspection was removed.) Once these repackaging effects are detected with thought experiments, they can be spotted in ordinary experiments. However, most of the illumination will flow from ordinary experiment to thought experiment; I shall mainly use the familiar to explain the obscure.

Chapter 1 motivates the study of thought experiments—in broad strokes. Their power is displayed by assembling influential thought experiments from the history of science. I then lay out my plan to understand philosophical thought experiments by concentrating on their resemblance to scientific relatives. Points of difference between philosophical and scientific thought experiments give us a preview of obstacles that must be overcome in the course of the campaign. Naive and sophisticated reservations about the philosophical cases are registered for the same purpose.

Detailed structuring of the issues begins by making chapter 2 a forum for sceptics. Doubts about thought experiment are given their most damning expression to date. Thus, the technique looks discredited by the time sceptics yield the floor to Ernst Mach in chapter 3. This Austrian philosopher–physicist was the earliest and most systematic writer on thought experiments (and, not coincidentally, mentor of the young Albert Einstein). Mach's views are of more than historical interest: they are insightful, fairly accurate, and fertile. My limited disagreements with Mach spring from his sensationalism—the view that everything worth saying is reducible to commentary on sense data. I argue that this extreme empiricism misled him into an overly narrow account of thought experiment, ill suited to thought experiments falling outside the natural sciences.

This criticism is followed by a chapter on armchair inquiry which will give us more breadth. Part of this chapter is deflationary; I trace a portion of our wonder about a priori enlightenment to modal fallacies. The positive part addresses curiosity about the mechanisms underlying armchair inquiry. Philosophers and psychologists have been influenced by several vague models of nonobservational enlightenment. Mach said that thought experiments draw from a storehouse of unarticulated experience. This answer to the question of how it can be fruitful to sit and think is an empiricist version of Plato's doctrine of recollection, which pictures knowers as rememberers. In the 1960s, ordinary language philosophers defended their appeals to "what we would say" with a linguistic descendent of Plato's doctrine: knowledge of *how* to speak is transformed into knowledge *that* the rules for the term are such-and-such. The homuncular model, made respectable by the rise of cognitive psychology in the 1970s, pictures people as composed of subsystems ("little people") that act within the larger system much as personnel behave within a firm. The armchair investigator shifts information from one subsystem to another, so that facts familiar to one part are news to another part. This intrapersonal communication, pooling, and delegation of questions produces a better-informed person even though no fresh data has been gathered. A fourth model focuses on how information can be made more informative by rearranging it in ways that facilitate its storage, retrieval, and deployment in inference.

Special use will be made of the cleansing model of armchair inquiry. It presents intellectual improvement as the shedding of intellectual vices, rather than the acquisition of virtues. Rationality is portrayed as analogous to health: just as health is the absence of disease, rationality is the absence of irrationalities. Thus, thought experiments make us more rational by purging us of bias, circularity, dogmatism, and other cognitive inefficiencies. All experiments work by raising the experimenter's status as an epistemic authority. Ordinary experiments confer authority mainly by improving the experimenter's perceptual abilities and opportunites. Thought experiments focus on nonperceptual improvements. But since an ordinary experiment can simultaneously provide perceptual and nonperceptual improvements, study of thought experiments can bring a neglected side of ordinary experiments into sharp relief.

Inconsistency is the most general and best understood of cognitive flaws. This invites a reduction thesis: all of the irrationalities eliminated by thought experiment can be formulated as inconsistencies. If this thesis is true (and I think it is), we need only standardize the format of thought experiments in order to apply standard logic directly. So while granting that all of the models of armchair inquiry have something to offer, I close the chapter with the conclusion that the reductionist version of the cleansing model offers the best chance for immediate elaboration.

Thomas Kuhn's work on thought experiments gives me a running start. In "A Function for Thought Experiments," he argued that they revealed a special kind of contradiction—a type of local incoherency. Standard logic has no room for this notion of a noninfectious, contingent contradiction. Hence, Kuhn's persuasive and insightful application of this feral concept startles conservatives like me. After showing why we should continue to side with standard logic, I try to salvage the insight driving Kuhn's heresy in chapter 5. The basic idea is that a group of tricky factors leads Kuhn to construe relative inconsistency as an exotic sort of absolute inconsistency. Crucial to the diagnosis is identification of the tricky thought experiments as *paradoxes*.

A paradox is a small set of individually plausible yet jointly inconsistent propositions. In chapter 6 I extrapolate to the thesis that every thought experiment is reducible to such a set. Indeed, I argue that they are all reducible to two highly specific forms of paradox—one targeting statements implying necessities, the other targeting statements implying possibilities. These two closely related sets are the standardized formats I craved in the discussion of the cleansing model. They are the molds into which raw thought experiments can be poured. They then enter the logician's mill. While in this admittedly artificial state, thought experiments can be systematically classified in accordance with which member of the paradox is slated for rejection. Both types of paradoxes have exactly five members, so the taxonomic system has a manageable scale.

The official role of thought experiment is to test modal consequences. The apparent narrowness of its function eases once we realize that there are many kinds of necessity: logical, physical, technological, moral. But the real flexibility of thought experiments wriggles up from the *indirect* uses of this official procedure. Just as jokes, metaphor, and politeness are conveyed through trick bounces off conventions governing literal conversation, thought experimenters use the standard format obliquely to transact a rich array of side tasks: concocting counterexamples to definitions and "laws," expanding the domain of theories, exhibiting modal fallacies, deriving astounding consequences, suggesting impossibility proofs.

The paradox analysis is further deepened in chapter 7 by a special application to the genre of thought experiments that fascinated Kuhn. This special class is powered by *conflict vagueness*, a linguistic property that appears to make a normally well-behaved concept "come apart." Thought experiments that probe this nerve of indeterminacy are apt to force conceptual revision. Since conflict vagueness can be neatly dissected, I complete the chapter with a classification of the ways concepts change in response to this provocative family of thought experiments.

Having taken Kuhn's insight as far as I can, I follow it with a chapter that returns thought experiments to their unregimented state. My goal here is to define 'thought experiment' and dig to its origin. I argue that thought experiments evolved from experiment through a process of attenuation. This builds inductive momentum behind the theme that thought experiments are experiments. My commitment to viewing them as limiting cases of experiment is solidified by defining thought experiments as experiments that purport to deal with their questions by contemplation of their design rather than by execution. But in the course of this analysis another reduction is endorsed: in addition to being experiments and paradoxes, thought experiments are stories. This brings one of the book's minor themes into prominence: many of the issues raised by thought experiments are prefigured in aesthetics and the logic of fiction.

The suspicion that 'thought experiment' is a systematically misleading expression is addressed in chapter 9. Since I put so much weight on the accuracy of the surface grammar, I itemize how 'thought experiment' is actually a systematically *leading* expression. This catalogue of hot tips raises a variety of issues ranging from how thought experiments differ from simulations to the ethics of fantasy.

My final chapter assesses the hazards and pseudohazards of thought experiment. Although I grant that there are interesting ways in which the method leads us astray, I attack most scepticism about thought experiment as arbitrary. Once we apply standards that are customary for compasses, stethoscopes, and other testing devices, we find that thought experiments measure up. They should be used (as they generally are used) as part of a diversified portfolio of techniques. All of these devices are individually susceptible to abuse, fallacy, and error. But, happily, they provide a network of cross-checks that make for impressive collective reliability. So if there are sides to be taken, I count myself among the friends of thought experiment.

1

Our Most Curious Device

It was not until I had attended a few post-mortems that I realized that even the ugliest human exteriors may contain the most beautiful viscera, and was able to console myself for the facial drabness of my neighbors in omnibuses by dissecting them in my imagination.

J. B. S. Haldane

Once in a blue moon, the field biologist spots a creature far from the established range of its species. A surprisingly large majority of these strays are young specimens. Experts on migration explain that animals go through an exploratory phase before settling into adulthood. The wanderings help new members find fresh territory and expand the range of the species as a whole. What holds for fish and birds and elk and seals holds for human beings. However, our exploratory drive has a richer array of effects, since human youth is prolonged and much of this drive is sublimated through our mental lives.

One of these effects is a gutsy attitude toward belief formation. Unlike most of the population, college students enjoy roving through exotic cognitive terrain and even court challenges to their central beliefs. Philosophy teachers are guides in the adventure and so must strike a balance between excitement and safety. They need to apply techniques of *controlled* speculation.

I. The Instrument of Choice

The philosopher's instrument of choice is the thought experiment. For example, I ask students to suppose that they are brains in vats on a planet orbiting Alpha Centauri, where playful neuroscientists electrify gray matter to maximize false beliefs. How would things seem under such conditions? Just as they do seem! We know all too well what it is like to be a brain in a vat. Hence, there is no test that can assure us that this possibility is not an actuality. The thought experiment shows that it is possible that you are mistaken in believing you have a nose, so there is room for doubt about your nosedness. But if you *know*

7

that you have a nose, then you must be certain. And if you are certain, then there is no room for doubt. But since the brain-in-a-vat scenario makes room for doubt, you are not certain, and therefore you do not know that you have a nose. But if you do not know a fact that is as plain as the nose on your face, what do you know?

At this point, many students feel they have got their money's worth. They leave class like they leave an absorbing matinee—a little disoriented, a bit preoccupied, depressurizing to everyday reality. But there is the occasional objection, "So what? What do those brains have to do with anything? They are just *hypothetical*." One could respond with correct but arcane allusions to counterfactuals and the general relevance of possible worlds. Tu quoque is also tempting. Doesn't any sensible student heed *possibilia* when crossing the street or practicing birth control? But the best response is to find partners in crime: scientists conduct thought experiments, so why pick on the philosophers?

II. Scientific Thought Experiments

Let's get down to cases. Galileo's fascination with pendulums led him to design an inclined plane that demonstrates the law of equal heights. Just as the pendulum's bob recovers its original height as it swings from its top-left point to its top-right, a ball rolled along a double inclined plane will recover its original height (see Figure 1.1). Of course, Galileo realized that the ball's track was not perfectly smooth and that air had to be pushed aside by the ball. In real life, the ball does not quite reach its original height because some of its energy is spent in the good fight against friction and air resistance. The law of equal heights only *directly* applies to an idealized counterpart of the physical setup, one in which there is no friction or air resistance. Finding a substantive sense in which it holds for familiar things is tricky business. Galileo's idealization lets us sidestep this mess and test the law with the behavior of an imaginary ball and plane. His line of inquiry brings us to a more dramatic kind of thought experiment in which we are doing more than stipulating away nuisance factors. Galileo asks us to suppose that one side of the plane is progressively lengthened, so that the ball must travel farther and farther to regain its original

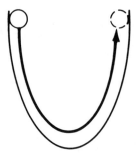

Figure 1.1

Figure 1.2

height (see Figure 1.2). In the limiting case of infinite lengthening, the ball never returns to its original height. Since the law of equal heights says the ball must continue until it does regain its original height, it follows that the ball will continue forever in a straight line (see Figure 1.3).

This thought experiment turned the theory of motion upside-down. According to Aristotle and common sense, it is natural for things to slow down and come to rest; continued movement is what needs explaining. But after Galileo's thought experiment, continued movement seemed natural and slowing required explanation. Stephen Toulmin calls this sense of what is in need of explanation an "'ideal of natural order'—one of those standards of rationality and intelligibility which . . . lie at the heart of scientific theory."[1] Toulmin rightly regards these ideals as important because they set the agenda of discussion. The key scientific questions are raised by anomalies—phenomena that defy expectation. Since our ideals of natural order provide basic reference points for what counts as strange (and thus what is in need of further inquiry), they control lines of inquiry. A change in these ideals will therefore alter inquiry by dropping old questions and introducing new ones.

This agenda-setting effect of Galileo's thought experiment illuminates a peculiarity of Isaac Newton's formulation of his first two laws of motion:

1. Every body continues in its state of rest or of uniform motion in a straight line unless compelled to change that state by impressed forces.

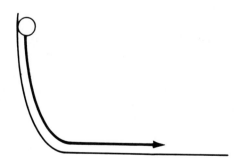

Figure 1.3

2. Change of motion is proportional to the impressed force and occurs in the direction in which that force is impressed.

Newton's first law of motion looks logically redundant; isn't it just a special case of the second law where the impressed force equals zero? Physics textbooks answer *yes* because they read the second law as $F = ma$. This is the simple answer, and Newton uses the formula as part of definitions 7 and 8 in the *Principia*. But some scholars are reluctant to attribute inelegance to Newton and so have attempted to break the entailment.[2] However, Toulmin's concept invites a distinction between logical and rhetorical redundancy. Think of how "That's all there is, and there ain't no more" achieves emphasis by repetition via the logical equivalence of its conjuncts. Hence, we can undercut the search for an independent message hidden in the first law by viewing it as inaugurating a new ideal of natural order. In other words, the real role of the first law is as a slap in the face of common sense and Aristotle. It is intended to jolt us into a new way of looking at nature. It expresses rejection of the belief that objects naturally move in circles and only continue to move when propelled.

The live rival to Newton's physics was René Descartes's. Like Aristotle, Descartes was subjected to hostile thought experiment. The Cartesian law of collision states that smaller objects rebound from larger objects with equal speed but that when larger ones hit smaller ones, they move together in a way that conserves the total quantity of motion. To refute Descartes, Gottfried Leibniz had us contrast a pair of hypothetical collisions.[3] In the first collision, ball A strikes an imperceptibly larger ball B. Since there is no restriction on how much larger the larger body must be, Descartes' principle implies that A would rebound while B remains stationary. But now suppose that A is slightly enlarged so that it is now imperceptibly larger than B. Descartes' principle implies that this second collision with B would send them both moving, each with about half the speed that A had prior to the collision. This discontinuity struck Leibniz as absurd. How could such a tiny change have such a dramatic effect? Nature does not make jumps!

By tripping Cartesian physics, Leibniz's thought experiment helped the rival Newtonian system in a relational way; some races are won by slowing down losers rather than speeding up winners. But it also invited intrinsic improvement by imposing an antiqualitative constraint on mechanics. To avoid Leibnizian absurdity, one must resist the temptation of thinking that the *size* of the colliding objects is *qualitatively* important. Instead, apparently qualitative differences should be explained away by a principle that makes do with quantitative differences. This paves the way for Newton's solution, embodied in his third law:

3. Reaction is always equal and opposite to action; that is, the actions of two bodies upon each other are always equal and directly opposite.

The third law does not discriminate by size and so explains away many of our intuitions about collisions as due to our insensitivity to small effects. The influence of big things on little things will seem to be one-way influences because much less energy is needed for a little thing to move about in a noticeable way. (Thus, it is commonly assumed that a person is gravitationally attracted to the earth but not vice versa.) The quantitatively minded third law goes on the offensive when it predicts some effects that are even more "qualitative" than those predicted by Cartesian physics. Consider the case where the moving ball *equals* the size of the stationary ball. The Cartesian would conclude that both balls move off together having exactly divided the momentum—for if only one moves, which will it be? But Newton's theory predicts that the moving ball would stop dead in its tracks while the second proceeds at the first ball's former speed. For when A hits B, B pushes back with the same force with which A pushes B, thereby canceling A's forward momemtum. Since B was stationary, it will have a net movement in the direction it was pushed. If you don't believe this, try it on a pool table. When the cue ball does not have much "English" (rotational velocity) and hits along the line of centers of the two balls, it comes to an abrupt halt.

III. The Bridge to Philosophical Thought Experiments

Our tour of scientific thought experiments shows that they are more than intellectual ornaments. Controlling extraneous variables and inverting ideals of natural order are real contributions. Since students respect scientific work as a paradigm of rationality, they are well on their way to accepting the legitimacy of *philosophical* thought experiments. Their only reservation is that there might be hidden differences between scientific thought experiments and philosophical ones.

The strong version of this worry is that all philosophical thought experiments are separated from *all* scientific thought experiments by a sea of disanalogies. This would prevent us from legitimating any philosophical cases by bridges to scientific cases. But I am bullish on the comparison. Further inquiry into science will reveal a great variety of thought experiments. Ditto for philosophy. Such double heterogeneity quashes fear that the two groups of thought experiments are logically isolated from each other. We will find that virtually all philosophical cases are analogically linked to some scientific thought experiment or other.

Certainly, there are *historical* links between science and philosophical thought experiments. Sydney Shoemaker's Lefty/Righty case was inspired by neurophysiological studies of patients who survived loss of half a brain. His extension of this research was first to suppose that a healthy person Smith has the left half of his brain transplanted into a debrained body and the right half destroyed. Given that the posttransplant person, Lefty, is psychologically similar to Smith, we are inclined to say that Smith survived the operation and so is identical to Lefty. But now suppose that the right half is simultaneously transplanted into another brainless body, resulting in Righty. Since Lefty and Righty are equally similar to Smith, it is arbitrary to identify either with Smith. Yet it seems that one must be Smith: a double success cannot be a failure!

The weaker version of the isolation fear is that *some* philosophical thought experiments are separated from all scientific thought experiments, thereby preventing a subset of philosophical thought experiments from being legitimated by their scientific brethren. For example, many thinkers are impressed with the fact that actual performance of Shoemaker's transplant would fail to resolve the question of whether Smith survives as Lefty or Righty. This suggests an asymmetry because scientific thought experiments seem completely superseded by their physical instantiations—the one you do is always better than the one you just thought about doing. Perhaps thought experiment functions in science as an ersatz empirical result but in philosophy as a kind of notational mannequin that can be clothed and reclothed with rival descriptive schemes. If so, philosophical thought experiments seem too distant from their scientific relatives to receive any help.

Philosophical thought experiments are more controversial than their scientific counterparts. Newcomb's problem illustrates this divisiveness. Suppose you are presented with two boxes. One box is transparent and contains a thousand dollars. The other is opaque and might contain a million dollars. It all depends on whether the Predictor put the money there two weeks ago. If he predicted you would take only the opaque box, he put the million in the box. If he predicted you would take both boxes, he put nothing there. You know he is pretty accurate. Perhaps past experience with other people has shown that he is right nine times out of ten. You also know there is no future penalty associated with taking both boxes and that no tricks are involved. Should you take only the opaque box or should you take both boxes? Two-boxers reason that since the contents of the opaque box were determined two weeks ago, you are sure to be $1,000 better off by taking both boxes regardless of whether there is any money in the box. One-boxers counter that your chance of being a millionaire is much higher if you only pick the opaque box. Given that the predictor is right nine times out of ten, the expected return on taking one box equals $9/10 \times \$1,000,000 = \$900,000$ as opposed to the $(1/10 \times \$1,000,000) + \$1,000 = \$101,000$ accruing from the two-box choice. These arguments persuade some people to switch sides, but wide controversy has persisted through twenty years of intense analysis.

The disanalogy with science is especially striking in philosophy's back-woods subfields of ethics and aesthetics; for *evaluative* thought experiments appear hopelessly marooned by the fact–value gap. Science tells us how things are, not how they ought to be. True, ethical thought experiments resemble scientific ones in that they often function as tests. For instance, to test the absolute prohibition against killing innocent people, ethicists suppose a fat explorer has become stuck in the mouth of a cave. The cave is flooding, so the remaining members of the party will drown—unless they use their emergency dynamite. If, as many think, the explorers are permitted to blast the fat man free, the universal prohibition against killing innocents is refuted. There are also generic ethical tests that can be used again and again. I may check whether my action is permissible by imagining everyone doing the same, or by determining the reaction of an ideal observer or by consulting the clauses of a hypothetical social contract. But "ethics by thought experiment" fails to yield a *description* of phenomena. The direction of fit is reversed; science tries to make words fit the world, ethics tries to make the world fit the words.

Although metaphysics is advertised as addressing the question "What is there?," scholars have suspected that its direction of fit occasionally flips. The doctrine of the eternal recurrence illustrates how a piece of ostensibly *descriptive* metaphysics can quietly begin to tell us how to run our lives. Friedrich Nietzsche reasoned that since there were only finitely many things, there were only finitely many arrangements of things and so only finitely many possible situations. But since time is infinite, every actual situation must repeat infinitely many times. Although Nietzsche believed this conclusion, his principal use of it was as a thought experiment by which to plot the course of your life:

> How, if some day or night, a demon were to sneak after you into your loneliness and say to you: "This life, as you now live it and have lived it, you will have to live once more and innumerable times more; and there will be nothing new in it, but every pain and every joy and every thought and sign . . . must return to you—all in the same succession and sequence—even this spider and this moonlight between the trees, even this moment and I myself. The eternal hourglass of existence is turned over and over—a grain of dust!" Would you not throw yourself down and gnash your teeth and curse the demon who spoke thus? Or have you once experienced a tremendous moment when you would have answered him: "You are a god and never did I hear anything more godlike!" If this thought were to gain possession of you, it would change you as you are, or perhaps crush you. The question in each and everything, "Do you want this more and innumerable times more?" would weigh upon your actions as the greatest stress. Or how well disposed would you have to become to yourself and to life to *crave nothing more fervently* than this ultimate eternal confirmation?[4]

Nietzsche is not *asserting* that the demon speaks the truth. He is only having you *suppose* the eternal recurrence ("the thought of thoughts") and drawing attention to how the supposition magnifies your choices: "Even if the repetition of this cycle is only probable or possible, even the thought of a possibility can deeply move and transform us, not just what we can perceive or definitely expect! What an effect the possibility of eternal damnation has had!"[5] The truth of the eternal recurrence is irrelevant as long as it functions solely as a decision aid. Nietzsche realized that his main point did not depend on the soundness of his argument for the reality of the eternal recurrence.

And that is fortunate because in 1907 Georg Simmel published a thought experiment that demonstrates the invalidity of Nietzsche's argument. Suppose that three wheels of the same size are supported by a common axle (Figure 1.4). Each wheel has a point marked on its circumference. The three marks are initially aligned along a thread stretched above the wheels. The wheels are then

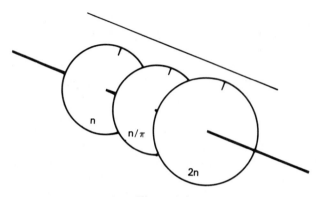

Figure 1.4

indefinitely rotated at speeds of n, $2n$, and n/π. Never again will the marks line up beneath the thread:

> If, therefore, there exist anywhere in the world three movements that correspond in their relationships to these three wheels, then the combinations among them can never return to their initial form. Finitude in the number of elements does not at all necessarily ensure, even if there is an infinite amount of time for their movements, that the situation of any moment is repeated unchanged. Naturally, the case could be different. The movements of the world could be so arranged that they run through an ever-repeating cycle of combinations. But the bare possibility just sketched suffices to show the putative proof for the eternal return to be an illusion.[6]

The mathematical character of Simmel's thought experiment sharply contrasts with its visionary target. We see metaphysical thought experiments at their extremes. This contrast makes the eternal return emblematic of the heterogeneity of philosophical thought experiments.

Even if the bulk of philosophical thought experiments can be grounded in scientific analogues, questions about the scientific thought experiments would still have to be answered. How could they really achieve what they appear to? After all, the results of the scientific thought experiments mentioned earlier are propositions about how the world operates; they are not merely conceptual propositions reflecting linguistic conventions. Logic alone does not dictate that Galileo or Newton is right. Their results are synthetic propositions. But how can we learn contingent facts by thought alone?

According to empiricism, everything we know about this world is through experience. Empiricists admit that reason alone can provide knowledge of tautologies such as 'A fortnight is less than a month'. But these are not informative truths about the world; tautologies merely reflect the rules we have for representing the world. Tautologies are a dime a dozen. Stipulate that 'splarg' means husbands wearing shorts. Now comes our bounty of truths: all splargs are men; no bachelor is a splarg; nude splargs do not exist. None of these are *substantial* truths, they just mirror notational resolutions. Scientific thought experiments, in contrast, give us truths about the real world. Thus, empiricism appears refuted by the thought experiments in physics, biology, and economics. Rationalists, on the other hand, say that we can learn about the world without experience, and so rationalists need not be as puzzled as the empiricists. Indeed, a rationalist could claim that thought experiments prove his point: there are synthetic a priori propositions. But even a rationalist has got to draw a line against magic. He has got to tell us how this extraexperiential technique works. The rationalist must do justice to the conviction that this world works without ad hoc spookery. His best story has been that the synthetic a priori knowledge is innate or that it is structured into experience. But few contemporary philosophers see much merit in either line. Since innate beliefs can be false, why would the fact that I am born to believe p justify p? Presuppositions of successful cognition appear equally vulnerable. Even if a proposition is "built into" inquiring minds as a precondition of experience, that would not force reality to fall within the grooves of the transcendental template.

Puzzlement about thought experiment can be placed within the traditional debate between rationalists and empiricists. The sophomore's scepticism about thought experiment is blunt but not sophomoric. Nor is it an atavistic issue displaced from some past stage of philosophy. Contemporary philosophers are also ill at ease with thought experiment. They, too, have doubts about how small-scale science fiction could prove anything. Why not avoid the problem by using methods other than thought experiment?

IV. Analytic Philosophy's Commitment to Thought Experiment

Analytic philosophers make heavy use of thought experiment because it is the natural test for the clarificatory practices constituting conceptual analysis: definition, question delegation, drawing distinctions, crafting adequacy conditions, teasing out entailments, advancing possibility proofs, mapping inference patterns. For instance, if I define 'man' as the featherless biped, contrarians envisage walking lizards. Their point is that the definition works only if it is necessarily true that all featherless bipeds are men. Since we can imagine featherless bipeds that are not men, it is possible for there to be featherless bipeds that are not men, and the definition is false. Of course, analytic philosophers are not the only ones who use imaginary counterexamples. Only mental cripples are barred from hypothetical reasoning. However, analytic philosophers are the characters who make the most extensive use of thought experiment. The reason is the premium they set on clarity. They differ from other philosophers only in what they are willing to exchange for freedom from ambiguity, vagueness, circularity, inconsistency, and sundry obscurities.

We gain evidence for the accuracy of this characterization by considering what critics of analytic philosophy aim at. Antianalytics prize clarity but castigate the analytics for sacrificing too much. Peace, health, and cleanliness are also good things; but their pursuit becomes excessive when other goals languish. Compare the analytic philosopher's pursuit of clarity with the germophobe's pursuit of sanitary surroundings. Many passions and phobias center on genuine goods and evils; they only acquire the status of disorder by virtue of what they interfere with. Some critics of analytic philosophy will say that the problem is more than interference. One criticism is that the obsession with clarity tempts the analytic into aping science. Thus, we find the accoutrements of notation, jargon, and specialization, as well as the cultivation of cordial relations with physics, biology, and psychology and parvenu indifference toward neighbor humanities. Another grievance is that the pursuit of clarity squeezes the life out of other values: neglect of other goals leads to their extinction. The imbalance creates a delusional fear that one's thinking is flabby and excessive along with a grim determination to slim down. The resulting academic anorexia is highlighted by parodies of analytic philosophy. The analytic is pictured as an eccentric who is so obsessed with sharpening his tools that he gives no further thought to what he is sharpening them for. Or he is compared to the drunk who loses his keys at his door but chooses to search for them under the street lamp because the lighting is better. Clarity is not enough!

Analytic philosophers deny that they pursue clarity beyond the point of diminishing returns. Indeed, hardliners like Saul Kripke complain that we do not press the demand for clarity vigorously enough. Most analytics have metaphilosophical beliefs that justify (or rationalize) their circumspection. Some appeal to the history of speculative failure, others to the high frequency of verbal illusion, and still others to the view that problems are philosophical just because other fields deem them hopeless. There is some temptation to extract an analytic creed from these metaphilosophical rationales. But there are too many different reasons for being careful. Also, many analytic philosophers do not provide any metaphilosophical reason for assigning great weight to clarity; some just have a taste for it. So trying to derive a creed from the love of clarity will be as unrewarding as the attempt to derive a creed from the love of health, safety, or cleanliness.

The absence of a creed makes analytic philosophy difficult to criticize. One can chase down some of the reasons for their attitude, but other far-flung motives will remain. The more effective objection to a passion is to show that it cannot be satisfied; for when an enthusiast is persuaded that he cannot do what he hoped, he stops trying and is willing to try something else. Hence, one way of criticizing analytic philosophy would be to show that crystal clarity is impossible. An absolute refutation of thought experiment would have this undermining effect because there is no substitute for this master test of conceptual analysis. More generally, analytic philosophy would be undermined in proportion to the severity of the refutation.

Compare this methodological commitment to the one phrenologists embraced in the nineteenth century. Phrenologists believed that the capacities of the brain were localized and were a function of size. Thus, the phrenologists believed that your mental capacities could be measured by mapping the shape of your skull. For instance, an aptitude for math could be detected in the form of a bulge on the temple. Phrenology was opposed by a variety of scientists, physicians, and dualist philosophers. Their most powerful objection was the discovery that the contours of the brain were poorly correlated with the contours of the head. (Bumps are better indicators of sinuses, bruises, and bone knits.) This discovery undermined the phrenologist's method of feeling heads; and since there was no alternative, this discovery ended scientific phrenology (although pseudoscientific phrenology still emits death rattles).

Antianalytics hope for the same pattern of demise from a successful refutation of thought experiment. Sure, there would be some irrational holdouts; but philosophers in the main (especially the younger ones) would reorient. Resources would be redirected to other philosophical goals, such as the general description of the universe and cultural critique. Since the critic is apt to view these as goals neglected by the analytics, he would view his refutation of thought experiment as liberating, a casting away of a false ideal.

And, indeed, we do find thought experiment criticized by outsiders. Some Marxists take the analogy with science fiction seriously and conclude that thought experiment is a genre of escapist literature.[7] One feminist criticism is that the use of hypothetical counterexamples constitutes a coercive, abstract, reductive, male-biased form of philosophy that needs to be balanced by a cooperative, concrete, holistic perspective. Richard Rorty contends that the method is circular because our beliefs determine what we find imaginable. Alasdair MacIntyre protests that thought experiment is naively ahistorical. He illustrates the point with a thought experiment of his own. Imagine that scientific practice is persecuted out of existence and then "revived" on the basis of a miscellany of writing fragments and relics from various stages of the old science. Since the theoretical context has been forgotten, the discourse of the reconstituted "science" would only bear a superficial resemblance to the original:

> The language of natural science, or parts of it at least, continues to be used but is in a grave state of disorder. We may notice that if in this imaginary world analytical philosophy were to flourish, it would never reveal the fact of this disorder. For the techniques of analytical philosophy are essentially descriptive and descriptive of the language of the present at that. The analytical philosopher would be able to elucidate the conceptual structures of what was taken to be scientific thinking and discourse in the imaginary world in precisely the way that he elucidates the conceptual structures of natural science as it is.[8]

But thought experiments that explore this jumble would give pointless results. Children who half-understand a collection of games sometimes blend them into a single, unstable "game." If we tried to apply thought experiment to determine its rules, we would get answers of a sort, but not edifying ones. MacIntyre believes that the current language of morality is in the same sorry state as the hypothetical scientific language. Hence thought experiment (at least within ethics) is a misconceived enterprise. The general problem is that the method can be blind-sided by past events that disorder the practice that the method is designed to articulate. More recently, Mark Johnston has portrayed the "method of cases" as the Achilles' heel of analytic philosophy and has speculated that recognition of its unreliability, artifices, and biases will inaugurate a new type of philosophy.

Within the last twenty years, we find more of these criticisms coming from the inside. Gilbert Harman has said that thought experiments only tell us what we believe, not the nature of reality. Other critics believe that ordinary language entombs folklore that the philosopher disinters as "conceptual truth." J. L. Mackie has ridiculed "ethics by thought experiment" as the cultivation of illusions seeded within moral language. Utilitarians warn that a steady diet of bizarre cases warps one's good sense. Eventually, the fantasist becomes more concerned with a theory's fidelity to storybook worlds than with its fit with the actual world. Like Madame Bovary, the otherworldly thought experimenter is doomed to sacrifice the good in vain pursuit of the perfect.

Daniel Dennett traces much of the persuasive effect of "intuition pumps" to imagery, making thought experiments seem more like poetry or rhetorical devices than instruments of rational persuasion. Dennett also cautions that thought experiments systematically oversimplify, that they are biased against complex processes that cannot be summed up in vignettes. W. V. Quine has long been wary of farfetched thought experiments (especially those in the personal identity literature) that try to extract more meaning from terms than has been infused by our practical needs. Back in 1964 Jerry Fodor elaborated Quinean scepticism about thought experiments into a critique of the appeal to ordinary language that is readily applied to any bizarre supposition. Many criticisms of the method of reflective equilibrium would also gut thought experiment as a corollary. Some argue that philosophical thought experiment rests on the false presupposition that our ordinary concepts are coherent. The purveyor of thought experiments is accused of pandering to one strand of inconsistency, while his rivals pander to other strands. Another sceptical theme rests on an instrumental view of ordinary language. If common sense only aims at the prediction and control of daily affairs rather than at describing reality, then folk concepts will behave erratically when pulled out of their workaday roles. More traditional is the criticism that thought experiment falsely presupposes that conceivability implies possibility. Yet other doubters debunk the concept of logical possibility itself. And there are critiques that weave these separate charges together, for instance, Kathleen Wilkes's *Real People: Personal Identity Without Thought Experiment*.

Most of these internal critics are not entirely sceptical of philosophical thought experiments. Some accept thought experiments that only suppose things to be a wee bit different from how we believe them to be. Some restrictions focus on the vocabulary of the supposition. Are its predicates entrenched in scientific practice? Do they designate natural kinds? Other restrictions focus on the relation of thought experiment to ordinary experiments, insisting the thought experiment be chaperoned by the possibility of physical instantiation. Still others allow it the role of prototype: there has to be a future in it, the possibility of cashing in on a real experiment.

But even these dispensations fail to save the moderate sceptics from tu quoque objections. The analytic sceptic about thought experiments tends to violate his own prohibition once he stops talking about philosophy and starts doing it. Insofar as he strives for clear discourse, he comes into the orbit of other possible worlds. Where else is he to turn when testing definitions, entailment theses, and feasibility claims? However, this hypocrisy is of only marginal comfort to the traditional analytic philosopher. Tu quoque is a fallacy, after all. The sceptics are free to recant and reiterate the hard line. But the fact that it is so hard to practice what you preach is evidence against what you preach. 'Ought' implies 'can'. Philosophers who wish to ban thought experiment need to articulate a feasible alternative.

The sceptics seldom elaborate an alternative way of doing analytic philosophy. The apparent exceptions are those who follow Quine's hints about naturalizing philosophy into a kind of scientific journalism. The idea here is that philosophers are perfectly entitled to presuppose standard science because they can hardly hope to improve upon the credentials of a scientifically established proposition. Rather than addressing questions that are not worth asking, the philosopher should systematize the scientific vision and perhaps kibitz along the conceptually confused frontiers. Thought experiment is thus to be replaced by careful historical, comparative, and logical analysis of science. Real-life intellectual work supplants the phantasmagoria of idle supposition.

I doubt that this type of metascientific study is a coherent *rival* to thought experiment. Thought experiment is part of accepted scientific practice. Therefore, if the Quinean naturalist wants to throw in with the scientists, he must see science as validating thought experiment. A naturalist might retrench to the position that thought experiment works in science but not philosophy. And indeed, we do find naturalists claiming that there are disanalogies between science and philosophy that prevent the spillover of legitimacy. However, naturalists also tend to follow Quine's metaphilosophical gradualism: philosophy only differs from science in degree, not in kind. A gradualist cannot hope to break the comparison with an important qualitative difference between science and philosophy.

Other critics of thought experiment are free to criticize the scientists and the analytic in the same breath. A Marxist, for example, can attack their effete retreat from practice. But the Marxist's alternatives, when consistently pursued, will be something alien to analytic philosophy. Hence, I do not think that the *analytic philosopher* has an alternative to thought experiment. He has, of course, other tools of analysis (as did the phrenologist). Nevertheless, he is committed to the adequacy of his most curious device, the thought experiment. Once he gives it up (not merely *says* he has given it up), his ardor for clarity cools; he warms to new desiderata.

My account of thought experiment is principally aimed at removing specific, internal doubts about thought experiment. Thus, I am not debating with those who derive scepticism about thought experiment only as a corollary of a much broader scepticism. For instance, nothing will be said in reply to those who reject all counterfactual reasoning; there is no *special* problem about thought experiments here. Moreover, since broad scepticism never constitutes a credible premise from which a position can be refuted, other threats merit more attention. Even the sceptics have a hard time believing what they finally wind up saying. Jonathan Dancy's attack on ethical thought experiments illustrates this widening spiral into unbelievably broad know-nothingism. As Dancy observes, ethicists frequently respond to a hard case by inventing an easier but similar case from which they can reason by analogy. He objects that this practice is inevitably futile:

> The original idea was that we found the actual case difficult, and so we derived help from a case which was easier. So there is a further difference between the two cases; in one we find it hard to discern what we ought to do, and in the other we find it easy. But this could only be so if there were other relevant differences between the two cases, in virtue of which one is hard and the other is easy. The one that is easy can only be so because it does not contain factors which complicate the issue in the hard case. And, this being so, it is odd to suppose that our decision in the easy case should be of help when we come to the actual, hard case.[9]

The obvious reply is that the difference between the easy and hard cases need not be a *relevant* difference; the easy case could just exhibit a crucial property more saliently or lack a property that causes distraction in the hard case. But Dancy argues that the notion of relevant similarity is viciously circular. How can we know which properties are relevant without already knowing whether they share the very property in question? Although Dancy only targets ethical thought experiments, he realizes that his reservation applies equally well to many nonethical hypotheticals. Indeed, any restriction to just ethics would be arbitrary. Moreover, his sceptical argument works just as well for comparisons between hard actual cases and easy actual cases. Although Dancy heroically welcomes this consequence, he cannot really believe that legal reasoning from precedent is fundamentally fallacious or that a chef is hasty when inferring the taste of his next spoonful of soup from his last.

Broad scepticism should be ignored even when the corollary is a restricted scepticism about a certain kind of thought experiment. For instance, sceptics about aesthetic knowledge put aesthetic thought experiment on a long list of discredited methods. But that says little about thought experiment even if thought experiment happens to be singled out as an entertaining whipping boy.

Nor do I address external doubts about the drive toward clarity that powers the proliferation of thought experiment. This book is not an apology for analytic philosophy. However, I will respond to most of the objections that antianalytics specifically target at thought experiment. External doubts can structure an issue as well as internal ones. But the benefits bestowed by a devil's advocate entitle him to his own chapter. So let us now give the devil his due.

2

Scepticism About Thought Experiments

Schoolmaster: Suppose x is the number of sheep in the problem.
Pupil: But, Sir, suppose x is not the number of sheep. (I asked Prof. Wittgenstein was this not a profound philosophical joke, and he said it was.)
John E. Littlewood

This chapter presents and motivates the issues surrounding thought experiments by assembling the case against their use. Since criticisms have tended to be off the cuff, I engage in rational reconstruction to organize the scepticism. From what I read between the lines, a core suspicion is that thought experiment is a disguised form of an already-discredited form of inquiry. I first explore the more specific charge that thought experiment is just introspection, then concentrate on the charge that it is merely an atavistic appeal to ordinary language. Even if thought experiment is distinct from either of these methods, it strongly resembles them. Hence, details of both introspection and the appeal to ordinary language will be discussed in the hope of illuminating thought experiment at least by analogy. Finally, I develop a more general problem about the informativeness of thought experiment. My policy will be to defer replies to later chapters.

I. Introspection on the Sly?

Despite a recent surge of psychological interest in mental imagery, introspection continues to be kept at the fringe of methodological respectability. Thus, thought experiment can be marginalized by assimilating it to introspection.

A. The Internal Horizon

Most definitions of 'introspection' resemble G. F. Stout's: "To introspect is to attend to the workings of one's own mind." For instance, when I look at a gnu, I observe a gnu; but when I focus attention on my looking at the gnu, I

21

introspect. Introspection is commonly pictured as an inward perception of one's own mental states. Thus, it raises the same questions as perception of external things. Nevertheless, introspection struck early theorists as the more fundamental process. For perceptual illusions are possible only if *judgments* are being made about the external scene. These inferences are from the contents of consciousness. Therefore, an understanding of perception should come from the study of the nature of these basic ideas and the processes by which they are combined into ideas of external things. Since introspection promises direct access to the contents of consciousness, it struck early theorists as crucial to perception. The traditional empiricists regarded perception as our keyhole into reality, so they took introspection to be crucial to *all* knowledge. Thus, philosophers such as David Hume undertook the exploration of the inner world in the same spirit that Isaac Newton undertook the study of the external world.

B. Complaints About Introspection

Allegiance to introspection frayed to bits at the turn of the century. Let's review the objections that caused the most wear and tear.

1. *Has our access to our own minds been overestimated?*   An early misgiving about introspection is that it endowed consciousness with miraculous knowability. Inner omniscience seemed too good to be true.

Yet special access does seem to mark the difference between external objects and mental states: our environment is independent of us perceiving it. Comets and omelets exist even when completely undetected. Mental states, on the other hand, are subjective things that depend on our awareness of them. This contrast was embraced by the first great British empiricist, John Locke. He insisted that propositions about our own mental states are strongly self-intimating: if you have mental state M, then you *know* you have M. This fountain of self-knowledge was also accepted by Thomas Reid, William Hamilton, and many other prominent eighteenth-century philosophers.

However, one of Locke's most distinguished adversaries, Gottfried Leibniz, objected that self-intimation implies an infinite regress. If having mental state M_1 requires another mental state, M_2, consisting of knowledge of M_1, then there must be a third mental state, M_3, consisting of knowledge of M_2—and so on for M_3 and the mental states that subsume it. In addition to this conceptual difficulty, the self-intimation thesis would force us to reject the possibility of empirical *discoveries* about mental states. This puts the match to literature on unconscious mental processes, inner illusion, and the collection of subtle insights that come from *extra* attention to your own mental processes.

The best-known expression of belief in the *incorrigibility* of introspection lies in René Descartes' *Second Meditation*. Descartes argues that even if there is an evil demon deceiving Descartes into believing that he is sitting in front of a fire, the demon cannot fool Descartes into thinking that he has visual *perceptions* as of fire and *sensations* of heat. A belief is incorrigible when believing so makes it so. For example, the belief that you exist implies you exist. Self-intimation and incorrigibility are converses:

p is self-intimating for A = p being true ensures that A believes p.

p is incorrigible for A = A believing p ensures that p is true.

A proposition can be self-intimating without being incorrigible. If you are omniscient, you know and therefore believe it. But if you believe you are omniscient, it does not follow that you really are. Conversely, a proposition can be incorrigible without being self-intimating. Anyone who believes that there are beliefs has got to be right. But it does not follow that anyone who has beliefs, believes there are beliefs. Some animals are sophisticated enough to have beliefs but are not sophisticated enough to be aware of the fact that they have beliefs. Although self-intimation and incorrigibility are independent, they are not mutually exclusive: pain is both self-intimating and incorrigible. Hume believed that all mental states are self-intimating and that our beliefs about them are incorrigible: "Since all actions and sensations of the mind are known to us by consciousness, they must necessarily appear in every particular what they are, and be what they appear."[1]

There are commonsense counterexamples to the thesis that our beliefs about our own mental states are incorrigible. An artist may be convinced he is imagining the home he is painting when actually remembering it. Experimental psychologists have collected evidence that introspection is heavily influenced by the same sort of interpretation we apply to other people. Market researchers have found that the position in which you display an item alters its likelihood of being chosen.[2] English speakers tend to prefer the item on the far right (presumably because they read left to right, giving the last item a recency advantage). However, the choosers never give position as their reason. They provide answers conforming to stereotypes of rational choice ("Because it tastes better"). Likewise, moral psychologists have found that the probability of an individual's aiding a person in distress varies inversely with the perceived number of potential helpers. But a bystander explaining his own behavior rarely cites this factor. He instead focuses on the plight of the victim and his personal ability or inability to render assistance.

2. *What is the organ of introspection?*   Perceptions can be traced to effects objects have on our sense organs. If we view introspection as inward perception, it is natural to inquire about the corresponding sense organ. Where is it?

This anatomical embarrassment is felt by other theories that assimilate a kind of knowing to perception. The assimilators include Kurt Godel, who said we have a mathematical intuition that enables us to "see" mathematical entities. Some ethicists have appealed to a moral sense by which we detect goodness. And memory theorists sometimes credit us with a means of "looking" at past events. But these perceptual theories all run into the disanalogy— no sense organ. Thus, we are not just picking on introspection. Maybe introspection is like our sense of humor—not really a mode of perception at all. Indeed, what would count as a sense organ for introspection? Suppose we find a small eyeball within the brain focused on a little screen. Wouldn't we still need to explain how the inner eye works? Wouldn't that require a faculty of introspection as well? Such questions inspire a more fundamental objection.

3. *Is self-observation even possible?*   Auguste Comte objected that introspection is self-observation and self-observation requires an impossible split of consciousness: "As for observing in the same way [as one observes something external to the organs of observation] *intellectual* phenomena at the time of their actual presence, that is a manifest impossibility. The thinker cannot divide himself into two, of whom one reasons whilst the other observes him reason. The organ observed and the organ observing being, in this case, identical, how could observation take place? This pretended psychological method is then radically null and void."[3] Many introspectionists accepted the unity of consciousness on empirical grounds. For example, William James cites a study of the French psychologist Paulhan who tried in vain to write a poem while reciting another aloud.[4] However, Comte had a priori grounds for believing that consciousness could not be divided. He thought a Leibniz-style regress follows from self-observation. The difficulty is that a self-observer would have to observe his own observing, and observe that second order observation, and so on ad infinitum.

John Stuart Mill responded by substituting *retrospection* for strict introspection.[5] The retrospector studies conscious states as suspended in short-term memory. Since past states are not part of current consciousness, there is no need to observe x and oneself observing x simultaneously. Thus there is no infinite regress. The main problem with this suggestion is that retrospection does not sufficiently resemble introspection. How does it differ from attentive memory?

4. *Is introspection constant and faithful?*   As long as introspection is used to test only properties of one's own current consciousness, it will be a hard method to discredit. But the introspective tradition was more ambitious. It used the method to investigate the general features of how we think. How well does it test these features?

A test is *constant* to the degree that it gives uniform results. For example, your weight scale is constant if it gives you the same readings when you step on and off a few times within a period of five minutes. A test is *faithful* if, and only if, it measures the property it was intended to.[6] If I step on and off the scale three times and it always reads fourteen pounds, then the test is constant but unfaithful. It is also possible for a faithful test to be inconstant; if I steadily gain weight, I should not get the same reading twice. (Indeed, highly sensitive weight scales oscillate about an ounce because your center of gravity shifts with the heart's cycle.) A test that is faithful within one's sample has internal faithfulness, but this does not guarantee that the test holds for the targeted population in general. My weight scale may work for my buddies while being inaccurate for people outside the narrow band of weights that my friends happen to occupy. We also wish to see how well the test matches other tests. One cross-check would be against other weight scales. If one scale always reads one hundred pounds and another always reads two hundred, both have high *intra*constancy but low *inter*constancy. A better check of constancy would be against a different *kind* of test—say, water displacement. When qualitatively different tests yield the same results, we gain a strong case for faithfulness because the best explanation of the uniformity is accuracy.

How does introspection fare? Comte harps on the poor interconstancy: "The results of so strange a procedure [introspection] harmonize entirely with its principle. For all the two thousand years during which metaphysicians have thus cultivated psychology, they are not agreed about one intelligible and established proposition. '*Internal observation*' gives almost as many divergent results as there are individuals who think they practice it."[7] These complaints were echoed at the turn of the century within the new field of psychology. T. Okabe's experiments on belief and disbelief illustrate the problem of inconstancy.[8] Okabe's procedure was to have the subjects introspect their response to statements designed to elicit strong agreement or disagreement, such as "Wife beating is good exercise." Many subjects gave trivial reports, like feeling that they were saying *yes* or *no* or nodding, but others reported bizarre images in which belief was represented "by a circle or a ball, of small size, two or three feet away from the eyes, which is very sharp in outline and seems very heavy. Doubt or hesitant belief may be represented by a larger, softer, vague, indefinite, and hazy ball."[9] Another subject said he felt belief as a pressure "just below the ribs, a little left of center." Other believers reported tingling sensations. In addition to the diversity among believers, there was diversity over time with the same subject.

5. *Is introspection complete?*   A test that fails to be faithful toward the targeted population leads us into *commissive* error when applied beyond their narrow domain. Tests are also criticized when they fail to give any response at all; these generate *omissive* error. For example, one flaw of early thermometers using colored water was that they failed to give readings once the temperature passed below the freezing point of water.

Introspection was intended to give answers to all questions about our current mental states because these states were supposed to be self-intimating. Hence, any omissive failure of self-intimation counts against the completeness of the method. Thus, introspectionist psychologists were alarmed by their subjects' inability to report the images, feelings, and acts of will associated with certain problem-solving tasks: "Given that: Joe is richer than Bill, and Joe is poorer than Ed, find the relation between Bill and Ed." Many subjects solved the problem with an imaginary diagram in which three dots representing Ed, Joe, and Bill were arranged vertically in order of wealth. However, they were not able to cite any image, feeling, or act of will that made the top dot represent Ed rather than Bill, or why any dot corresponded to any of the characters in the problem scenario. The thought that the top dot represented Ed was called an *imageless thought*. These imageless thoughts show that there are mental events (the representing of Ed by the top dot) that are not revealed by introspection.

The woes of introspection are relevant to thought experiment because it is difficult to separate the two methods in a principled fashion. Indeed, one might suspect that 'thought experiment' is just a twentieth-century euphemism for the currently reviled term 'introspection'.

C. The Parallel Plight of Thought Experiments

Whether or not thought experiment is just clandestine introspection, it draws similar criticisms.

1. *Is the subject matter of thought experiment accessible?* Hypothetical examples are intended to reveal attitudes toward propositions (i.e., whether we believe them or desire them to be true) and to determine whether those propositions frame genuine possibilities. Sceptics caution that attitudes and possibilia are much more elusive than they appear. For instance, Annette Baier argues that our moral convictions are masked by processes of self-dramatization. The storytelling setting of thought experiment brings out the ham in us. Instead of reporting bland truths, we embellish. Since even real-life decisions tend to be confabulated in retrospect, Baier's scepticism extends to actual cases: "I see no nonsuspect way, by interviewing people about other people's actual or hypothetical decisions or even about their own past actual ones, to gauge what are or were their *effective* moral beliefs."[10]

We are also apt to overestimate the ease with which possibilities are ascertained. The mere impression that p is possible is enough to make most people assert that p is possible. Indeed, the possibility of an event is normally *presupposed* as soon as the speaker begins to talk about the event. That is why folks are stumped by "What is the sound of one hand clapping?" Explicitly asking about the possibility of p overcomes this presumption. Impatience is responsible for error in the opposite direction. When explicitly asked whether p is possible, we quickly infer impossibility from any strain in envisioning p. For example, it is easy enough to imagine boring through a steel cube to form a hole through which a slightly smaller cube could pass. But what about boring a hole through the slightly smaller cube to allow the larger cube to pass through? Frustrated visualizers quickly conclude the feat is impossible. Their reason: they cannot conceive of a bigger cube fitting inside a smaller one. But safe passage can be arranged by cutting the tunnel *diagonally* rather than through a face of the cube.

The sceptic can supplement this appeal to the record with an important conceptual problem about our knowledge of *possibilia*. If we picture modal truths as facts about possible worlds, we appear to be cut off from modal reality. Every possible world is causally isolated from every other possible world, so our sense organs cannot detect what is going on "out there."

2. *Does thought experiment commit us to voodoo epistemology?*   The sceptic can press his attack further by charging that the method forces us to accept a modal sense akin to Godel's mathematical sense. Platonistic defenders of thought experiment tolerate this and even revel in it.[11] But the vast majority of philosophers insist that the postulation of a sense must satisfy the standards of physiology. Scientists have proven to be open-minded about perception. They have expanded the list of senses beyond the five endorsed by common sense. They have not required senses to be directed at the external environment. For instance, awareness of body temperature has been traced to an internal sense. Nor do they require the perceiver to be aware of the sense; physiologists allow for the possibility that we have a weak magnetic sense that helps us navigate. But for all this liberality, physiologists insist there be a corresponding sense organ and that it operate in accordance with causal laws.

3. *Is 'thought experiment' a contradiction in terms?*   Experiments are deeds, not thoughts. Real experiments are a posteriori, thought experiments are a priori. Thus, the phrase 'thought experiment' runs afoul of the contrast between conceptual and empirical inquiry. We can look and see, or we can sit and think; but we cannot do both simultaneously. Experiments are designed to answer questions by actions that produce empirical data; but thought experiments cannot deliver empirical data. In short, the objection alleges that 'thought experiment' is a contradiction in terms and should be grouped with oxymorons such as 'tiny giant' and 'married bachelor'.

4. *Is thought experiment constant?*   Locke himself noted, "Everything doth not hit alike upon every man's imagination." Divergence in cultural background is one source of inconstancy:

> A party of white men and Indians were amusing themselves after the day's work by attempting to throw stones across a deep canyon near which they had encamped. No man could throw a stone across the chasm. The stones thrown by the others fell into the depths. Only Char, the Indian chief, succeeded in striking the opposite wall very near its brink. In the discussion of this phenomenon, Char expressed the opinion that if the canyon were filled up, a stone could easily be thrown across it, but, as things were, the hollow or empty space pulled the stone forcefully down. The doubts of European Americans as to the correctness of this conception, he met by asking "Do not you yourselves feel how the abyss pulls you down so that you are compelled to lean back in order not to fall in? Do not you feel as you climb a tall tree that it becomes harder the higher you climb and the more void there is below?"[12]

Divergence is also possible among natives of the same culture. Charles Schmitt presents the medieval debate over the possibility of a vacuum as an internal conflict of intuitions.[13] Fellow medievalists deployed the same thought experiments in support of diametrically opposed theses. For instance, vacuists claimed that a void could form within a collapsible container; just empty it, make it airtight, and force it back into its larger uncollapsed shape. Vacuists admitted that this container could not be an ordinary one (such as a wine pouch) because it would leak. The force could not be the tugs of ordinary hands because they are not strong enough. So the vacuists just stipulated the container leak-proof and the forces strong enough. Plenists would use the same scenario to demonstrate the *impossibility* of a vacuum! They invited readers to picture the comical futility of trying to seal the container *and* enlarge it. If you succeed in enlarging the container, you can be sure of a leak; so you reinforce it. Now you need more enlarging force. You get it; but your success in enlarging the container shows you have a new leak. Thus, the vacuum maker is doomed to an endless spiral of ever stronger sealants and ever stronger enlarging forces. Thomas Hobbes ridiculed Robert Boyles's attempt to build an air pump as just such a misconceived enterprise—the pump was bound to leak![14] In any case, Schmitt concludes that the plenists and vacuists of the Latin Middle Ages should have been more suspicious of a method that gives incompatible results when applied by different people.

We need not travel back in time or to alien cultures to find divergence. Most philosophy professors have a stable of thought experiments that foment controversy within otherwise homogeneous lecture audiences. Some thought experiments are even diachronically inconstant: the same thought experimenter need not get the same verdict when he repeats the thought experiment. Sometimes the verdict is reversed. On other occasions, the switch is from neutrality to partisanship. Consider Hilary Putnam's feline robots.[15] Putnam has us imagine that all ostensible cats are actually robots, whose every move is controlled by Martians sending messages to a receiver in the cat's pineal gland. Would these robots be cats? Putnam initially observed that opinions varied widely and said that he himself lacked any answer. In his later writings, he stoutly defends a positive verdict.

Some kinds of constancy are better than others. We can distinguish them by borrowing the physicist's distinction between stable, unstable, and neutral equilibria. A marble inside a bowl is in stable equilibrium because it returns to its equilibrium point when disturbed (Figure 2.1). Balancing the marble on top of the bowl yields an unstable equilibrium because it rolls far away from its equilibrium point at the slightest nudge: the marble's potential energy "points away from," instead of toward, the original position. When the marble is on a floor, it is in neutral equilibrium, because a disturbance will move it away from its equilibrium point but its potential energy does not point toward, or away from, this point. Each of the equilibria correspond to a kind of constancy once

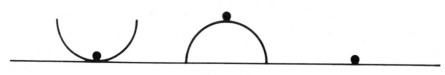

Figure 2.1

we conceive the marble as our *credence*, spatial positions as *propositions* "in logical space," and disturbances be *redescriptions*. We only prize the stable sort of constancy. Indeed, our satisfaction grows with the degree of stability: we like the bowl tall and narrow.

The distinction illuminates thought experiments that produce uniform results under one mode of presentation but not under a slightly different mode. Many hypothetical claims owe their plausibility to slanted storytelling. The method of counterdescription cancels this effect with the help of a description that is slanted in the opposite direction. Bernard Williams applies this technique to "body transfer" thought experiments that support the dominant view that personal identity is more a matter of psychological, than of bodily, continuity.[16] The typical scenario features two people, A and B, who enter a machine that swaps their skills, memories, and character. Now suppose that A knows he will enter the machine. He is given a choice as to which person will receive a reward (say $100,000) and which will be penalized (say, by torture). It seems in A's self-interest to assign the penalty to the person emerging with his old body and the reward to the person emerging with B's old body. This shows that we believe A and B have changed bodies rather than their psychologies; hence, (just as the dominant view says) our psychology is more important to personal identity than physical features. Williams objects by redescribing the situation as one in which amnesia is induced, skills erased, and character is blanked out, so that new "memories," skills, and character can be written into the erased person. Under this description, it seems that A's self-interest is served by rewarding the person emerging with the A-body and penalizing the B-body person. We think it is bad enough for A to lose his memory, character, and skills—that it is bad enough to have a new psychology foisted upon him. Then to be tortured is to suffer *further* harm.

Mark Johnston argues that Williams's counterdescription undermines thought experiment ("the method of cases"): "Taken at face value, the intuition associated with the second presentation: that A would suffer pain in the A-body despite intervening psychological discontinuity of the most radical sort, threatens the dominant view. And on the face of it, the two intuitions taken together threaten the method of cases. For how can intuition be reliable if we can be got to react so differently to the very same case?"[17] Prior to Williams' redescription, our intuitive judgments are uniform in the sense that most people agree that A and B swap bodies. The hidden instability of the equilibrium is exposed by a shift in the mode of presentation. Thus, the uniform results of some thought experiments is a pseudoconstancy masked by a failure of ingenuity. Johnston goes on to note that the results of thought experiments vary with their order of presentation. In his experience,

> it is difficult to get the uninitiated to come up with the body-swap intuition in response to the first presentation of Williams's case when the first presentation comes immediately after cases that emphasize the importance of bodily continuity. Such inconstancy in our intuitive reactions itself suggests that our ordinary capacity to make correct judgments about personal identity is not well

engaged by such bizarre cases. We are bewildered, and in our bewilderment we opt for one or another partial extension of our ordinary practice of reidentification.[18]

Peter Unger also emphasizes that thought experiments can be affected by their proximity to each other.[19] He further argues that they are influenced by social context. Unger illustrates the point by supposing Putnam's thought experiment conducted in the presence of a cat lover. We are to assume the cat lover has no affection for machines. Given his sentiments, the cat lover would deny that the robots are really cats. Our sympathy with the cat lover would incline us to agree with his verdict. But once away from the cat lover, our opinion may fall under the sway of another setting.

5. *Is thought experiment complete?*   A method is incomplete when silent on questions it was designed to answer. For instance, sextants are useless on cloudy nights. Thought experiments also draw blanks: "I do not know what to say in that case"; "I have no intuitions here"; "I cannot get a purchase on the situation." Many people are unresponsive toward whole groups of thought experiments—especially bizarre ones. Perhaps Quine had this in mind when critiquing the debate between Sydney Shoemaker and David Wiggins over personal identity:

> Later, [Shoemaker] examines Wiggins on personal identity, where the reasoning veers off in familiar fashion into speculation on what we might say in absurd situations of cloning and transplanting. The method of science fiction has its uses in philosophy, but at points in the Shoemaker–Wiggins exchange and elsewhere I wonder whether the limits of the method are properly heeded. To seek what is "logically required" for sameness of person under unprecedented circumstances is to suggest that words have some logical force beyond what our past needs have invested them with.[20]

Read Quine as objecting that thought experiment fails to deliver verdicts. Then Derek Parfit's reply suffices:

> This criticism might be justified if, when considering such imagined cases, we had no reactions. But these cases arouse in most of us strong beliefs. And these are beliefs, not about our words, but about ourselves. By considering these cases, we discover what we believe to be involved in our own continued existence, or what it is that makes us now and ourselves next year the same people. We discover our beliefs are revealed most clearly when we consider imaginary cases, these beliefs also cover actual cases, and our own lives.[21]

However, Quine is more charitably read as troubled by the lack of *faithful* answers to our questions. Hence, this aspect of test quality must come off the back burner.

6. *Is thought experiment faithful?*   You cannot measure the soundness of an automobile by kicking its tires or criminality by facial characteristics. I am

concerned with the charge that thought experiment belongs in this group of pseudotests, that they do not measure their targeted property.

I can distinguish the charge of infidelity from a constancy criticism by focusing on thought-provoking questions that *systematically* produce mistaken answers. Imagine repeated folding of a sheet of tissue paper, originally .001-inches thick. How thick would it be after fifty folds? Most people agree the answer is less than five feet. In fact, multiplying 2^{50} by .001 inches yields 17,770,000 miles, a quarter of the distance to Venus. Our weakness with large numbers is responsible for the myth that "six monkeys, set to strum unintelligently on typewriters for millions of millions of years, would be bound in time to write all the books in the British Museum."[22] In fact, the age of the universe is far shorter than necessary for even the production of *Hamlet*. Now suppose there is a ribbon running all around the equator. Lengthen the ribbon by one foot. How far up from the equator would this new ribbon be? Most people estimate that the distance would not be detectible—quite nearly 0. But in fact, the distance would be almost two inches.[23] The riddles teach a lesson: a thought experiment could fail to yield the correct result even though it persuaded everyone to accept the result.

a. *Do thought experiments merely measure a lookalike property?*   Often, the infidelity of a test is masked by its reflection of a property that is apt to be confused with the intended property. Gilbert Harman alleges that this confusion occurs with thought experiments. Rather than reflecting reality, thought experiments reflect *beliefs* about reality. Harman plays this theme against ethicists who maintain that moral theories are tested by the use of imaginary cases:

> The moral philosopher who develops a normative theory by using the method of reflective equilibrium does "thought experiments," not real experiments. Thought experiments are not confined to ethics, of course. They have been used in physics and other sciences in order to bring out the implications of a theory and to test the theory against a particular scientist's sense of what ought to happen in certain circumstances. But scientific theories can often be tested against the world and not just against a particular scientist's sensibility. That is what makes them empirical theories. The issue I am concerned with is whether moral theories can be tested in this further way, against the world.[24]

Harman agrees that the study of "commonsense physics" is feasible and may be of psychological interest. It might be discovered that the average person believes that a car driven off a cliff goes straight out and then drops straight down, as in cartoons. But this would be irrelevant to physics:

> A moral philosopher who tries to find general principles to account for judgments about particular cases, hoping eventually to get principles and cases into "reflective equilibrium," is studying commonsense ethics. The results of such a study are results in moral psychology. Just as a study of commonsense physics is not a study of physics, the study of commonsense ethics is not an investigation

of right and wrong, *unless* what it is for something to be right and wrong can be identified with facts about moral psychology.[25]

But since Harman denies that this identification is possible, he concludes that thought experiments are irrelevant.

b. *An impure measure?*   Some tests track the intended property but only in an adulterated form. An example is Galileo's thermometer. It was made from a narrow twenty-two-inch tube that was closed at the top end and stood in a bowl of colored water. You take the tube out and hold it in your hand to warm it. Then you stick it back into the bowl and wait for the glass to cool. Since the air inside the tube contracts as it cools to the surrounding temperature, the water will rise in the tube and thereby indicate the temperature. Amused physicists point out that Galileo's thermometer is as much a barometer as a thermometer. It really responds to two things: the volume of trapped air (which does depend on temperature) and the atmospheric pressure pushing on the water in the bowl.

Thought experiments are intended to reveal beliefs and preferences or (more ambitiously) to measure the modal status of certain propositions (whether the proposition is possible, impossible, or necessary). Critics grant that the thought experiment is partly sensitive to this targeted property but allege that other properties inevitably adulterate the measure. Commonly cited impurities include the proposition's charm, credibility, desirability, familiarity, fashionability, salience, and simplicity.

c. *Bias?*   Thought experiment has been alleged to be biased *against* complex processes, theories that stress origins, and flexible thinking. It is said to *favor* aesthetic values, familiar facts, and the theoretician's own pet theories. A couple of these charges have been lodged against the use of hypothetical dilemmas in moral psychology. Lawrence Kohlberg has shown that these stories push subjects toward more abstract principles.

> In Europe, a woman was near death from cancer. One drug might save her, a form of radium that a druggist in the same town had recently discovered. The druggist was charging $2,000, ten times what the drug cost him to make. The sick woman's husband, Heinz, went to everyone he knew to borrow the money, but he could only get together about half of what it cost. He told the druggist that his wife was dying and asked him to sell it cheaper or let him pay later. But the druggist said, "No." The husband got desperate and broke into the man's store to steal the drug for his wife. Should the husband have done that? Why?[26]

The forced choice prompts many to resolve an inconsistency between the prohibition against theft and the duty to save the lives of loved ones. Subjects tend to adopt an overarching principle that ranks the two considerations. Further hypotheticals lead to further sophistication. This progression leads Kohlberg to postulate stages of moral development. Carol Gilligan complains that Kohlberg's methodology winds up praising those who model moral dilemmas on mathematical puzzles.[27] This ends up as a sex bias because males are more apt to accept the situation as "given" and apply impartial principles. Females tend to rewrite the dilemma and apply a partialist ethic of care. Instead of accepting the given, as one must in math, females change the story so that there is a third alternative. (Maybe there is another cure. Maybe Heinz can get a government loan, etc.) Kohlbergians dismiss this inventiveness as immature evasion. But Gilligan applauds the tendency to challenge the problem definition. She grants that there may be merit in making hard choices but insists that there is also merit in questioning the premises that seem to force the hard choice.

Absolutists also balk at exotic *What if?s*. G. E. M. Anscombe, a well-known defender of Catholic ethics, avers that it is always wrong to wittingly punish the innocent. At one of her lectures, a member of the audience asked whether we should stick to this rule even at the cost of catastrophe. Suppose, for example, that the only way to avoid a threatened nuclear war was to try, sentence, and execute an innocent man. Would punishing the innocent not be permissible in these extreme circumstances?

> It would seem strange to me to have much hope of so averting a war threatened by such men as made this demand. But the most important thing about the way in which cases like this are invented in discussions, is the assumption that only two courses are open: here, compliance and open defiance. No one can say in advance of such a situation what the possibilities are going to be—e.g., that there is none of stalling by feigned willingness to comply, accompanied by a skillfully arranged "escape" of the victim.[28]

Anscombe rejects the question. Others accept the question but grumble. Bernard Williams, for instance, lists the artificial restrictiveness of moral hypotheticals as the first of two ways they beg important questions: "One is that, as presented, they arbitrarily cut off and restrict the range of alternative courses of action. . . . The second is that they inevitably present one with the situation as a going concern, and cut off questions about how the agent got into it, and correspondingly about moral considerations which might flow from that."[29] Williams thinks that we just have to eat these problems because any discussion of ethics requires consideration of how one *would* think or feel. His main reason for raising the second difficulty is to post notice that thought experiment has a bias in favor of a view he is in the midst of attacking: utilitarianism. This doctrine defines 'right' as the act that maximizes good consequences and is thus a forward-looking view. It says that the past is relevant only insofar as it helps to predict the future. The method of thought experiment inadvertently reinforces this tendency to ignore the past by making questions of origins irrelevant (as they are in fiction: Who was Sherlock Holmes's great-grandfather?).

d. *Is thought experiment an intrusive measure?*   Years ago, a physician built a reputation as a diagnostician, especially in detecting the onset of typhoid fever.[30] His method was to feel the patient's tongue with thumb and forefinger. Colleagues and students would follow him on rounds in amazement because his large collection of typhoid diagnoses would nearly all be vindicated about a week later. But awe boiled off when they learned that the secret of his success was contaminated fingers.

The story illustrates an *intrusive test*—one that significantly affects the very property it was designed to measure. The results of a faithful test should be an *effect* of either a cause of the property or of the property itself. When the tested property is caused by the test, the test is intrusive and thereby an unfaithful recorder.

Perhaps thought experiments tamper with what they are intended to measure. An individual hypothetical often leads people to form opinions on previously uncrystallized issues. If its purpose is to *reveal* our beliefs about a certain topic, then we court the danger of *creating* the very beliefs the test was intended to detect. The impact of thought experiments is magnified when they are presented collectively. Students staggering down a gauntlet of hypotheticals reverse earlier judgments to avoid inconsistency. Numbed but still nimble, they volunteer generalizations and corollaries of their increasingly elaborate system of beliefs. As pollsters and Socratic teachers appreciate, questioning affects the beliefs of the questionee.

I have also heard some philosophers grouse that thought experiments alter what is possible. Their idea is that some statements express *borderline* possibilities that can be precisified one way or another. When an opinion leader such as Putnam publishes a thought experiment with the verdict that it is indeed possible for cats to be robotic spies, other philosophers jump on the bandwagon. Since the meaning of a word depends largely on what the linguistic community believes it to mean, the bandwagon effect changes the borderline possibility of robot cats into a clear possibility. Hence, according to this criticism, Putnam's thought experiment winds up with the correct verdict—but only because it causes robot cats to become possible.

Servility is not the only cause of our sheeplike behavior. The principle of charity tells us to interpret people in a way that maximizes their rationality— hence our agreement with them and therefore the truth of their beliefs. Thus, when someone says 'France is hexagonal', we adopt loose standards that make the utterance come out true. If he had instead said 'France is not hexagonal', stricter standards would have been adopted. So by giving speakers the benefit of the doubt, we grant them a degree of discretion over borderline cases. This enables a thought experimenter, says the sceptic, to parlay a borderline modal claim into a widely acclaimed insight. Of course, the thought experimenter has to worry about competition from rival thought experimenters who try to precisify the case in the opposite direction. Whether you prevail in your attempt to colonize the indeterminacy depends on the social dynamics of the confrontation. Did you get in the first word? How big was your audience? Can you browbeat opponents into silence? Such is the politics of modality.

The bandwagon effect can work even if perceived as such by rational and well-informed speakers. Believing that the others are persuaded that the meaning is x justifies my belief that the meaning is x regardless of whether I think the others were converted by fashion or bravado. There might even be a scientific role for thought experiments of the sort Thomas Kuhn assigns to dogma.[31] Scientists need to limit their search space and coordinate their inquiry. Otherwise, they become as impotent as a scattered army. A thought experiment that makes scientists think alike will provide the needed organization. The important matter is not so much *how* the borderline cases get decided as *that* they get decided. If bluster is needed to keep the troops in line, then so be it.

The sceptic about thought experiments can grant that dictatorial thought experiments might have these Kuhnian virtues. After all, his doubts concerned its power to *justify* beliefs. The sceptic was always free to concede that they might have prudential value or that they might help science in the way sociologists say a Protestant work ethic helps science. Such contributions are cold comfort to the friends of thought experiment because they need to show that their method delivers beliefs with *intrinsic* epistemic value, not ones that are merely useful as a means to the production of other valuable beliefs.

e. *Are we falsely assuming that conceivability implies possibility?* Traditional scepticism chiefly targets the ambitious sort of thought experiment that purports to be more than a device for revealing hidden preferences or beliefs. The targeted thought experiments are designed to demonstrate possibilities and necessities. To power them up to these modal conclusions, we must add premises linking conceivability with possibility. Since the debate about the adequacy of these premises is well developed, I need to follow the dialectic down pretty deeply to reach the major contributions to the issue. Indeed, I need to go so far that I might as well break policy and state my position toward the end rather than lead everyone through the catacombs again.

Many philosophers assume that conceivability and possibility entail each other. Hume endorses the equivalence: "'Tis an establish'd maxim in metaphysics, *That whatever the mind clearly conceives includes the idea of possible existence*, or in other words, *that nothing we imagine is absolutely impossible.* We can form the idea of a golden mountain, and from thence conclude that such a mountain may actually exist, We can form no idea of a mountain without a valley, and therefore regard it as impossible."[32] Hume's maxim is a conjunction of two conditionals:

H1   If p is possible, then p is conceivable

H2   If p is conceivable, then p is possible.

A philosopher might affirm one and deny the other. Belief that imagination is boundless commits you to accepting H1 but rejecting H2. The reverse position beckons to philosophers who picture 'conceive' as a success word, as meaning the grasping of a state of affairs. Since a state of affairs is associated with a possibility, your ability to grasp it would establish possibility, hence H2. Other states of affairs might lie beyond reach of any conceiver but nevertheless exist, thereby refuting H1. An intermediate position warrants rejection of H1 *and*

H2. Suppose that our imagination is limited, but not by consistency. Then there will be things conceivable but impossible and things possible but inconceivable.

i. *Appeals to 'Possibility implies conceivability'.*   Appeals to H1 tend to be *modus tollens* impossibility arguments: if p is possible, then p is conceivable; but p is inconceivable, hence p is impossible. For example, H1 has figured as a premise in arguments to show that spatial properties are essential to perception. Although colorless and odorless perceptions are easy to imagine, it is impossible to visualize objects without assigning them position and extension. So perception theorists conclude that to perceive is to perceive spatially. The freewill debate provides a second instance. After reviewing disagreements about whether deliberation is an activity, Peter van Inwagen adds:

> But all philosophers who have thought about deliberation agree on one point: one cannot deliberate about whether to perform a certain act unless one believes it is possible for one to perform it. (Anyone who doubts that this is indeed the case may find it instructive to imagine that he is in a room with two doors and that he believes one of the doors to be unlocked and the other to be locked and impassable, though he has no idea which is which; let him then attempt to imagine himself deliberating about which door to leave by.)[33]

The conservation of the center of gravity is demonstrated in textbooks with an appeal to physical inconceivability. Imagine a man in the middle of a frozen lake whose surface is perfectly smooth. Wind, magnets, and all other external forces are absent. Although it initially seems easy for the man to leave the lake, one eventually realizes that if he moves part of his body in one direction, the rest will move in the opposite direction (by Newton's Third Law).

ii. *Appeals to 'Conceivability implies possibility'.*   The favorite use of thought experiment is to establish a possibility. This emphasis on fresh possibilities gives thought experiment a reputation as a liberating enterprise; we discover more degrees of freedom. Love of liberty confers a diffuse positive aura over the imaginative exercise. But since the possibility is usually marshaled as a counterexample to some promising definition or entailment thesis, the effect tends to be the negative one of refutation. Consider how dualists attack behaviorists. Behaviorists insist that mental discourse reduces to claims about actual or possible behavior. This implies that you must have a body to exist. However, we can conceive of disembodied existence. Imagine leaving your body and observing events from a perspective within the Kremlin. Thus, concludes the dualist, it is *possible* to exist without a body, so behaviorism is false.

iii. *Objections to 'Possibility implies conceivability'.*   John Stuart Mill attacked the principle by citing how the task of imagining a round earth stretched their limits. Obviously, the caps on their thinking did not constrain the shape of our planet. Our own thinking caps are just less obvious to us.

Thus, many thought experiments seem cogent just because of our lack of ingenuity. For example, the hypothetical about the man in the middle of the lake looks decisive until a wiseguy points out that the "stranded" man can blow his away across.

A second set of objections springs from the thesis that our imagination is determined by perceptual opportunities. Since even the most well rounded life will be experientially incomplete, there are sure to be pockets of gratuitous inconceivability. The point can be made more dramatically by considering animals with qualitatively different perceptual abilities. Fish have lateral lines along the length of their bodies that detect water pressure gradients. Since we do not, we cannot imagine what is it like to be a trout. So there probably are possibilities that lie beyond our ken because we do not have the requisite perceptual abilities.

A third class of counterexamples to 'Possibility implies conceivability' turn on conceptual presuppositions. We are now able to ask questions that our intellectual ancestors could not ask.[34] The reason is that we have concepts they lack. Ptolemy could not ask—let alone answer—the question of whether black holes exist. He did not know enough to ask. Likewise, we are not in a position to ask questions our intellectual descendents will ask. Since every stage of inquiry has its unaskables, it follows that there will always be ineffable questions. If we cannot ask the question, we cannot conceive the possible answers. Since these possible answers are possibilities, it follows that there will always be (at least temporarily) inconceivable possibilities.

Hume's best reply to these objections would be a charge of equivocation. Some of the charges, such as Mill's, confuse conception with other propositional attitudes such as belief and visualization. It was a cinch for the ancients to *conceive* that the earth was round but hard to *believe* it. Hume was willing to grant that a great many things are far more incredible than this. For instance, none of us can believe that bricks exist only when perceived or that the future will not resemble the past. But there is no problem in *conceiving* of these possibilities. Confusion between conception and visualization is responsible for another "counterexample" to 'Possibility implies conceivability'. This one rests on our alleged inability to imagine the inside and the outside of a barn simultaneously. Since we are highly visual animals, we try to perform the task by visualizing the scene. When we find ourselves unable to do this, we hastily conclude that the task is impossible. But conception and imagination does not have to be by visualization. Visualization is the technique of conceiving a state of affairs by imagining how it would look. Although we do not have a word for it, we also conceive of states of affairs by imagining how they would sound or feel or smell.

We can illustrate how conceivability need not involve visualization with Peter Strawson's acoustical thought experiment.[35] The ultimate question he wishes to answer is whether there could be particular things without there being material things. His thought experiment's purpose is to establish a lemma, namely, that nonspatial experience is possible. Imagine what it would be like to have hearing but no other senses. Unlike touch or sight, sound has no intrinsic spatial character. Of course, correlations can endow it with *extrinsic* spatial character. For example, the Doppler effect lets us infer whether a train is coming or going on the basis of sound alone. But we must learn this by correlations with what we see or touch. A purely auditory person would not be able to exploit the Doppler effect or any correlation linking sound and spatial position. Since the person would lack the notion of spatial position, he would not have the notion of spatial objects. Since material things are extended in space, any notion of particular things would have to be obtained without the help of thoughts about material objects. So by showing that this sound-only person could nevertheless derive the concept of particular things, one can show that material objects are not basic to all conceptual schemes that contain the notion of particulars.

Some of the other objections to "Possibility implies conceivability" recognize the difference between *impossible to conceive* and *difficult to conceive*. It is difficult to conceive of a one-inch cube serving as a tunnel for a slightly larger cube. But the principle only says that if p is possible, then it is also *possible* to conceive of the possibility. Other objections equivocate between 'possible for anyone to conceive' and 'possible for someone to conceive'. Hume himself granted that a person who had never tasted pineapple could never conceive its exact taste. Mental handicaps may curtail the range of conceivability for a particular person. The defender of 'Possibility implies conceivability' could even grant that some possibilities can only be grasped by elite groups of thinkers or a single individual. Perhaps that one person is epochally precocious, like Evangelista Torricelli, who in 1641 showed how a solid of finite volume can have an infinite length. But one is enough. Indeed, the weakest interpretation of the principle only says that for every possibility there is a *possible* conceiver of that possibility, formally, $(p)[\lozenge p \supset \lozenge (\exists x)Cxp]$. (For the $\lozenge$, see p. 135).

This weakest interpretation of "Possibility implies conceivability" makes any specific counterexample self-defeating. If I tell you 'p is possible but inconceivable', you can point out that if either of us understands the sentence, it is false. To understand the first conjunct, 'p is possible', I must conceive of the possibility of p. On the other hand, if neither of us understands the sentence stating the counterexample, how can we know it is a counterexample? The difficulty is analogous to the one faced by those who believe that there are ineffable truths: if they are right, they cannot state the example.

Although stressing 'Possibility implies *the possibility* of conception' has the virtue of disarming many counterexamples, it weakens the principle's value as a *test* for impossibility. Failure to conceive can be explained away as merely due to a deficit of ingenuity. Even so, most opponents of 'Possibility implies conceivability' are not content to accuse it of being true but useless. They want to show it is false. Since they cannot refute it by counterexample, they must rely on other strategies.

It is indeed possible to refute a universal generalization without giving a specific counterexample. Recall Mark Twain's proof that not all of Europe's reliquaries contain genuine fragments of Jesus' cross: when added together, the reliquaries have enough wood to make more than one cross. Again, if I say 'All the books on this library shelf are in order', you can refute me by pointing out that two copies of *Breads of France* sandwich *American Baloney*. You need not know that copy 1 is out of order or that copy 2 is out of order to know that at least one of them is. A third way of refuting a universal generalization is by analogy. You might point out that the past ten library shelves have had some books out of order.

One of the most popular analogical arguments against 'Possibility implies conceivability' is evolutionary. Humans conceive of possibilities that no other animals on earth can conceive: the constancy of the speed of light, reincarnation, Madonna as First Lady. Hence, all of these animals have imaginative limits; for each of them, there is a possibility of which they cannot conceive. By analogy, therefore, human beings have imaginative limits. The only way to avoid this conclusion is to suppose that humans have passed a cognitive threshold at which all possibilities become conceivable. But to suppose that we alone are "over the hump" is arbitrary. Evolutionary biology everywhere emphasizes our continuity with other animals. Thus, an unblinking application of naturalism requires us to admit pockets of darkness, maybe even vast regions uninhabitable by the human intellect.

Objections to 'Conceivability implies possibility'. Most criticism has been directed at H2. Contemporary sceptics of thought experiments in studies of personal identity, reference, and science argue that imagination overflows the banks of logic. Witness our comprehension of sloppily told time-travel stories and incoherent religious convictions (such as the doctrine of the trinity and transubstantiation). Examples are readily exhumed from the history of mathematics. Antoine Arnauld agreed that Descartes conceived the mind as distinct from the body but asked him to consider the possibility "[that your] conception is inadequate when you conceive yourself as a thinking but not extended thing."[36] Arnauld pointed out that a man could conceive of a right triangle and yet firmly deny that the sum of its sides is equal to the square of its hypotenuse.

Intellectuals are not the only ones who have overestimated what is logically possible. Sam Lloyd offered fabulous prizes for anyone who could achieve an exact reversal pattern with the famous *Fifteen Puzzle*. The puzzle consists of fifteen blocks in a square tray. The object of the game is to slide the blocks into a designated sequence. Since one can easily manuever the blocks into a formation close to the one specified by Lloyd, excited players thought that his prize offer was recklessly generous. But Lloyd was risking nothing. No sequence of moved blocks leads to the winning arrangement. From the starting position, you can get close, but never all the way there (Figure 2.2).

Psychologists have performed experiments that claim to detect a variety of inconsistencies in our spatial imagery. For example, town A may be judged as northeast of town B but B judged as not southwest of A.[37] Philosophers who accept 'Conceivability implies possibility' would have to condemn this research on a priori grounds.

This commitment may seem hopelessly high-handed, but there are two ploys worth examining. The first distinguishes between *basic* conceivability

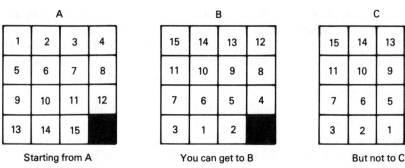

Figure 2.2

and *derived* conceivability so that 'Conceivability implies possibility' applies directly to some propositions but only indirectly to the rest. For example, Albert Casullo has argued that we should apply Hume's distinction between intuition and demonstration.[38] In the case of propositions knowable by intuition, the mind can "see" the relationship between the ideas. The mind lacks this ability for other propositions and so must learn them demonstratively through intuitive steps. But how do we know which propositions are knowable by intuition? Without a selection principle, the revision will not be a practical test.

The second defence of 'Conceivability implies possibility' relies on conceptual paternalism: you deny that I ever imagined the impossible scenario and assert that I instead imagined a highly similar situation. At first, this sounds presumptuously sophistical: who are you to tell *me* what I imagined? Our indignation smacks of lingering allegiance to a corollary of the incorrigibility of introspection: if I believe that I am imagining that p, then I am imagining that p. But the principle is refuted daily in college classrooms. Students listening to philosophy lectures on the problem of other minds often report that they can imagine feeling another person's pain. The instructor counters that the student is only imagining feeling a qualitatively similar pain, not the same particular pain. Other students claim to be able to imagine the falsehood of the law of noncontradiction. They invite the instructor to imagine a wife who both loves her husband (because he defends her honor) and hates him (because he has broken her jaw). But the instructor replies that the imagined wife loves and hates different aspects of the husband.

For a more controversial counterexample to the incorrigibility of beliefs about the content of your imagination, turn to the debate over whether resurrection is possible. Christians claim to be able to imagine themselves deconstituted and then reconstituted after a long period. Unbelievers say the Christians only imagined the formation of a duplicate person from the old parts. A similar debate occurs between those who deny the possibility of time travel and those who assert that they can imagine a time traveler sprinkled through a "block" universe in which time is regarded as the fourth dimension. Lastly, consider disputes over whether a person can try to do something he knows to be impossible. One side will say that joggers who warm up by pushing against buildings are trying to move the building even though they know the feat is impossible. The other side redescribes the exercise as that of *pretending* to try moving the building.

Since conceptual paternalism is compatible with the view that imagination can extend beyond possibility, we have a choice of interpretations. If p is impossible and a thinker believes he has imagined p, we can blame either his belief or his imagination. However, champions of 'Conceivability implies possibility' must be extreme paternalists and always blame the belief about what we imagine. This policy has been motivated with an analogy to perception.[39] We don't think the world could be inconsistent, so we pin the blame on what we believe we see, rather than what we see. If p is impossible but a perceiver believes he is seeing p, he is simply wrong about what he sees! So if imagination stands to *possibilia* as perception to external objects, one cannot imagine impossibilities.

The price of defending 'Conceivability implies possibility' with extreme conceptual paternalism is paid at the level of application. If my beliefs about the content of my supposition are fallible, I cannot make easy use of the principle I sought so hard to defend. I now need extra, noncircular tests to confirm my beliefs about what I have imagined. Casullo's solution also complicates application because I need assurance that my imaginings are basic ones. In any case, we can see that 'Conceivability implies possibility' is open to serious challenge at the level of application, if not truth. At best, apparent conceivability establishes a prima facie case for possibility. It is not a sufficient condition. We may open a modal inquiry with a casual appeal to what we can imagine, but we cannot close it.

My opinion is that conceivability and possibility are only linked by a statistical law. People conceive of states of affairs that are impossible, but these errors are far less common than their correct beliefs. A sceptic might say that I have no grounds for my optimism. How do I know that any of my modal beliefs are correct? Compare this question with the challenge to memory: How do I know that any of my beliefs about the past are correct? If the range of his doubt is sufficiently comprehensive, then the sceptic "wins" because he deprives the opposition of premises. The sceptic does not achieve much of a victory: he himself finds his thesis incredible. Furthermore, this kind of scepticism is too wide to be deep; no interesting evidential connections are illuminated. This is not to say that we have to take the reliability of our modal judgments on faith. Once the sceptic eases up, justifications flow (as we shall see in later chapters).

II. A Repackaged Appeal to Ordinary Language?

In addition to its disquieting likeness to introspection, thought experiment is suspiciously reminiscent of the appeal to ordinary language. However, this resemblance is not so alarming, because the appeal to ordinary language is viewed with less contempt. Nearly all philosophers assign it some weight.

Nevertheless, there are widely varied opinions as the appeal's proper weight, its scope, and the way it should proceed. The suspicion is that philosophers who describe what they are doing as a "thought experiment" are just smuggling an appeal to ordinary language past our distracted scruples. Whenever an argumentative move becomes encumbered by accumulating methodological reservations, enterprising thinkers reclothe the gambit in new terminology. The makeover breathes new life into the move because it now passes unrecognized through the sequence of checkpoints that have arisen as quality control measures against the old, familiar move. In a nutshell, the charge is that *thought experiment* is a euphemism for *appeal to ordinary language*.

A. How the Appeal to Ordinary Language Is Supposed to Work

Centuries of philosophers have made appeals to what we would say. But the frequency of the appeal and the weight attached to it dramatically increased with the rise of ordinary language philosophy after World War II.

The appeal to ordinary language is designed to distinguish between empirical regularities and logical implications. Statements such as 'Owl eyes are bigger than human eyes' report perfect correlations that are not logical implications. Others, such as 'Every wife has a husband' are clearly expressing logical relationships. But in other cases, it is hard to tell whether the connection between the properties is logical or empirical: 'Every man is a son', 'Every echo is started by another sound', 'We see with our eyes'. The appeal to ordinary language is a test for these hard cases. In essence, the method amounts to asking your informant whether he would ever be willing to apply F in the absence of G. For instance, to test 'Every man is a son' we may ask the informant whether he would call Emo Phillips a man if he learned that Emo was grown in a vat from inorganic ingredients.

B. Strong Scepticism About the Appeal to Ordinary Language

Sworn enemies of this method insist that linguistic intuitions are no more trustworthy than grassroots sentiments about other matters. Just as we do not ask ordinary people for their intuitions about mending bones and tracking satellites, we should not ask for their opinions of English. Folk linguistics is no better credentialed than folk medicine. Linguists warn of a vast reservoir of myths and fallacies about language. For example, native speakers underestimate their language's degree of conventionality, reverse the priority of oral language over written, and chauvinistically adhere to easily refuted rules such as "Never end a sentence with a preposition." The man in the street confuses the properties of words and their referents and arbitrarily regards upper-class dialects as setting the standard for correct usage. Some of the more repressive members of the linguistic community go on to propose bans against words such as *ain't*, *hopefully*, and *irregardless*, while syntactic Stalinists press the attack to punctuation marks such as the comma and semicolon. Why ask these yahoos for their opinion of linguistic rules?

Fluency alone does not enable one to assemble clear examples, to form promising linguistic hypotheses, or to test semantic generalizations. Nevertheless, the distinction between metalinguistic competence and linguistic competence should not be overdrawn. Cooks and musicians are also prone to myths and fallacies about their specialities. But their experience makes them more qualified than an outsider to describe the structure of their activities. Similarly, the speaker's intuitions are the beliefs of an insider. He has mastered the rules constituting the language; hence his judgment is a (sometimes distorted) reflection of those rules. Since the rules determine the meanings of the terms in question and the meanings determine logical implication, the appeal to ordinary language is a good (but, of course, not perfect) test for logical implication. Compare the evidentiary value of the appeal to ordinary language to the information obtained from people who learn dances, etiquette, and chess by immersion. Although such insiders have problems articulating mastery of the rules they have absorbed, they are fairly reliable sources. Philosophers forewarned of linguistic fallacies are forearmed against miscodifications. Hence, we should heed their appeal to ordinary language.

C. Moderate Scepticism About the Appeal to Ordinary Language

Many critics of the appeal to ordinary language are only sceptical about a subset of the appeals. William Labov rejects appeals that emanate from the theoretician's own intuitions because of the danger of bias.[40] Labov is willing to count the judgments of naive subjects as data and so is sympathetic to attempts to develop well-controlled linguistic interviews.[41] Another worry is people's accuracy in predicting their own speech. So language elicitation experiments strive to prompt natural speech.

A more popular division between acceptable and unacceptable appeals is by bizarreness. For example, Jerry Fodor accepts appeals having mild suppositions but is sceptical about appeals to ordinary language that require informants to suppose central beliefs to be mistaken.[42] The double standard is promoted by an application of Quine's *meaning holism*. Meaning holism is the view that statements lack independent meanings, that they only have meaning within a large group of statements just as *syncategorematic* terms (*or, not, if, then, some*) only have meaning within a full sentence. For example, 'Don remarried Meg' depends on the meaning of 'divorce' which is in turn linked to 'marriage' and thereby to 'promise'. If the intentional element of 'promise' faded away, so that promises were just predictions about one's own behavior, there would be a corresponding shift in the meaning of 'marriage', 'divorce', and therefore in 'Don remarried Meg'. Although each statement affects the meaning of all the rest, they vary in influence. Quine describes our "web of belief" as being composed of virtually irrevisable convictions at the center and progressively more negotiable beliefs near the periphery. Changes at the outer edge do not upset the web's stability; but changes close to the center are disruptive, requiring a complicated set of readjustments to regain equilibrium. It is easy to see what follows from a mild supposition, because we are imagining what happens when a peripheral belief is changed. But it is virtually impossible to see what follows from a radical supposition, because we are imagining a revision of a central belief.

The use of strange scenarios has also been attacked on the grounds that it creates interference between our semantic competence and other elements of our psychology. For instance, Alvin Goldman contends that our linguistic judgments are only partly governed by our grasp of the term's meaning.[43] Speakers are also influenced by a host of psychological factors that enhance linguistic efficiency: stereotypes, contextual salience, general heuristics, and so on. In most contexts, these facilitators work together with our semantic knowledge. But in peculiar cases, they get in each other's way. For instance, normal discussion of bachelors is guided by an idealized model of bachelors as men who can get married but choose to lead a self-sufficient, solitary life. This prototype tends to exclude marginal bachelors such as priests, homosexuals, and eunuchs even though they conform to the definition of 'bachelor'. The prototype also gives us a bum steer by including as bachelors men who have been separated from their wives, men with marriages of convenience, and so on. Thus, Goldman concludes that fringe cases elicit garbled judgments because they break up the collaboration between our understanding of meaning and our nonsemantic means of understanding.

Since Goldman and Fodor allow mild suppositions, there is still room for many of the best-known appeals to ordinary language. J. L. Austin's illustrates the appeal in "A Plea for Excuses." To refute the claim that 'mistake' and 'accident' are synonymous, Austin told his famous donkey story:

> You have a donkey and so do I and they graze in the same field. The day comes when I conceive a dislike for mine. I go to shoot it, draw a bead on it, fire: the brute falls in its tracks. I inspect the victim, and find to my horror that it is *your* donkey. I appear on your doorstep with the remains and say—what? "I say, old sport, I'm awfully sorry, etc. I've shot your donkey . . . *by accident*? or *by mistake*"? Then again, I go to shoot my donkey as before, draw a bead on it, fire—but as I do so, the beasts move, and to my horror yours falls. Again the scene on the doorstep—what do I say? "by mistake"? or "by accident"?[44]

A second well-known illustration from Austin is directed against the thesis that succumbing to temptation implies losing control of yourself.

> I am very partial to ice cream and a bombe is served divided into segments, corresponding one-to-one with persons at the High Table: I am tempted to help myself to two segments and do so, thus succumbing to temptation and even conceivably (but why necessarily?) going against my principles. But do I lose control of myself? Do I raven, do I snatch the morsels from the dish and wolf them down, impervious to the consternation of my colleagues? Not a bit of it. We often succumb to temptation with calm and even with finesse.[45]

Fodor would also allow suppositions that involve merely improbable events. For instance, Putnam has us imagine an ant crawling on a patch of sand.[46] By pure chance, the path in the sand forms a recognizable caricature of Winston Churchill. Is this a *depiction* of Churchill? Since most people answer *no* on the grounds that the ant does not intend the lines to be taken as representing Churchill, Putnam concludes that resemblance is insufficient for depiction. This inference is acceptable to the holist because a merely unlikely event does not violate a natural law or alter important contingencies, so it does not overburden the questionee with the superhuman task of predicting the right recovery from a cognitive cataclysm.

D. Semantic Descent to Thought Experiments

What is the difference between an appeal to what we would say and a thought experiment? The critic suggests the following answer: the mechanics of the appeal to hypothetical speech are clearer. One ascertains the rules of the language by checking how a native would talk in a particularly telling situation. This harmonizes with the analytic's hard-core passion for semantic ascent: the policy of reformulating questions about things into questions about talk about things. Questions about the result of a thought experiment are questions as they stand prior to the application of this illuminating strategy. Hence, a thought experiment, in the eyes of a stout linguistic philosopher, is just an unduly obscure ancestor of an appeal to what we would say. Hence, any problem for an appeal to hypothetical speech is a problem for thought experiment.

The critic supports his allegation that old wine was poured into new scientistic bottles by showing that it is just as rancid. In particular, meaning holism is just as devastating to bizarre thought experiments as it is to bizarre appeals to ordinary language. In its Fodor form, it amounts to an infeasibility thesis. Fodor grants that we have revised central beliefs in the past. However, their consequences were worked out by battalions of intellects over decades, not by one philosopher in an afternoon. So the problem with bizarre thought experiments does not directly concern their hypothetical status. Instead, they are defective in the way of verification procedures that give rise to a "combinatorial explosion" of steps. Thus, Fodor could grant that the thought experiment has an answer, even an answer knowable in principle. His take-home message need only be that the answer is too expensive.

Other meaning holists say that radical suppositions are *meaningless*, rather than unworkable. Wittgenstein discourages fascination with mental freak shows:

> It is only in normal cases that the use of a word is clearly prescribed; we know, are in no doubt, what to say in this or that case. The more abnormal the case, the more doubtful it becomes what we are to say. And if things were quite different from what they actually are—if there were for instance no characteristic expression of pain, of fear, of joy; if rule became exception and exception rule; or if both became phenomena of roughly equal frequency—this would make our normal language-games lose their point. The procedure of putting a lump of cheese on a balance and fixing the price by the turn of the scale would lose its point if it frequently happened for such lumps to suddenly grow or shrink for no obvious reason.[47]

Wittgenstein boosts this operationalist theme with the claim that concepts depend on extremely general facts of nature. (Presumably, the connection is indirect: general facts prompt shared beliefs, which provide the necessary backdrop for coordinated rule following.) The very generality of facts such as the stability of memory, emotion, and artifacts and the distinguishability of human bodies puts them beyond discussion. Language requires this fundamental, unspoken consensus: "It is as if our concepts involved a scaffolding of facts—that would presumably mean: If you imagine certain facts otherwise, describe them otherwise, than they are, then you can no longer imagine the application of certain concepts, because the rules for their application have no analogue in the new circumstances."[48] This conservatism about thought experiment appears to be at odds with Wittgenstein's practice. Unlike Austin, who sticks close to home in his hypotheticals, Wittgenstein regularly supposes that things are radically different. He was a Marco Polo of the imagination, not a J. Alfred Prufrock. Wittgenstein imagines surgeons discovering him to be brainless, a man whose expressions of sorrow and joy alternated with the ticking of a clock, people who look alike but have their features migrate from body to body. Consistency could be achieved if we suppose that Wittgenstein only had a soft-hearted scepticism about thought experiment. For instance, James Broyles says that the function of Wittgenstein's bizarre thought experiments is to promote an awareness of "scaffolding facts."[49] This interpretation fits nicely with Wittgenstein's career-long conviction that some kinds of nonsense are illuminating. Bizarre thought experiments would then *show* important facts without *proving* anything, because only a meaningful supposition can entail consequences.

But the hard-headed sceptic will see no need for nonsense. According to him, the appeal to what we would say at least had the advantage of honesty. Thought experiment is just a bad-faith substitute.

III. Thought Experiments and the Dilemma of Informativeness

Thought experiments raise a problem of informativeness akin to that raised by analysis and theoretical terms. The dilemma for analysis is that either the analysans are synonymous with the analysandum (making the analysis trivial) or are not synonymous (making the analysis false). The dilemma for theoretical terms is that either they serve their purpose of indirectly establishing connections among observables (in which case they can be dispensed with in favor of laws and interpretative statements that directly establish those connections) or they do not serve this purpose (in which case they are just as dispensable).[50]

The dilemma for thought experiments is rooted in their evidential inferiority to public observation and experiment. Consider again the medieval vacuum debate. Franciscus Toletus deployed the following thought experiment against the void:

A very strong jar full of hot water is taken, and let it be well closed; then let it be in a very cold place. Either the water will freeze or it will not. If not, it seems absurd since the natural agent should necessarily perform its action. If it does freeze, then the water will occupy a small place because of its condensation and the rest of the jar will remain void.

. . . I admit that the water should become cold and should freeze, nevertheless this could not be without the evaporation of subtle vapors, which would fill the place left by water.[51]

Other plenists were willing to say the water fails to solidify because they thought that coldness is as good at condensing vapor as it is at solidifying water. Still others insisted the jar would implode. Domingo de Soto averred that implosion would occur even if the jar was replaced with a perfectly spherical iron container. Vacuists such as Bernardino Telesio and Francesco Patrizi, on the other hand, maintained the container could be reinforced and that the cold would ensure both solidification of the water and condensation of any vapor. As ingenious as these thought experimenters were, both sides overlooked the fact that freezing water *expands*. We should not sit and think when we could look and see!

Galileo eloquently expresses the primacy of observation in a letter:

Oh, my dear Kepler, how I wish that we could have one hearty laugh together! Here at Padua is the principal professor of philosophy, whom I have repeatedly and urgently requested to look at the moon and the planets through my glass, which he pertinaciously refuses to do. Why are you not here? What shouts of laughter should we have at this glorious folly? And to hear the professor of philosophy at Pisa labouring the Grand Duke with logical arguments, as if with magical incantations, to charm the new planets out of the sky.[52]

Carl Hempel, the closest philosophers of science have to a patriarch, provides a modern endorsement of the primacy of observation. Hempel defines "theoretical thought experiments" as ones that rest on explicitly stated principles and strict deductions. These are just as good as any derivation of a scientific theorem. However, such completeness and explicitness is rare. Intuitive thought experiments lie at the opposite extreme: underlying principles are left unarticulated, and the inferences are made casually. Hempel locates most thought experiments between these two extremes of the theoretical and the intuitive. Since nearly all thought experiments are significantly intuitive, they only have heuristic value. Hempel acknowledges that thought experiment is a fruitful method of *discovery* but emphasizes that

of course, intuitive experiments-in-imagination are no substitute for the collection of empirical data by actual experimental or observational procedures. This is well illustrated by the numerous, intuitively quite plausible, imaginary experiments which have been adduced in an effort to refute the special theory of relativity; as for imaginary experimentation in the social sciences, its outcome is liable to be affected by preconceived ideas, stereotypes, and other disturbing factors. . . . These could evidently affect the outcome and defeat the purpose of intuitive thought experiments in sociology. Such experiments, then, cannot provide evidence pertinent to the test of sociological hypotheses. At best, they can serve a heuristic function: they may *suggest* hypotheses, which must then be subjected, however, to appropriate objective tests.[53]

By placing all but a limiting case of thought experiment outside the context of justification, Hempel relegates their study to the history and psychology of science. Those who believe that thought experiments *justify* and *test* hypotheses face a dilemma that we can formulate in terms of their connection with public experiments:

1. If a thought experiment can be checked through public experimentation, then not actually checking leaves the results unverified, and an actual check would render the thought experiment redundant or misleading.
2. If a thought experiment cannot be checked through public experimentation, then its results are unverifiable.
3. Any experiment having results that must be either unverified, redundant, misleading, or unverifiable is without scientific value.
4. No thought experiment has scientific value.

Those sceptical of thought experiments can concede that they are fun or beautiful or revealing in the way that one's fantasies and dreams are. But these sorts of value are not cognitive values and so do not secure a scientific role for thought experiment.

The spirit of scepticism about thought experiments comes through in slogans such as "Science is about the real world, not fictional ones." Behind the slogans dwell doubts about reliability, replicatability, and relevance. Many scientists view preoccupation with thought experiments as a sign that one has been sidetracked. Thought experiments, they would say, belong to the *rhetoric* of science. A scientist, like anyone seeking influence, must maximize the accessibility of his work. So he may clothe his strict arguments with colorful stories. But one should not mistake the glove for the hand that does the real work. Thought experiments are eliminable: none need appear in science.

Some historians of science are convinced by this line of reasoning and so try to explain away the popularity of thought experiments. According to Pierre Duhem, the glut of thought experiments in physics is a legacy of Newton's inductivism.[54] This methodology maximizes the importance of experiment and minimizes the role of theory. However, inductivism's demand that every scientific claim be backed by an empirical demonstration is impossible to satisfy. So the physicist tries to meet his obligation with a "fictitious experiment." His prediction about the hypothetical outcome of this experiment is based on his theoretical expectations, rather than observation. But these expectations are based on the very principle that is to be tested by the imaginary experiment! The only way to break out of this circle is actually to perform the experiment. This seals the fate of "ideal experiments" involving an unperformable procedure. The market for these "experiments" will collapse once physicists recognize the heavy role of theory. Since even the simplest experiment takes off from a hive of background assumptions, there is no hope of finding atheoretical adjudication. Once we outgrow the picture of experiments as neutral touchstones of truth, we can stare down the inductivist's demand that every scientific claim be backed by an experiment. Hence, Duhem thinks that thought experiments should disappear from physics in the way that nominal payments disappear from an economy that no longer outlaws gifts.

More recently, we find Ron Naylor dismissing the uncanny accuracy of Galileo's thought experiments with the hypothesis that they relied on secret executed experiments.[55] Galileo knew that unadorned experimental reports would not achieve his goal of converting a large number of thinkers to his new science. Readers would be reluctant to take so much on authority and would be easily bored in their passive role as information sponges. Hence, Galileo peppered his published work with do-it-yourself mental exercises that seemed to necessitate the propositions of the new science. Although you cannot really prove contingent facts in this manner, you can rig the pseudodemonstrations so that they have true conclusions. Thus, Naylor portrays Galileo's thought experiments as veridical propaganda.

A follower of Alexius Meinong might say that scepticism about the cognitive value of thought experiments rests on a bias in favor of the actual. Meinong believed that in addition to existing things, there are subsisting things. A key argument for this position appeals to the possibility of true, negative existential propositions such as 'Santa Claus does not exist'. If there are only existing things, how do we manage to refer to Santa Claus? Since we do refer to Santa Claus and since he does not exist, Santa must have another sort of being: subsistence. Unicorns, phlogiston, and the bumblebee on your nose lack existence; but they do have subsistence. Meinong believed that this new realm of being ought to be explored. Traditional metaphysicians scoffed at Meinong's forays into the ontological nightlife. But he dismissed their ridicule as betraying a deep prejudice in favor of existence and against subsistence. Perhaps Meinong would detect the same metaphysical provincialism among those who criticize thought experiments as being mere fantasy. Perhaps thought experiments constitute a broadening of science beyond the confines of existence to the neglected realm of subsistence. We have heard reports of other cultures who take their dreams as seriously as waking experiences. Westerners discount dreams. This dismissive attitude is largely due to the meager results obtained by those who do take their dreams seriously. But this could be due to bad observational and experimental techniques. Recall the trouble the ancient Greeks had in getting a fruitful account of the physical world. Maybe if scientists take their methodological sophistication out onto the dreamscape, we could claim a whole new domain for scientific study.

But regardless whether thought experiments could be justified in terms of the information they convey about nonpublic realms, they are in fact used to learn about the same world with which physical experiments are concerned. It is this relevance to a world of blood and steel that makes thought experiments puzzling.

The natural reply to the irrelevance objection is to claim a resemblance between the two worlds. But what two things could be more different than the stuff of thought experiment and the stuff of nature? Hans Hahn, in his attack on the rationalist notion of grasping facts about the world by thought alone, underscored the vanity of airy-fairy dualism:

> The idea that thinking is an instrument for learning more about the world than has been observed, for acquiring knowledge of something that has absolute validity always and everywhere in the world, an instrument for grasping general laws of all being, seems to us wholly mystical. Just how should it come to pass that we could predict the necessary outcome of an observation before having made it? Whence should our thinking derive an executive power, by which it would compel an observation to have this rather than that result? Why should that which compels our thoughts also compel the course of nature? One would have to believe in some miraculous pre-established harmony between the course of our thinking and the course of nature, an idea which is highly mystical and ultimately theological.[56]

3

Mach and Inner Cognitive Africa

Plato . . . says in Phaedo that our *"necessary ideas"* arise from the preexistence of the soul, are not derivable from experience—read monkeys for pre-existence.

Charles Darwin

The natural response to Hahn's two-world objection is to synchronize the worlds. Argue that just as the experimental results obtained in Peking hold for Peoria, experiments from the province of thought hold in the public realm. Concede a lower degree of resemblance. But insist that it is high enough to support fruitful analogies.

I. Instinctive Knowledge

Ernst Mach (1838–1916), who coined the term *gedankenexperimente*, grounded the similarity between our inner private world and the outer public one in the biological necessity of conforming thought to environment. Our unconscious compulsion to mimic natural patterns ensures that "iron-filings dart towards a magnet in imagination as well as in fact, and, when thrown into a fire, they grow hot in conception as well."[1] The "instinctive knowledge" tapped by thought experiments apes reality with tropistic inevitability:

> Everything which we observe in nature imprints itself *uncomprehended* and *unanalyzed* in our percepts and ideas, which, then, in their turn, mimic the processes of nature in their most general and most striking features. In these accumulated experiences we possess a treasure-store which is ever close at hand and of which only the smallest portion is embodied in clear articulate thought. The circumstance that it is far easier to resort to these experiences than it is to nature herself, and that they are, notwithstanding this, free, in the sense indicated, from all subjectivity, invests them with a high value.[2]

Out of this inner cognitive Africa come mostly negative judgments, such as "[that] heavy bodies do not rise of themselves, that equally hot bodies in each other's presence remain equally hot and so on."[3] Rather than announcing what must happen, instinctive knowledge brutely waves off what cannot happen, "since the latter alone stands in glaring contrast to the obscure mass of experience in us in which single characters are not distinguished."[4] Our raw sense of the absurd is a touchstone of truth because it blots out alternatives.

Mach illustrates how we metamorphose animal faith into explicit scientific principle with a thought experiment conducted by Simon Stevin (also known as Stevinus). In 1605 this former bookkeeper and military engineer published a book, *Hypomnemata mathematica*, on idealized machines whose parts do not stick or bind. One of his problems is to determine the force needed to keep a ball on an inclined plane from sliding downhill. Two extreme cases are easy. In Figure 3.1, the ball is at rest on a perfectly horizontal plane. Since the ball is fully supported by the plane, zero force is needed to hold it steady. In Figure 3.2, the ball is flush against a perfectly vertical plane. Since it receives no support from the plane, the ball requires its weight to be fully matched. Figure 3.3 has one of the many intermediate cases in which the plane partially supports the ball. We know that some force is needed and that the force need not match the ball's weight, but exactly how much force is required?

Stevin's procedure has three steps. First, imagine a triangular prism. Second, lay a circular string of fourteen balls over the prism (Figure 3.4). Now either the balls are in equilibrium or they are not. If they are not in balance, we could give the string a little tug sending the balls into motion—forever. Since this perpetual motion is absurd, we conclude that the balls are in equilibrium. The third step is to cut the string simultaneously at the lower corners (Figure 3.5). The balls won't slide off the prism because this bottom strand must have been pulling each side equally. (If the balls slid, which direction could they slide?) This makes it apparent that the short portion of the stringed balls balances the long portion. The original question can now be answered. The force needed to keep a ball on an inclined plane from sliding downhill varies inversely with the length of the plane (when the distance to the floor is constant).

Stevin was proud of his thought experiment and placed an illustration of it on the title page of his book (Figure 3.6). The inscription means "The wonder is that there is no wonder." There is nevertheless a feeling of magic to the result.

Unquestionably in the assumption from which Stevinus starts, that the endless chain does not move, there is contained primarily only a *purely instinctive*

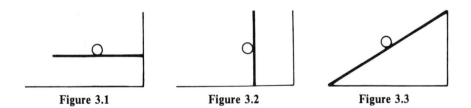

Figure 3.1   Figure 3.2   Figure 3.3

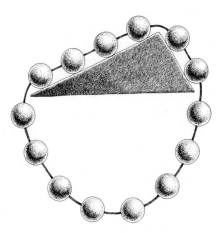

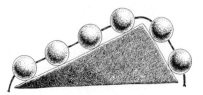

Figure 3.5

Figure 3.4

Figure 3.6

cognition. He feels at once, and we with him, that we have never observed anything like a motion of the kind referred to, that a thing of such a character does not exist. This conviction has so much logical cogency that we accept the conclusion drawn from it respecting the law of equilibrium on the inclined plane without the thought of an objection, although the law, if presented as the simple result of experiment, otherwise put, would appear dubious.[5]

Had we physically placed a string of balls on a prism, the lack of motion could be due to interfering factors such as friction. So one reason for greater confidence is the superior control over extraneous variables. A second reason is the avoidance of observer error. "We feel clearly, that we ourselves have contributed *nothing* to the creation of instinctive knowledge, that we have added to it nothing arbitrarily, but that it exists in absolute independence of our participation. Our mistrust of our own subjective interpretation of the facts observed, is thus dissipated."[6] These two immunities from error make thought experiments attractive to a rationalist. A rationalist is someone who believes that there are synthetic a priori propositions: propositions about the world (not just our way of describing the world) that can be known without experience.

Mach, however, was an empiricist. In fact, he was an extreme one—a sensationalist. Empiricism commits you to the claim that all of our knowledge is learned from experience. Sensationalism goes further by saying that all knowledge is reducible to judgments about sensations. Thus, talk of the moon and the stars, the birds and the bees, boils down to talk about feelings, images, and inclinations. Likewise, the physicist's talk of atoms, forces, and ether amounts to even more roundabout discourse about sensations. Allusion to theoretical entities is just a handy way of summing up and predicting the flow of experience. Science is stuffed with bookkeeping concepts: potential energy, cash flow, breaking points. When a physicist states that a doughnut's center of gravity is in the hole, do not ask what holds it in place! Just as it is a mistake to infer that the average man exists from the helpfulness of the expression *average man*, it is fallacious to infer the existence of atoms, forces, and ether from the utility of thinking in their terms. Mach lamented his colleagues' runaway reification. His career was marked by a running battle with materialists who were inclined to believe that atoms really existed. Mach thought it perverse to inquire whether feelings can be explained by the motions of atoms.[7] He deemed sensations far more familiar than atoms. Any attempt to reduce sensations inevitably reverses the order of explanation; we must explain the obscure in terms of the familiar, not vice versa! Turning the reduction inward also guarantees the unity of science; although the various fields have different approaches, they have sensations as their common subject matter.

Since empiricists deny the existence of synthetic a priori propositions, Mach denied that thought experiments provide synthetic a priori knowledge. He believed that whatever knowledge we obtain from thought experiment is synthetic a posteriori. When asked whether a proposition is a posteriori, we tend to answer by imagining a well-planned empirical investigation that could settle the matter either way by observation and experiment. This optimal way of determining whether a proposition is a posteriori biases us toward an intellectualist conception of empirical knowledge. We expect order, deliberation, explicitness. Instinctive knowledge is murkier. It is formed by experience but not well formed by it. Thought experiments draw from this dark well of half-treated experience. Hence, although thought experiments do not directly appeal to observation and experiment (and so seem a priori), thought experiments are indirectly based on observation and experiment. The a priori knowledge provided by thought experiment is a *relative* a priori; no new observations are required, but the old ones are. Crick and Watson could discover the double helix "without dirtying a test tube" because they could draw from data in which many test tubes were dirtied. Likewise, thought experimenters can make discoveries from their armchairs, but only because they can draw from a reservoir of past experience.

Mach's term "instinctive knowledge" suggests innate knowledge, a type of knowledge widely regarded as taboo for the empiricist. John Locke had formulated empiricism as a denial of innate knowledge. Locke stressed that we enter the world as blank slates. All we know can be traced to our individual experiences through our own sense organs. David Hume later dismissed the issue of innate beliefs as a trifle.[8] Hume insisted that the existence of such beliefs is compatible with empiricism because empiricism only says that all *knowledge* is derived from experience. Innate beliefs lack justification and so are not knowledge.

For good or ill, Mach accepted innate beliefs. He maintained that children fear the dark and fear ghosts instinctively.[9] Innate belief that dark places are dangerous can be traced to the truth of the belief. Humans are highly visual animals and so are greatly disadvantaged by the dark. Mach noted that other highly visual animals, such as sparrows, become more irritable and defensive at night. But the fear might be especially useful to humans because it curbs their great curiosity. Children without a fear of darkness would be less likely to survive to adulthood than those with the fear. Fear of heights, diseased people, and corpses may also be due to the damage and infections those fears prevent.

The chief reason for interest in innate knowledge among contemporary epistemologists is that innateness serves as a point of divergence between two conceptions of knowledge. According to the justificationist conception, you know p only if you have acquired strong evidence for p. Since a baby is born without evidence for anything, it has no knowledge (regardless of whether it believes anything). So the justificationist empiricist is dead set against innate *knowledge*. However, an empiricist who subscribes to *reliabilism* would be less hostile to innate knowledge. The reliabilist conception of knowledge only requires that one's belief in p arise from a reliable means such as perception, testimony, or reasoning. To this standard list of reliable belief-producing processes, we might add the generate-and-eliminate process described in Darwinian biology. Just as testimony allows us to borrow from the experiences of contemporaries, the evolutionary filtering mechanism allows us to inherit from our distant, uncommunicative ancestors. But unlike testimony, the lessons may be "hard-wired" into our brains. In one sense, the innate knowledge would be synthetic a priori. For the propositions would be known without the benefit of the knower's experience. However, in the more important sense, they would be synthetic a posteriori. For although the knower's experience was not involved, the experiences of his ancestor were necessary. Peter Unger's duplicate man provides a parallel case.[10] Suppose we make an exact duplicate of a man. Since the copy has the same "memories" as the original, he would be able to answer questions and make his way through the world just as well as the original. Thus, the duplicate satisfies all behavioral and introspective criteria for knowledge even though his apparent knowledge does not depend on his own experience. A tolerant empiricist will stress that the duplicate's knowledge is based on the original man's experience and so is really a posteriori.

In Mach's day, the reliabilist conception of knowledge was "in the air" because of Darwin.[11] Before Darwin, innate beliefs were couched in Kantian terms. Even Heinrich Hertz took the laws of thought to be an innate endowment of the mind. However, in 1904 we find Ludwig Boltzmann casting innate beliefs in an evolutionary framework: "Our innate laws of thought are indeed the prerequisite for complex experience, but they were not so for the simplest living beings. There they developed slowly, but simple experiences were enough to generate them. They were then bequeathed to more highly organized beings. This explains why they contain synthetic judgments that were acquired by our ancestors but are for us innate and therefore a priori, from which it follows that these laws are powerfully compelling but not that they are infallible."[12] Darwin was alive to the epistemological implications of evolution. He even suggests that imagination lets us bob above the cold generate-and-eliminate process that sustains innate knowledge; for imagination lets us conduct some of the trial-and-error process in thought so that "reason, and not death rejects the imperfect attempts."[13] Although Mach did not have access to Darwin's unpublished notes, he writes as if he did. Mach even makes the same kinds of mistakes. Like Darwin, Mach reacted to his ignorance of the mechanism of inheritance by leaning farther and farther toward Lamarck's theory of acquired traits. Mach was fond of Hering's theory of "unconscious memory" and so thought it plausible that children inherited technological techniques. In 1915 Mach suggested in Culture and Mechanics that a study of how children make and repair toys would reveal how primitive people made the first tools.

In summary, then, Mach did take innate beliefs to be part of our instinctive knowledge. Hence, Mach implicitly allowed for the possibility that some thought experiments draw upon experiences that the thought experimenter did not personally have.

Given the ultimately empirical origin of instinctive knowledge, we should expect it to falter when faced with unfamiliar phenomena. Mach's illustration involves an experiment that was viewed as an almost magical violation of the principle of symmetry by nineteenth-century physicists. This principle states that a symmetrical arrangement of causes produces an equally symmetrical effect. Archimedes illustrates the principle by having us imagine a symmetrical beam suspended from its center. Can the beam tilt? Only if we introduce an asymmetry such as an intruding fly. A corollary of the symmetry principle is the parity principle. It states that nature is indifferent between left and right; all physical processes are ambidextrous. An interesting consequence of the parity principle is that no experiment can discern the difference between left and right. But in 1820 Hans Christian Oersted stumbled into an experiment that seemed to do just that. According to one version of the story, Oersted was lecturing to his students and wished to give them a quick demonstration of the independence of electricity and magnetism. We can see why he wound up doing the reverse with the help of some diagrams. In Figure 3.7, a dead wire runs north beneath a compass, so the compass needle points north. In Figure 3.8, current is sent from south to north and the needle swings counterclockwise and points west. In Figure 3.9, the current is reversed and the needle points east. Nineteenth-century physicists concluded that there was an asymmetry in the laws of nature. Mach used this episode to underscore the fallibility of instinctive knowledge:

> Even instinctive knowledge of so great logical force as the principle of symmetry employed by Archimedes, may lead us astray. Many of my readers will recall, perhaps, the intellectual shock they experienced when they heard for the first time that a magnetic needle lying in the magnetic meridian is deflected in a definite direction away from the meridian by a wire conducting a current being carried along in a parallel direction above it. The instinctive is just as fallible as the distinctly conscious. Its only value is in provinces with which we are very familiar.[14]

To this we might add the astonishment caused by Pasteur's demonstration that microscopic organisms kill huge animals such as horses and Rutherford's demonstration that atoms were emptier than the universe. Our appreciation of the depth of these disruptions is clouded by their acquired familiarity. At the time they happen, they grate against the primal grain. Since science marches

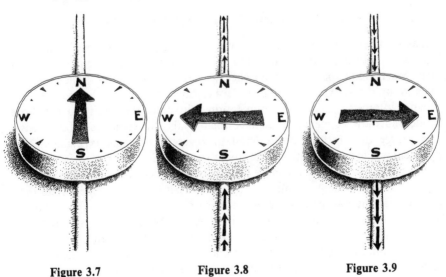

Figure 3.7   Figure 3.8   Figure 3.9

on and its results are only slowly assimilated, there are many contemporary examples that enable us to feel the violation first-hand. For instance, physics students are taught that boiling is a *cooling* process. The point is driven home with a demonstration of water simultaneously boiling and freezing (with the help of a vacuum pump).

Mach contends that we overestimate the reliability of instinctive knowledge because the obscurity of its origins cloaks its rough-and-tumble growth. Seedlings are magical to children because they seem to spring spontaneously from lifeless dirt. A standard school project is to grow seedlings in glass jars with moist blotter paper. When the kids see the seeds bloat with water, sprout roots, and only eventually break toward surface, their belief in a radical jump from earth to life is snuffed out. The children come to view the soil as a concealer of natural processes rather than as a magical source of life. A parallel demonstration (initiation?) would be instructive for instinctive knowledge. The subterranean development of instinctive knowledge makes its verdicts appear as if they well up from an impersonal, independent, inerrant oracle: "No longer knowing *how* we have acquired it, we cannot criticize the logic by which it was inferred. We have personally contributed nothing to its production. It confronts us with a force and irresistibleness foreign to the products of voluntary reflective experience. It appears to us as something free from subjectivity, and extraneous to us, although we have it constantly at hand so that it is more ours than are the individual facts of nature."[15] But this inner voice, this empirical conscience, is only an echo of past bumbling mumbles. Since all knowledge claims should be subject to scrutiny, thought experiments perform an important service by raising bits and pieces of instinctive knowledge to the surface. Once exposed to the light of criticism, the truths sprout into explicit knowledge, while falsehoods are sterilized.

II. The Continuum of Cognitive Bargain Hunters

Thought experiments are a natural outgrowth of primitive experimentation. The Darwinian necessity of adapting thought to fact applies equally to animals. Mach credits people with a richer mental life than animals, one that suits them to rapid and extensive changes. In addition, Mach notes that people have broader interests, greater communication skills, and the ability to use more subtle and indirect means to achieve their goals. Nevertheless, Mach insists that people only differ from animals in degree.[16] Animals experiment at a rudimentary level, instinctively varying a situation in response to frustration. Trial and error reinforces the more productive responses thereby improving the animal's fit with its environment. Children are better experimenters than their pets because human beings are better at recognizing and sharing similarities and so form more fruitful generalizations. As the child is progressively enculturated with theories that stoke this talent for noticing likenesses, it grows beyond spontaneous experimentation and begins to conduct planned experiments. The next stage is prompted by the costs of physically varying one's environment. The cognitive bargain hunter substitutes thought for things, converting his mind into a private laboratory. Instead of expensive physical variation of the facts, he opts for cheap mental variation of the facts. The idea is that one's habits of thought reflect the environmental dispositions that caused those habits. Hence, the pattern of thought follows the pattern of nature, enabling one to predict nature from thought. The utility of thought experiments is evident from the fact that (albeit tame) ones are used by hardheaded merchants and engineers. The quest for the predictive power conferred by generality will lead some thinkers to thought experiments that are further and further removed from any physical situation. Mach thus places thought experiment on a continuum of experimentation running from the instinctual variation of animals to the abstruse speculation of theoretical physicists.

Aye, Pythagoras was right: human beings are microcosms.[17] Like Leibniz's monads, minds are natural mirrors. Indeed, the Romantics recommended the study of natural history on the grounds that it revealed an ennobling unity: "Do you want to know nature? Turn your glance inwards and you will be granted the privilege of beholding nature's stages of development in the stages of your spiritual education. Do you want to know yourself? Seek in nature: here works are those of the self-same spirit."[18] But unlike the Romantics, Mach is no friend of unfettered imagination. Imagination is something that must be exposed to experiential pressure. It must be cut and molded to fit the contours of observation and experiment. It is in shape only when shaped; and the more shaping, the better. This developmental theme puts Mach in sympathy with Comte's notion of a hierarchy of thinkers:

> The spontaneous play of ideas and the changing conjunctions of thoughts, divorced from sensation or immediate needs, indeed far beyond them, these are what raise man above other animals. Phantasies about what has been seen and experienced, poetry, is the first step up from the everyday burdens of life. Even if such poetry at times bears evil fruit if applied uncritically to practice, . . . it is the beginning of spiritual development. If such phantasies make contact with sense experience and seriously aim at illuminating it while learning from it, then we obtain one after the other religious, philosophic and scientific ideas (A. Comte). Let us therefore look at this poetic phantasy that busily completes and modifies all experience.[19]

Mach's continuum of thinkers is in bad odor with egalitarianism about thought experiments. Those with greater experience tend to have more accurate thought experiments because their veteran minds have been brought into finer tune with the environment. Perhaps others have an innate gift for thought experiment in the form of brains particularly well suited for current conditions.

Many scientists credit Michael Faraday with an instinctive genius for physical experimentation. Faraday worked himself up from the slums of nineteenth-century London. With almost no formal education, he managed to conduct the best experimental work of his day. He just seemed to have a nose for what was important and how nature works—great "physical intuition" (as the physicists say). Maybe Faraday had a lucky match with his universe. As a further extrapolation beyond Mach, we might conjecture that the standard-bearers of the perennial parade of pseudoscientists, cranks, quacks, and kooks are just the many losers of a natural lottery. One can see how it might profit a cognitive creature to "hard-wire" a minority of its progeny in eccentric ways, prejudicing them in order to secure a running start in the mastery of nature. Most of these cognitive gambles would fail. They are the misfits. But a few would be superfits. These lucky winners will have an uncanny feel for the rhythms of nature. For the rest of us, it is "Monkey see, monkey-do." Being substantially less committed than the lucky partisan, we must learn the hard way; we need to methodically eliminate alternatives he instinctively spurns. But slowly, his gains become our gains; and we come to appreciate the value of tolerating eccentrics.

Speculations about hard-wired scientists run counter to the ideal of open inquiry. There is deep allure in the picture of the mind as being completely unfettered, as being as likely to strike out in one direction as any another. But this ideal is overwhelmed by the sheer number of possibilities. An open mind would be lost in logical space. Hence, even as babies, we attend to certain things (milk, faces, loud noises) and are oblivious to most everything else. There is mounting anthropological and psychological evidence that we have cognitive predilections for structures that are linear, spatial, discrete, hierarchical, balanced, componential, and sharable.[20] You say you do not *feel* constrained? Neither does a spider when it spins its characteristic web. Of course, the *degree* of mental predetermination can vary. Mother Nature might only *lightly* load the dice in favor of some hypotheses. We can disagree on the structure of the necessity as long as we agree on the necessity for the structure.

If there are superfit physical experimenters, there will be superfit thought experimenters. This fitness could be a matter of absolute cognitive power. Nikola Telsa (inventor of fluorescent lighting and the self-starting induction motor) claimed that he would determine which parts of a machine were most subject to wear by inspecting a mental model he had "run for weeks."[21] Contemporary psychology provides less fantastic but better corroborated differences. For example, experiments have established individual differences in rates of mental rotation and in the number of novel patterns one can construct from a stock of simple shapes.[22]

Extra fitness could also be a relative matter. Galileo, Newton, or Einstein may have owed some of their success at thought experiment to a lucky, tight *match* between genes and environment. Individuals born with a good inner laboratory and the opportunity to improve it through experience may form a recognizable thought experiment elite. Their capacity to recognize absurdity might be due to a single gene like the perfect pitch of gifted musicians. Or maybe our ear for theoretical coherence is composed of independent subabilities that can be coached into effective collaboration.

Such speculations lead naturally to pedagogy. How do we detect good thought experimenters? Can thought experiment be decomposed into educable skills? The idea of training thought experimenters has a precedent in Wilhelm

Wundt's insistence on seasoned introspectors. He required his subjects to master elaborate guidelines to prevent overinterpretation and to encourage attentiveness. These precautions were intended to meet Brentano's objections to direct introspection (which he called "inner observation").[23] The direct introspector focuses on his mental life, while the indirect introspector only notices mental phenomena out of the corner of his mental eye. Brentano favored the indirect variety ("inner perception") because it was less intrusive. Focusing the mental eye would drain off the attention needed to sustain one's first-order mental life. Mach's thought that the experimenter needed to use both sorts of introspection. He had to be direct when turning the mind toward the experimental setting but indirect when he then let his inner nature take its course.

A further consequence of Mach's view is that thought experiments should be more accurate and reliable as their subject matter becomes more familiar. For example, thought experiments about middle-sized objects should be more trustworthy than those about tiny things and huge things. (This should make us wary of thought experiments in quantum mechanics and relativity theory.) Thought experiments should be most reliable when about phenomena at near normal temperature and pressure. (This produces caveats about thought experiments concerning superconductivity and vacuums.)

III. Mach's Response to the Problem of Informativeness

Mach's account of thought experiments would lead him to challenge the first premise of the sceptical argument advanced in the previous chapter (see p. 48).

1. If a thought experiment can be checked through public experimentation, then not actually checking leaves the results unverified, and an actual check would render the thought experiment redundant or misleading.

Sometimes a thought experiment is forceful enough to make public replication unnecessary. Mach cites Galileo's criticism of Aristotle's principle that the natural velocity of a heavier body is greater than that of a lighter one. Aristotle's principle seems vindicated by everyday observation; if you simultaneously drop a brick and a feather, the brick lands first. Aristotle also noted that when a faster object joins a slower one, the faster one slows down. For example, if a swift charioteer grabs a lumbering oxcart, he slows down. Galileo sets these commonsense principles against each other by having us imagine that a big stone becomes connected to a little stone in the middle of their fall to earth. The principle that heavier objects fall faster predicts the composite stone will speed up. The opposite prediction follows from the principle that says attaching a faster object to a slower one slows the faster one. Since it is impossible for the composite stone to simultaneously fall faster and slower, Aristotle's theory of motion is refuted without the need to perform the experiment.

Measurement puzzles also illustrate the occasional evidential autonomy of thought experiments. As Mach notes, "vivid phantasy" suffices to solve the problem, "given three vessels of capacities 3, 5 and 8 units respectively, the first two empty and the third full, to divide it into two equal parts without any further tools."[24] Hence, some unverified thought experiments have scientific value.

Further, when we do check a thought experiment, replication need make it no more redundant than physical replication of other physical experiments. Two experiments are better than one. Indeed, the qualitative difference between a thought experiment and an executed experiment may make mixed replication better than pure replication. Redundancy is also an unfair charge in that it overlooks the preparatory value of thought experiments:

> Our ideas are more ready to hand than physical facts: thought experiments cost less, as it were. It is thus small wonder that thought experiment often precedes and prepares physical experiments. Thus Aristotle's physical investigations are mainly thought experiments, in which he uses the store of experience kept in memory and above all in language. Thought experiment is in any case a necessary precondition for physical experiment. Stephenson may be familiar, from experience, with carriages, rails and steam engines, but it is by first combining them in thought that he can next proceed to build a locomotive in practice.[25]

This use of thought experiments as pilot studies follows from the economy of thought. Businessmen make the cheapest commitments first so as to minimize their exposure to misfortunes and to maximize their freedom to switch to more lucrative ventures. Scientists also lead with their cheapest investment— thought—and so work out the theoretical end before making the more expensive commitment of an executed experiment.

Some laboratory settings resemble the strange scenarios concocted by thought experimenters. When psychologists first suspected that perception presupposes physical regularities, they tested the hypothesis by imagining how our visual system would fare in a world in which those regularities failed to hold. For example, if objects frequently expanded and contracted like balloons, change in image size would not be the good clue to distance that it is in our rigid world. In the elastic world, the object's growing image on the retina could mean that it is coming nearer or it could mean that it is stationary and expanding. This thought experiment can lead to an executable experiment because psychologists are able to build miniature elastic worlds in their laboratories. For instance, one experiment has the subjects judge the relative distance of two partially inflated balloons.[26] Although the balloons are both at ten feet from the viewer, inflating one makes it look nearer. Varying its inflation makes the balloon look like it is moving back and forth.

Mach's point is that many executed experiments owe their existence to preliminary thought experiments. Criticizing superseded thought experiments as inevitably redundant is like chiding car manufacturers for constructing prototypes on the grounds that the final version renders the prototypes obsolete. If the results of the thought experiment are disconfirmed by regular experiments, then our assessment should not end with the conclusion that the thought experiment misled us. A more complete assessment would include ways in which it was also positively leading and the remedial value of correcting an unfruitful tendency in our thinking.

The propaedeutic value of thought experiments is Mach's pet pedagogical theme. He encouraged students to form educated guesses about the outcomes of class demonstrations and to discuss those guesses critically.

> Most people will go for what is most obvious by association: just as the slave boy in Plato's *Meno* thinks that doubling the side will double the area of the square, a primary pupil will readily say that doubling the length of the pendulum will double the period of oscillation, while the more advanced will make less obvious though similar mistakes. However, such mistakes sharpen one's feeling for the differences between what is logically, physically or associatively determined or obvious, and in the end one learns to discriminate between the guessable from that which is not.[27]

The point of the exercise is to improve the inevitable process of forming expectations. According to Mach, we are always busy trying to anticipate future experiences. The well-educated thinker is distinguished by his *controlled* speculation and his refined sense of absurdity. The question for Mach, was not *whether* we should do thought experiments: we have no choice. The question was how to do what we must do, well.

IV. Appraisal of Mach

Mach was the first to study thought experiment systematically, and he remains the most thorough and successful. Much of what Mach says is quite right, leaving me little to do but register points of agreement. Points of disagreement will require more space because the objections require stage-setting.

A. What Mach Got Right

Several problems are elegantly solved by placing thought experiment within an evolutionary framework. Why should events in the world of thought indicate the course of events in the physical world? Answer: there are pressures selecting minds that mimic patterns of nature. Hence, the connection between the worlds is biological causation, not an occult bond. Biologists could detect physical patterns of an unexplored planet by studying thought patterns of an alien from that planet. Our prowess for imitation is due to a combination of opposite strategies. The first is impressionability: design a cognizer so it resembles putty. In this way, a large variety of possible regularities can be quickly and precisely impressed upon the mind. The second strategy is hardwiring. Sometimes we cannot wait for Mother Nature to teach us her lessons. The solution here is to design a creature with innate beliefs. Since the mechanism by which these beliefs are formed is the reliable generate-and-eliminate process detailed by evolutionary biologists, one's innate endowment of truths constitutes a poor man's synthetic a priori. Unfortunately, the continuing nature/nurture controversies will quickly ensnarl any discussion of hardwiring. For instance, it is an open question whether genetic determination can be fine-grained enough to preset a particular belief. Maybe heredity can only mildly affect the probability with which a belief is formed. It is one thing for Mother Nature to present you with a menu and another for her to pick your dish. Since I have neither the ability nor the need to develop the consequences of the poor man's synthetic a priori for thought experiment, I will simply record my optimism about its future development. The part of Mach's account of imitation that I do need, his major theme of learn-or-die impressionability, should be widely acceptable. However, we should bear in mind that selective forces need not be as heavy-handed as early Darwinians assumed. Geneticists now teach that a gene need only confer a very slight advantage in order to spread through the population. Other modernizations will be introduced in chapter 4.

Mach's account also escapes the narrowness of the appeal to ordinary language. Conventionalists reasoned as follows. Speakers know the rules of language. These rules are what make tautologies true. Hence, the philosopher's linguistic competence puts him in a position to identify hidden tautologies; and thought experiments help by converting implicit knowledge of the rules into explicit knowledge. However, this explanation does not account for the reliability of thought experiments that aim to establish *synthetic* statements. Nor was it intended for this task: conventionalists believed that philosophers were only competent with analytic statements, and conventionalists were only interested explaining philosophical and mathematical methodology. Hence, the inability to explain scientific thought experiments was deemed irrelevant.

Those who deny that all philosophical remarks aim to be tautologies will be less sanguine about the gap. Happily, Mach can explain synthetic thought experiments; for he sketches out a psychology of thinking that gives us rough but serviceable intuitions about physical possibility. Synthetic thought experiments use these intuitions to test lawlike generalizations.

In addition to establishing the relevance of thought experiments, the evolutionary framework grounds their function in economic terms. We learn that most thought experiments do what ordinary experiments do—test hypotheses about the world. The difference is price. Unfortunately, thought experiments also tend to be "cheap" in quality—they are not as reliable as executed experiments. When a correct answer is important, we are usually obliged to follow up with a *real* experiment just as, when you need reliable transportation, you get a *real* car, not a cut-rate model. Nevertheless, thought experiment has a place among the scientist's other "quick and dirty" techniques. Those who insist on always going first class do not go far. These insights position Mach for a partial solution to the problem of informativeness. He correctly points out that the execution of a thought experiment is sometimes superfluous because it can be decisive in its own right. Thought experiments that are physically replicated at least have preparatory value and may even lend autonomous support.

Mach also pinpoints several sources of illusion about thought experiment. Our ignorance about how it works, combined with its a priori, involuntary feel, make it fuel for rationalism and even occultism. The errors of this inner oracle are jarring. The disappointment sours some thinkers into a resentful rejection of thought experiment.

Mach correctly identifies familiarity as a key factor of the thought experimenter's accuracy. The point is corroborated by recent experimental work on the four-card selection problem.[28] In this conditional reasoning task, subjects were shown four cards (see Figure 3.10, task 1) and told that each card had a letter on one side and a number on the other. Which cards must you turn over to determine the truth of the task 1 sentence? The correct answer (D and 7) was only given by 5 of 128 people in the course of four experiments. The popular response was D and 3, suggesting that people try to verify rather than falsify the conditional. The experimenters suspected that the abstract nature of the material was partly to blame for the poor performance and so tried more realistic conditionals such as "If I go to Manchester, I travel by car." Performance rose from 2/16 correct to 10/16 correct. An even greater improvement was reported by an Italian experiment in which 0/24 were correct for task 1 but 17/24 were correct for task 2. However, University of Florida students were found to perform equally badly at tasks 1 and 2.[29] The experimenters conjectured that the Italians had an advantage because postal rates for sealed and unsealed letters are different in Italy. Since the college students were familiar with the Florida law requiring that beer drinkers be over nineteen, the experimenter were able to confirm their conjecture with task 3 (which specified that each card represented a person, with what the person was drinking on one side and the person's age on the other). Here, students imagined they were police

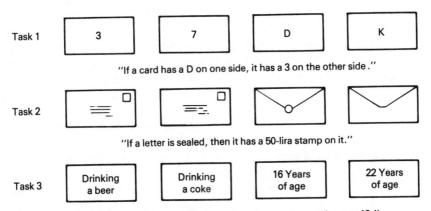

Task 1: "If a card has a D on one side, it has a 3 on the other side."

Task 2: "If a letter is sealed, then it has a 50-lira stamp on it."

Task 3: "If a person is drinking a beer, then the person must be over 19."

Figure 3.10

officers enforcing the regulation. Twenty-nine of the 40 students gave correct answers. In telling contrast, no student gave the correct answer to the abstract question in task 1.

As suggested by the implicit role reversal in this last hypothetical, exact precedents are not necessary. People can draw on general experience. In a fourth variant of the selection task, the subject imagines he is a manager of Sears store.[30] One of his duties is to check whether the sales receipts conform to this rule: "If a purchase exceeds $30, then the receipt must be approved by the department manager." When presented with four receipts corresponding to the cases in Figure 3.10, nearly 70% of the subjects gave correct answers even though none had specific experience as Sears managers.

In another study, the postal problem was run in Great Britain, which had, until 1968, one rate for sealed and another rate for unsealed letters.[31] Subjects younger than forty-five did nearly as poorly as the Florida students, but those over forty-five did nearly as well as the Italians. The difference was that the older people had experience with the defunct postal system.

Reasoning buffs stress that familiarity does *not* make the subjects more logical.[32] Success with the homey versions of the task never improves performance with the abstract version. This absence of transfer suggests that a familiar setting only helps by cueing a relevant memory that can then be applied directly or by analogy. So whereas the original, abstract, task 1 is clearly a logical task, its content-enriched descendents look more like memory tasks.

The literature on the selection task has several implications and nonimplications for thought experiments. First, it confirms Mach's idea that hypotheticals tap our instinctive knowledge and that our judgment improves as the situation becomes more familiar. However, the research does not show a universal incompetence with abstract scenarios. Normally, a small percentage of the subjects get the answer right. Moreover, increased familiarity should lead to *declining* performance in cases where memories are inaccurate. Hence, when we have reason to think that our past experience is misleading, we should defamiliarize the thought experiment by either draining its content or running it against inexperienced subjects.

If thought experiment is primarily memory-driven, there is the danger that subjects can easily fake the procedure. People often just ignore difficult instructions and answer on the basis of what they do understand. This lessens the credibility of conservative thought experiments that seek to reinforce widely held beliefs.

Finally, note that familiarity is not the same as realism; familiarity is relative to cognizers, reality is not. Indeed, the prohibition against "far out" thought experiments is partly undermined by the experiments just discussed. The underlying factor affecting performance is familiarity—not the distance to other possible worlds. The restriction to our own modal neighborhood only seems vindicated because of the correlation between modal proximity and familiarity. And the correlation is only rough: we can be unfamiliar with real things (radiation, the deep sea, the sun's core) and familiar with unreal things (Santa Claus, lucky numbers, miracle cures).

The pedagogical value of thought experiments has been vindicated by research on systematic physical misconceptions.[33] Psychologists have shown that students enter physics courses with naive but hardy principles such as "Motion implies a force." Simply exposing students to correct theories only marginally erodes allegiance to the old theory. Students just half learn: they hybridize the new theory with the old, or only apply the new when cued, or wind up confused at the points where the new and old diverge. So the psychologists recommend that teachers actively root out the old theory by using the techniques favored by Galileo. Embarass the opposition by extracting absurd consequences. Use contrast cases that maximize the differences between the two theories. Encourage your audience to work out the problems themselves by using entertaining scenarios.

B. What Mach Got Wrong

1. *Thought experiments need not involve introspection.* My first objection to Mach is that his reliance on introspection makes his theory of thought experiment too narrow. If thought experimenters must be introspectors, all thought experiments fall within the historical, developmental, and cultural boundaries of introspection. But the earliest clear use of the concept of introspection only dates back to Augustine's fourth-century *De trinitate*. Kathleen Wilkes has even argued that the ancient Greeks lacked the concept.[34] If so, it is unlikely that introspection was common enough or sophisticated enough to constitute the thought experiments of Plato and Aristotle. Indeed, thought experiments would have had to have arrived on the scene over a hundred thousand years earlier in order to play the evolutionary role Mach suggests. Even those who doubt Wilkes and believe that introspection was abundant in the days of yore can readily appreciate the danger of making thought experiment wait for the emergence of introspection. Developmental psychologists say that children first acquire awareness of their own thought processes and the ability to verbalize them at about the age of eight.[35] Yet they are able to absorb and invent simple thought experiments as soon as they learn the grammar of subjunctive conditionals and the conventions governing storytelling. On the cultural front, consider reports of the amazingly apsychological Balinese, Maori, and Ifaluk.[36] These folks have little interest and littler vocabulary for inward reflection. But they get the point of thought experiments.

Indeed, thought experiment is universal to all cultures. This is evident from the pervasiveness of the golden rule: "Do unto others as you would have them do unto you." The role reversal that constitutes the application of this moral test is a form of thought experiment. The agent need not recognize his role reversal as such. Indeed, he may lack the concept of thought experiment. Just as one does not need the concept of sublimation to sublimate, one does not need the concept of thought experiment to thought-experiment.

2. *Mach overestimates the subjectivity of thought experiment.*   Mach makes the initial stage of thought experiment incorrigible and too subjective. Unlike Comte, Mach was a firm believer in introspection and regarded beliefs about sensations as infallible. Since Mach pictured thought experiments as experiments within our own private laboratories, they amounted to reports of sensation sequences that are then extended to the public world. Although the reported sequence need not match the sequence produced by physical realization of the thought experiment, Mach's sensationalism prevents him from suspecting the report is erroneous. He has to say that all sincere reports of thought experiments are true. Should two thought experiments report different types of sequences, there would be no conflict. Everyone has his own inner reality. But the differences in subject matter that preclude disagreement will also preclude agreement. If the thought experimenters report the same types of sensations, there is no genuine agreement, just as there is no genuine agreement when I say that my eyes are blue and you say your eyes are blue; for I am talking about my eyes, and you, yours. Mach can only allow the thought experimenter to err at the stage where he infers facts about outer reality from the inner one. But subjectivism is not the normal attitude toward the first stage of thought experiment. Philosophers and scientists frequently accuse each other of botching the procedure and insert safeguards against bias and inattention. (Recall the examples of conceptual paternalism from chapter 2.) In short, thought experimenters resist divergent results even at the first stage.

3. *Mental imagery is not essential to thought experiment.*   Sensationalism also leads Mach to overstate the imagistic aspects of thought experiments. If thought experiments are constituted by sensations, people conducting the same thought experiment would have the same imagery. Yet people manage to agree on their thought experiments without agreeing on their internal experiences. Some are vivid visualizers but others think with little psychic color. (Most people cannot form a mental image of the gothic letter *G* but most of them can recognize it.) Even the same person thinking the same thought experiment at different times will often lack the same inner events. Moreover, people with the same images can be conducting different thought experiments. I imagine a black box and then a black box with a mouse in it. My images are the same. What makes them different imaginings is my intention. Just as two artists can depict different scenes with paintings that happen to be qualitatively indistinguishable, scientists can conduct different thought experiments with qualitatively indistinguishable mental imagery. Any introspectionist account of thought experiment will be vulnerable to this variant of the imageless thought objection.

4. *Mach's account is indiscriminate.*   Mach does not draw systematic distinctions between thought experiments and related phenomena such as fantasy, dreaming, hallucination, storytelling, and meditation. One can make out the vague outlines of how the taxonomy would proceed. Sensationalists generally base the distinction between dreams and waking experiences on degrees of

coherence. Dream experiences are disorderly, unpredictive, and lack intersubjective agreement. Hallucinations are almost as disorganized as dreams. The difference is that parts of the experience—those having to do with background—have a limited degree of reliability even though the foreground (featuring those exciting pink rats and devilish voices) is poorly connected with the rest of your experience. Following cues from Hume, the sensationalist can also distinguish perception, memory, and imagination along a scale of decreasing vivacity. But as we have learned from Hume's troubles with the classification of ideas, sensationalism provides us with meager taxonomic resources. There is not enough to finely discriminate between related mental states and activities. Thus, the suspicion is that Mach will not be able to atone for his sin of omission.

5. *Evaluative thought experiments are left unexplained.* Mach says little about morality and does not discuss any ethical, prudential, or aesthetic thought experiments. Hence, we lack guidance as to how he would account for evaluative thought experiments. Further, his severe empiricism makes any adequate explanation of evaluative thought experiment hard to imagine. A hedonist defines 'good' as pleasure and 'evil' as pain. So a Machian hedonist could say that good and evil are sensations that figure into thought experiments just as any other experience. However, hedonism is almost universally rejected. Indeed, many ethicists reject all attempts to identify goodness with a natural property such as self-development, harmony with nature, or preference satisfaction. Many even deny that moral utterances *describe* anything. Thus, they regard experimental investigation of moral questions as fundamentally misguided. What is the scientist to look for when measuring goodness? Is it green? Heavy? Loud?

Even if Mach could reconcile his empiricism with the possibility that thought experiments faithfully track goodness and rightness, there would remain the epistemological problem. Why should ethics be part of instinctive knowledge? The reliability of instinctive knowledge is underwritten by the savage principles of natural selection. Why should this brutal guarantee encompass moral knowledge? Is moral ignorance *maladaptive*? Critics of social darwinism complained that the law of the jungle actually favors cutthroat egoism.

Happily, this gaping hole in Mach's account can be partly filled with recent developments in biology. First of all, the theory of kin selection has made room for some altruistic behavior. The "selfish" agents are my genes, so that although they will program me to look after my own welfare, they will also program me to sacrifice for those who share my genes. Kin selection can be supplemented with an idea from information theory.[37] Simple rules are easier to encode, store, and follow. Hence, wide altruism may be more efficient than narrow altruism. This may explain why male baboons defend *any* member of their herd, not just their progeny. We can also appeal to the inaccuracy of reproductive behavior. In particular, some altruism may be misplaced parenting. Individuals who are too egoistic or too altruistic are displaced by individuals who strike a better balance. Thus, there is a tendency for moral beliefs and desires to become universal.

These points are reinforced by Leda Cosmides' analysis of the four-card selection problem.[38] According to Cosmides, what little capacity we have for formal reasoning is due to our ancestors' need for social contract reasoning. Hunter–gatherer bands depend on a system of favor exchange. The menace of exploitation puts pressure on group members to separate cheaters from good faith participants. Hence, our ancestors evolved mechanisms for tracking reciprocity, for making allowances for accidents and other excusing conditions, and for calculating the outcomes of agreements. Cosmides argues that our consequent aptitude for social reasoning is domain-specific and so leads to the fallacies discovered in the four-card selection task. If Cosmides is right, humans could be excellent at ethical thought experiments and poor at logical ones. This would not be a blank check for ethics. The domain-specificity of social reasoning raises a question about scenarios that involve nonhumans or large populations; for our moral competence may be specially suited to the thirty or so people that make up a hunter–gatherer band.

But even if these speculations pan out, the supplemented theory of kin selection will have only provided a partial solution. Its success is at the level of moral *psychology*. Kin selection explains the broad pattern of our moral concerns; but it does not *justify* them. Morality could be a useful myth. After all, even emotivists can accept the sociobiological explanation of morality. Emotivists maintain that moral utterances merely express or evoke emotions like 'Boo!' and 'Hurrah!' and so are neither true nor false. However, emotivists deny that emotions are *biologically* arbitrary. Attitudes conducive to prosperity spread. Hence, the emotivist expects a certain amount of moral "agreement" and welcomes psychological and sociological discoveries of these attitudinal uniformities. But the same goes for etiquette and fashion.

6. *How are open-ended thought experiments possible?*   Some experiments cannot be executed because they involve too many steps. Carrying out the steps "in one's head" is still possible if the mental operations are faster. But quick thinking cannot explain our success with thought experiments involving an open number of steps.

A standard problem for evolutionary theorists is to explain the existence of apparently useless intermediate steps. For instance, the eye gave Charles Darwin "a cold shudder" because it could not develop all at once and only appears useful when complete. Darwin's solution was to tell a "just so" story in which the eye is functional at each step of its development. Long sequences of enzyme reactions present biochemists with the same problem. The chain of reactions is only useful when complete, so how could the chain evolve? N. H. Horowitz applied Darwin's strategy of constructing a hypothetical history that assigns a past function to each of the intermediate steps.[39] Suppose life begins with a primitive organism that depends on a substance A, which is readily available in its environment. Some of its descendents mutate and acquire an enzyme that can synthesize A from two other plentiful substances, B and C. This "preadaptation" becomes a crucial asset because the amount of natural A is being depleted by the expanding population of primitive organisms. As A becomes scarcer, the mutants prevail because they can exploit the environment's B and C. The subsequent depletion of B leads in turn to the ascendency of a new "hopeful monster" that has an enzyme capable of synthesizing B from D and E. This leads to the depletion of D and another addition to the lengthening enzyme sequence. Eventually we have organisms that puzzle biochemists because of their apparently prescient enzyme chains. Thus, the solution is that the individual enzymes were acquired in a sequence opposite to the one used by current organisms.

The Horowitz scenario reconciles evolutionary theory with enzyme chains of *variable* length. Hence, contrary to the "inner laboratory" picture, the thought experimenter is not always carrying out a fixed number of operations "in thought." Infinite thought experiments dramatize the point. Simmel has us imagine his three wheels spinning forever. We do so without performing infinitely many checks "in thought" of whether the wheels realign.

7. *Inference rules are not part of instinctive knowledge.* A third gap in Mach's account is revealed by logical and mathematical thought experiments. This shortcoming can be traced to his allegiance to psychologism, the thesis that logic is reducible to psychology. There are many serious objections to psychologism, but let us concentrate on one pertinent to thought experiment: logical and mathematical principles are the *means* by which we conform our thinking to our environments, not the *results*. The means/result distinction is made vivid by Lewis Carroll's dialogue beween Achilles and the Tortoise.[40] The Tortoise challenges Achilles to convince him that Z follows from A and B:

A. Things that are equal to the same are equal to each other.
B. The two sides of this triangle are things that are equal to the same.
Z. The two sides of this triangle are equal to each other.

Actually, we can let A and B be the premises of any valid argument for a conclusion Z. The Tortoise is just using the geometrical propositions for the sake of illustration. Since Achilles sees that A and B imply Z, he thinks his task is easy. He just asks the Tortoise whether he agrees that

C. If A and B are true, then Z must be true.

The Tortoise is willing to grant C but still refuses to accept Z. So Achilles then asks for the Tortoise's assent to another hypothetical:

D. If A and B and C are true, then Z must be true.

Once again, the Tortoise agrees but still refuses the conclusion. So Achilles asks for yet another hypothetical:

E.   If A and B and C and D are true, then Z must be true.

Yet again, the Tortoise generously agrees but still refuses the conclusion. By now we can see that Achilles and the Tortoise are off on an infinite regress of ever larger hypotheticals.

What the Tortoise taught us is that we must distinguish between *premises* and *rules of inference*. We infer conclusions *from* premises *by* inference rules. In order to learn anything from experience, we must have the means to process the data. Hence, the rules by which we learn from experience are not themselves learned from experience and are not examples of thoughts conforming itself to empirical regularities. Thus, they are not part of the stock of instinctive knowledge. Thought experiments concerning rules of inference therefore disconfirm Mach's belief that all thought experiments are drawn from instinctive knowledge.

Lewis Carroll's Tortoise thought experiment is itself an example of the recalcitrant phenomenon. It is a logical thought experiment because it reveals the role of inference rules by quietly supposing the rules do not exist. We see that if arguers only had premises, they could not draw conclusions. Other logical thought experiments try to show the necessity of logical laws by our inability to conceive of illogical worlds. The same has been done with mathematical laws.

Another illustration will be of use in later chapters—W. V. Quine's task of radical translation. Suppose a linguist must translate the language of a completely foreign culture without the benefit of translation manuals, bilinguals, or knowledge of similar languages. Quine's main goal is to demonstrate the indeterminacy of translation: "manuals for translating one language into another can be set up in divergent ways, all compatible with the totality of speech dispositions, yet incompatible with one another."[41] But he is also out to reveal the incoherency of a "prelogical mentality." Some anthropologists have reported cultures that do not abide by logical laws such as noncontradiction. But could our imaginary linguist gain evidence that the people he is translating are a deeply illogical folk?

Evidence that he could not discover their illogicality can be gleaned from our trouble making a charge of illogicality stick to speakers of English. The rules of natural languages are never laid down in a constitution; and even if they were, actual usage would override it. We must form hypotheses as to what the rules are and test these hypotheses against the data of ordinary speech. One of the key principles of interpretation is the principle of charity, which tells us to maximize the rationality of those we are interpreting. A corollary of this principle is that we should minimize the inconsistencies we attribute to interpretees. In Shakespeare's *Julius Caesar*, Brutus defends slaying his friend by saying "not that I loved Caesar less, but that I loved Rome more" (act 3, scene 2). Suppose a smart aleck points out that if Brutus loved Rome more than Caesar, then he jolly well did love Caesar less than Rome. Champions of Brutus will reply that he must have meant that his assassination of Caesar was not out of hatred but because his love of Rome overwhelmed his ample love of

Caesar. Consider a balder example. You ask a friend whether he wants ice cream and he says "I do and I don't." You should not interpret his reply as a contradiction. Instead, try hypotheses such as 'He has a desire for ice cream but not an all-things-considered desire' or 'His desire for the ice cream conflicts with his desire to lose weight'. If these interpretations fail, check whether you misheard him, or he misspoke, or was speaking metaphorically, or was joking. The inconsistency hypothesis is only a last resort. Indeed, it is so far down the list that we are never entitled to stoop that low. Even if the belief is indeed a patent contradiction, we cannot recognize it as such. Rival explanations will always look at least as good. Compare this methodological point with Hume's epistemological scepticism about miracles: the essential improbability of a miracle ensures that we could not recognize such an event. The sensible explanation would hold on to core science and blame appearances on fraud or error.

Epistemological scepticism about belief in obvious contradictions leads to epistemological scepticism about inconsistent languages. Languages are inconsistent only if the rules constituting them are inconsistent. But you can show that the rules are inconsistent only if you can show that the vast majority of speakers are disposed sincerely to assert obvious contradictions. But the inevitability of stronger, rival explanations of the speech behavior of individual speakers ensures that we cannot justifiably attribute this disposition to the vast majority of speakers. Hence, we are never in a position to interpret the speakers as followers of inconsistent rules. This conclusion is reinforced by the radical translation scenario. A translation that portrays the natives as deeply inconsistent will inevitably count as a mistranslation.

The point can also be made by supposing that the radical translator reports that natives mean by 'and' what we mean by 'or'. According to the translator, their word for 'and' is 'gloz'. When a person says "p gloz q," he only means to exclude the possibility that both components are false. Hence, their truth-tabular definition is as follows.

p q	p gloz q	p or q	p and q
T T	T	T	T
T F	T	T	F
F T	T	T	F
F F	F	F	F

Instead of accepting the translator's claim that there is a logical difference between us and them, we will say he has simply mistranslated 'gloz' as conjunction. It really means 'or'. After all, the truth tables exhaust the meaning of the logical connectives.

Quine uses this point to impale deviant logicians on a dilemma. They say they reject standard logic and offer alternative definitions of the logical connectives and rules of inference. But if the rules constitute the very meaning of our logical vocabulary, any dispute with the deviant will be purely verbal. Thus "the deviant logician's predicament: when he tries to deny the doctrine, he only changes the subject."[42]

8. *Thought experiments sometimes "trump" executed experiments.*   Notice the understatement in Mach's point about the superfluousness of physical verification in the three-vessel thought experiment. If we fail to get physical replication, we would question the physical experiment, not the thought experiment. Notice the role reversal of thought and physical experiment in mathematics; thought experiment dominates, and physical experiment plays an ancillary, heuristic role. If a teacher wishes to illustrate mathematical induction with dominoes, he does not take the failure of the dominoes to all fall as refutation of the principle.

The robustness of our mathematical convictions shows that Mach overestimates the relative strength of physical experiment and thought experiment. Mach talks as if physical experiment *always* prevails over thought experiment. But even thought experiments outside of mathematics and logic can put up a spirited struggle and even prevail over their physical rivals. Consider Mach's own example of the magnetic deflection of the compass needle. This did not really overturn the principle of symmetry. Subsequent analysis of the phenomenon led physicists to the conclusion that the apparent asymmetry of the magnetic field was due to a conceptual confusion. Indeed, the symmetry principle was well entrenched until 1957. Only then did microphysicists at Columbia University detect a decisive counterexample.[43] Mach is, of course, correct to emphasize the fallibility of the deep intuitions that drive thought experiments. And no doubt, we usually give precedence to the executed version. My point is that Mach should not convey the impression that thought experiments *always* lose to physical experiments.

9. *Experiments are not always preceded by thought experiment.*   I now wish to attack Mach's claim that thought experiment is a necessary prelude to regular experiment. In addition to being warmly received by those with a casual interest in thought experiments, the claim is embraced by commentators on the topic. Sheldon Krimsky incorporates the theme in his definition of 'thought experiment': "A thought experiment is an experiment which is planned in thought."[44] Erwin Hiebert opens his article on Mach with the statement "Experimentation in thought is an indispensable precondition for the execution of any physical experiment."[45] My objection is that the only way to make this claim true is to stretch 'thought experiment' so wide that it means any kind of forethought about experiment.

This distortion of ordinary meaning is motivated by misguided benevolence. The protector of thought experiment would like to show the hardheaded experimentalist that they are on the same side; just as medieval scholars shielded philosophy by presenting it as the handmaiden of theology, contemporary guardians of thought experiment hope to hire it out as the handmaiden of experiment. But this purchases security at the price of triviality. "Think before acting"—like "Look before you leap"—is a platitude in science, as well as ordinary life. The sceptic about thought experiments readily concedes the value of planning one's experiments. But he cannot be appeased by merely redescribing this planning as thought experiment. Planning a meal is not thought-cooking and planning a trip is not thought travel. At most, thought experiment is a technique used in planning an experiment, not the planning itself. Even if we suppose that it is a useful technique (that it leads to better-executed experiments), we still need to show that it is useful in the right way.

Think of how a psychiatrist might use the technique of visualization to treat a chemist who has developed a phobia about experimentation. The psychiatrist hopes to condition the fear away by pairing relaxed states with visualizations of the anxiety-provoking scenes. The first step down the slippery slope is simply to have the chemist imagine walking by the laboratory. Once that can be done comfortably, the therapist has the chemist imagine walking into the laboratory and then handling some of the apparatus. Eventually, the chemist can calmly visualize performing formidable experiments and returns to a productive career. Although the patient's visualized experiments lead to actual, scientifically valuable experiments, the visualizations are not thought experiments. As therapeutic exercises, they do not seek to answer or raise any questions. Perhaps confident colleagues find that "positive thinking" visualization helps them facilitate excellent experimention, just as expert divers claim that mental practice helps them execute dives. But even these highly particularized visualizations fail to be thought experiments. Other psychological studies of experimentation may reveal that crossword puzzles, aspirin, and a big breakfast help the design of physical experiments. But finding that thought experiments help in this extrinsic way would satisfy neither the sceptic nor the traditional upholder of thought experiment. They are interested in whether thought experiments can help via the sort of inferences scientists make in their capacity as scientists. The problem of thought experiments is only solved once a place is found for them in the context of justification.

Mach's mistakes can be traced to his sensationalism and a one-sided diet of examples. His sensationalism led him to overemphasize the mentalistic aspects of thought experiment and to throw away tools needed to explain its genuinely a priori features. Perhaps because of this, Mach's attention gravitated toward thought experiments he could explain (natural science cases) and away from the recalcitrant normative examples (in aesthetics, ethics, logic, and mathematics). To get a complete account of thought experiments, we should reject sensationalism and consider how armchair inquiry works in general.

4

The Wonder of Armchair Inquiry

> I have now led you to the doors of nature's house, wherein lie its mysteries. If you cannot enter because the doors are too narrow, then abstract and contract yourself into an atom, and you will enter easily. And when you later come out again, tell me what wonders you saw.
>
> Thomas Hariot (to Kepler)

Thought experiment appears to open an enchanted portal. It has the feel of clairvoyance and so excites awe in some, suspicion in others. But the wonder of thought experiment is just a special case of our vague puzzlement about how a question could be answered by merely thinking. There is no mystery when investigators look, measure, and manipulate. Their answers come from the news borne by observation and experiment. But if you just ponder, then the information you have leaving the armchair is the same as the information you had when you sat down. So how can you be better off?

I shall first argue that part of our wonder is based on a modal fallacy, then survey theories of armchair inquiry that promise answers to the legitimate portion of our wonder, and finally I develop a cleansing model of rationality that sets the stage for a detailed analysis of how thought experiments work.

I. The Pseudoanomaly

Three men pay $30 as a group to rent a room. The clerk discovers he has overcharged the group $5 and so gives the bellhop the money to return to the three men. The bellhop decides that five is too hard to divide by three and so gives the men a dollar each and keeps $2 for himself. The three men have paid $9 each, the bellhop has kept $2 making $29. Where did the extra dollar go?

The way out of this conundrum is to challenge the question. Why should there be an extra dollar? This forces an explicit justification for the transition from "The three men have paid $9 each and the bellhop has kept $2" to the conclusion that this money should *add* up to $30. Once the inference is out in the open, we are more apt to notice that the bellhop's two dollars must be *subtracted* from the payment of $30 − $3 = $27 to account for the $25 they paid for the room and the $2 that has been stolen by the bellhop.

Sometimes exposure of a fallacy behind a question does not wholly dissolve it. For the question can be motivated by a variety of inferences, some good, some bad. Challenging the question can still help by giving us a more realistic idea of what is needed to answer it. Even if the answer still eludes us, the challenge helps by straightening out the rationale behind the question. Consider the way Darwin cools the overheated problem of transitional species. The objection is "Why, if species have descended from other species by insensibly fine gradations, do we not everywhere see innumerable transitional forms? Why is not all nature in confusion instead of the species being, as we see them, well defined?"[1] The positive aspect of Darwin's response furnishs information. The negative aspect criticizes the objector's degree of curiosity. Why should evolution imply that our environment teems with intermediates? Although the challenge does not show that the question has a false presupposition, it does lighten Darwin's burden of proof. For it unearths sampling fallacies that make the problem of transitional species look worse than it is.

Likewise, much of the wonder expressed by 'Why should thought experiment justify beliefs?' is dispelled with the retort 'Why shouldn't it?'. The counterquestion flushes out a fallacy; for people articulate their gut sense that thought experiment is preternatural by stressing variations of the theme that contingent facts about our thoughts can only imply other actualities. Thought experiment, they note, promises an exit from the circle of actual facts: it aims to show how things *could* be or *must* be. Thus, thought experiment violates a natural limit on knowledge.

A. Modal Gap Illusions

Two gap theses are afoot. The less deceptive one is an is/could gap: you cannot reason from how things are to how things could be. But obvious counterexamples flow from the principle that whatever is actual is possible: "Someone walked on the moon; therefore, it is possible for someone to walk on the moon." The is/could gap is more charitably construed as barring inferences to *uninstantiated* possibilities.

If we had to travel to other possible worlds to learn what goes on there, we could know nothing about uninstantiated possibilities or necessities; for possible worlds are not foreign countries or planets. Rather, they are alternative complete universes and so are physically inaccessible. Nevertheless, there are standard patterns of inference entitling us to infer facts about ways things could be—about other possible worlds. From the fact that lightning struck my neighbor's house, I infer by analogy that lightning can strike mine. Mach pictured the appeal to conceivability as a similar analogical inference: from the fact that identical triplets can be born in thought, I infer that they can be born in the real world. There are also valid *deductive* arguments from actual premises to uninstantiated possibilities: Donald Trump knows it is possible to make a bowling ball out of gold, therefore, it is possible to make a bowling ball out of gold. Once we get our initial stock of possibilities, we can derive new possibilities. Some of the more staid inferences are codified in modal logic. Others use more freewheeling metaphysical principles. The principle of recombination tells us that whatever is separately possible is possible together (plaid elk) and whatever is possible together is possible separately (furless elk). The principle of nonarbitrariness lets us pass to neighboring quantities.[2] For example, the first evidence of neutron stars was good news to believers in black holes because a neutron star has a radius that is only a few times the critical value needed for transformation into a black hole: "If a star could collapse to such a small size, it is not unreasonable to expect that other stars could collapse to even smaller size and become black holes."[3]

The more widely held is/must gap can be refuted in the same manner as the is/could. According to this thesis, a contingent premise never implies a necessary conclusion.[4] The converse of this principle is true; what is implied by a necessary truth must itself be a necessary truth. Nevertheless, a necessity need not be derived from another necessity: "All of Victoria Jackson's beliefs are true. She believes there are infinitely many prime numbers, therefore, there are infinitely many prime numbers." Indeed, any argument having a necessary truth as a conclusion is trivially valid. Since many of these arguments have no false premises, we can further say that there are *sound* arguments which reason from contingent premises to a necessary conclusion.

Perhaps the real spirit behind the tempting requirement centers on *cogency*: contingent premises are never key players in a rationally persuasive argument for a necessary conclusion. But there are several kinds of cogent *inductive* arguments for necessary conclusions. Some of these proceed from necessary premises. For instance, mathematicians infer that primes become rarer and rarer by simply measuring their frequency along the number line.[5] Logicians often reason by analogy. G. H. von Wright's first system of deontic logic was explicitly based on the resemblance between the deontic notions of obligation and permission and the modal concepts of necessity and possibility.[6] At the applied level, the technique of refutation by logical analogy consists of comparing the argument in question with one that is plainly invalid.

Other inductive arguments for necessary truths proceed from contingencies. Since similar things are governed by similar laws, the discovery of a hidden similarity invites an extension of the laws governing one thing to its newly recognized analogue. For example, Darwin's evolutionary theory encourages us to extend laws about human physiology to gorillas. Discoveries overturning apparent differences will also invite nomic colonization. Frederich Wohler's production of synthetic urea in 1828 weakened the relevance of the organic/inorganic distinction, inviting the extension of chemical principles into biology. Indeed, breaking a disanalogy sometimes opens the way for a grand unification. By showing that the earth revolves around the sun, Copernicus cast doubt on the Aristotelian separation between terrestrial and celestial laws and cleared the ground for Newton's universal physics. An inductive argument for a necessity can be viewed as an enthymeme with a statistical premise that bridges the gap between what is possible and what is actual:

1. No mammal has ever grown green fur.
2. For many of these animals, green would be the best coloration.
3. For most species, if the species can evolve the best trait, it has.
4. No mammal can grow green fur.

Another formulation treats nondeductive arguments for necessities as inferences to the best explanation. The conclusion is then regarded as explaining the premise. Under this view, we reason directly from the absence of green fur to a mammalian incapacity.

Since stronger necessities explain weaker ones, best explanation inferences let us power up from weaker impossibility results to stronger ones. The appeal to inconceivability can be read as an instance of this nomic amplification: "It is impossible for me to think of a counterexample to 'All Fs are Gs'—hence, necessarily, all Fs are Gs." The argument is driven by the explanatory power of the conclusion; the failure to perform a task is always neatly explained by the task's impossibility. But as emphasized by critics of the appeal to inconceivability, the conclusion has to outperform rival explanations of our cognitive failure: bias, distraction, impatience, incompetence, ignorance. This pattern of criticism is prefigured in critiques of the appeal to ignorance because the appeal to inconceivability is analogous to this fallacy. An argument from ignorance infers the falsehood of p from lack of knowledge that p. This is an acceptable effect to cause inference when our evidence supports the counterfactual 'If p were the case, then p would be known'. The problem is that people are prone to overestimate their epistemic access to the world and so overlook rival explanations of their ignorance. Like the appeal to ignorance, the appeal to inconceivability is a slippery fish because people overestimate their imaginations. One source of the illusion is that there is no direct method of ascertaining your own limits; you cannot learn these boundaries by surveying possibilities that you cannot imagine. It is not like learning of your limited ability to lift or eat. The limits of your own imagination are learned in slow and fitful measures, principally by analogy with the vacancies of others and memory of your own past narrow-mindedness.

B. How Thought Experiments Yield Modal Conclusions

Happily, the sort of inductive modal argument of greatest interest to us is one of the simplest, the appeal to authority:

1. Psychological premise: Peter van Inwagen believes a deliberator cannot view his contemplated action as inevitable.
2. Reference class identification: Peter van Inwagen is a philosopher who failed to think of a counterexample after a diligent attempt.

3. Accuracy claim: Most philosophers can think of a counterexample to a false generalization about deliberation if they try hard.

4. No one can deliberate about a perceived inevitability.

Since the argument is inductive, addition of further premises can affect the degree to which the conclusion is supported. Learning that van Inwagen specializes in the freewill issue enhances the argument. Learning that he formed the belief while ill would be bad news for the argument. Since the appeal to a single authority only lends modest support to a conclusion, a conscientious investigator wants further authorities. So van Inwagen volunteers the generalization that all of those who have thought about deliberation agree with him. Information about how such a consensus forms is also relevant to gauging the weight of authority, so van Inwagen leads his reader through a sample thought experiment (see p. 36).

The customary provisos also extend to the arguments seeking to establish *possibilities*:

1. Psychological premise: Georg Simmel believes it possible for an endless universe of finitely many objects to never repeat the same situation.
2. Reference class identification: Simmel is a philosopher who soberly imagined such a universe in all relevant detail.
3. Accuracy claim: A proposition that can be soberly and precisely imagined to be true by a philosopher usually frames a genuine possibility.

4. It is possible for an endless universe of finitely many objects to never repeat the same situation.

Simmel also specifies his procedure: his three-wheel thought experiment. Some of his readers lack the mathematical background to appreciate how the wheels invalidate Nietzsche's argument for the eternal return. But these half-informed readers are nonetheless, rationally reassured by the presentation of the thought experiment. Mathematical innocents defer to Simmel's expertise and gain further confidence from the fact that no one has challenged the mathematical aspect of his thought experiment. As in the publication of ordinary experiments, detailed specification of your procedure enhances the persuasiveness of your testimony. By making the thought experiment common knowledge, you enlist the cognitive authority of the community of inquirers; for their silence is itself evidence that no error exists.

Silence is not always golden. Listen to the argument from the cross: if there were anything wrong with the death penalty, Jesus would have condemned it from the cross; but Jesus did not broach the topic. Other Christians offer rival explanations of his silence: maybe Jesus had other topics on his mind, perhaps his agony diminished his lucidity, or maybe the evil of judicial execution was too obvious for words. A good appeal to silence must top all counterexplanations. Happily, academia provides a forum for an important subset of these appeals to silence; for there are rewards for exposing interesting falsehoods. Hence, the failure of interested researchers to challenge a claim is evidence in favor of that claim. On the other hand, little comfort can be taken from the quiet surrounding dull works or those soundproofed by censorship.

1. *Supporting the psychological premise.* The most straightforward revealer of an authority's position is the direct question, Do you believe p? If the authority reports that he does and if he is sincere and competent, then we have good evidence for the psychological premise. Unfortunately, respondents frequently lie about their own opinions and preferences in order to please or frustrate the interviewer. Others answer randomly out of fatigue or cussedness. Respondents are also fallible. Attitudes are not self-intimating. The mistake may be due to distraction, confusion, or self-deception. Another weakness of direct questioning is nonresponse. Questionees are often unable or unwilling to answer a direct question about their beliefs and preferences.

Survey researchers cope by carefully designing their questionnaires to reduce insincerity and error while increasing the response rate. Lying is minimized by concealing the researcher's preferences and adding related questions that detect inconsistency. Mistakes are prevented by using short, clear questions. Researchers increase the response rate by framing questions delicately to avoid embarrassment, by guaranteeing anonymity, setting deadlines, begging for answers, and offering incentives for completion. Survey research has become increasingly sophisticated and professionalized over the past fifty years. Each nuance of the question is the subject of many research articles with titles such as "Trends in Nonresponse Rates," "Effects of Question Order on Survey Responses," and "What We know about 'Don't Know'."

The closest analogues to thought experiments in the literature on questionnaires are *vignettes*—"systematically elaborated descriptions of concrete situations."[7] These little stories are used to cope with a particular subrange of survey problems. The first is that respondents have trouble with abstract questions that leave their habits of judgment uncued. A second obstacle is that the researcher is often interested in variables that are highly correlated. The few actual cases in which they are apart tend to be hard to find. The high correlation also makes control difficult. A third problem is that direct questions are more susceptible to bias because the point of the question is more salient. Direct questions also require the questionee to be insightful about the factors that influence his own judgment. Vignettes surmount these four problems by using hypothetical, detailed, scenarios:

> Vignettes are short descriptions of a person or a social situation which contain precise references to what are thought to be the most important factors in the decision-making or judgment-making processes of respondents. Thus, rather than allowing or requiring respondents to impute such information themselves in reacting to simple, direct, abstract questions about the person or situation, the additional detail is provided by the researcher and is thereby standardized across respondents.[8]

Vignettes have been used in the studies of jury decision making, social status, and the assignment of responsibility for crime to victims. Here is a sample from a vignette study intended to measure social status:

> Marvin Silver, 34, and his wife Sheila, 31, have been married two years. They have just moved into their first home in Wellesley, Massachusetts. They met as students in business school. Mr. Silver has just been appointed assistant administrator of a small Boston hospital. The Silvers have a one-year old daughter (who is a victim of cerebral palsy). Mrs. Silver does not work, but she does volunteer jobs for the Crippled Children's Fund.[9]

This story was followed with two groups of questions. One contained questions about which types of people the Silvers were most apt to associate with: An architect? A barber? An electrician? The second concerned reputational information: their shopping habits, recreational preferences, politics.

Thought experiments share many of the virtues of vignettes because both methods are intended to solve similar problems of psychological measurement. However, their problems are not equally severe. The audience for a thought experiment is more motivated, educated, and cooperative. They are addressing the question because they think it may provide an insight into the issue at hand. People who receive vignette surveys, on the other hand, lack any antecedent interest in the topic and do not expect enlightenment. This makes them less likely to consider the story carefully. One disadvantage of thought experiment is that the questionee usually understands and values the import of the available answers and so is apt to be biased by theoretical commitments. The casual presentation of thought experiments also gives wider play to interviewer bias; the questioner is apt to ask the question in a way that gives some answers an illicit boost.

Sometimes our interest in the thought experiment only extends to the support it gives to the psychological premise. We ignore the support it gives to the remaining premises in the way a sound engineer ignores most of a debater's performance when listening for the acoustic properties of the auditorium. For example, Jean Piaget uses thought experiments to plot the intellectual development of children. To show that they pass from an objective stage (where they blame in accordance with consequences) to a subjective stage (where they go by intentions), Piaget told five-to-thirteen-year-olds a story in which one boy, John, accidentally broke fifteen cups and another story in which another boy, Henry, irresponsibly broke one cup.[10] Who was naughtier? At about seven, children say John deserves more punishment. But at about nine, they disregard the extent of the damage and blame Henry. Obviously, Piaget is not treating the children as authorities. Moral psychologists do not care whether the thought experiment reveals that a deed is really right; they are interested moral beliefs, not morality itself.

Philosophers also make truncated use of thought experiments. In "Toward a Psychology of Common Sense," Unger just brackets the question of truth when conducting thought experiments intended to show that our existential beliefs are stronger than our property beliefs (i.e., when forced to choose between 'Some Fs exist' and 'All Fs are Gs', we tend to side with 'Some Fs exist'.)[11] Thought experiments also work in this way when used as an interpretative tool. Thus, to check what an aesthetician means by "Beauty and ugliness are interdependent," one might have him suppose that all the beautiful things are removed from the world. Would any ugly things remain? What about the reverse case? The point of these reconnoitering hypotheticals is to locate a position, not to attack or defend it.

Some philosophical positions restrict you to the truncated use of thought experiments. For instance, a noncognitivist about a field (ethics, aesthetics, politics) denies that utterances within that field have truth values; therefore, insofar as he seeks truth, he limits thought experiment to the role of revealing beliefs and desires. (Note, however, that a noncognitivist need not always seek truth. An emotivist dedicated to famine relief may feel free to use thought experiments as propaganda.)

Of course, psychologists also use thought experiments to back full-fledged appeals to authority. For example, they use Buridan's ass to refute the psychological principle that every choice expresses a preference. In these *untruncated* cases, they take a lively interest in the degree to which the exercise supports the premises ascribing accuracy and assigning reference classes. Even in the truncated case, the thought experimenter himself will aim at providing an authoritative conclusion. Like the debater, he cannot give half a performance.

Thought experiments can provide psychological evidence in unintended ways. One interrogation strategy is to tell the suspect that you are persuaded of his innocence and then coax him into outlining how he *would have* performed the crime. The suspect's answer is frequently self-incriminating. In the movie *Bladerunner*, robots passing as human beings are routinely detected by their emotional reactions to a standardized sequence of hypothetical questions. Just as a logician's proof can give away information about his penmanship, a thought experimenter's demonstration can give away information about his past activities and personality traits. But this is not what I mean by the *untruncated psychological use* of thought experiments. I want the phrase to cover thought experiments that *aim* at justifying the attribution of a psychological state.

The philosopher's most common psychological use of thought experiments is as fuel for the furnace of reflective equilibrium. You attain reflective equilibrium when your theoretical principles cohere with your atheoretical judgments. To reach this state, you must remove conflicts between them. Sometimes the conflict is resolved by giving up the principle and sometimes by giving up the intuition. The idea is to work back and forth, adjusting principles to intuitions and intuitions to principles until you get a tight fit. Your initial stock of intuitions has to be supplemented in the course of deliberations because the original corpus often lacks good test cases for proposed amendments. Thought experiments fill the breach by quickly and cheaply eliciting beliefs of just the crucial sort. Nevertheless, thought experiment is not *essential* to the method of reflective equilibrium because there are many revealers of propositional attitudes: bets, confession, election, hypnosis, insurance sales, legislation, polls, polygraphs, riddles, torture.

One criticism of reflective equilibrium is that it provides no guarantee of the truth of the resulting system of beliefs. My internal harmony could be neat but nutty. The operative bromide could be "Garbage in, garbage out." But do not blame the thought experiments! If they are only intended to reveal propositional attitudes, then they have accomplished their mission even if the attitudes are asinine.

2. *Supporting the reference class premise.* Most thought experiments purport to be more than devices for revealing propositional attitudes. The typical thought experiment is designed to underwrite all three premises of the appeal to authority; the question is posed in the hope of eliciting a *reliable* opinion. The power of a thought experiment to reveal attitudes is typically viewed as only a means to the ultimate end of obtaining knowledge. Thus, the poser of a thought experiment is untroubled by the influence his question exerts on the questionee's belief as long as the questionee is not changing to a less reliable belief.

Who does the thought experimenter have to be? The reflex egalitarian answer is "No one in particular." The reflex elitist answer is "An expert." But when we examine the interviewing techniques of people leading a thought experiment, we see that they are aiming for the sort of categories used to qualify judges and scientists. Thus, the thought experimenter should be sane, sober, attentive, calm, careful, honest, impartial, informed, and mature. These traits are favored because common sense assures us that they are conducive to sound judgment. The point of posing a thought experiment in a special way is to place the thought experimenter in this esteemed reference class—at least for *this* type of question and in *this* type of setting. Critics can lower confidence in the thought experiment by disproving this premise about the thinker's credentials—hence, the faintly ad hominem litany of complaints about bias, distraction, and self-deception. This attack on the thought experimenter is no more fallacious than inquiries into an eyewitness' character, motive, and perceptual capacities and opportunities.

The need to anticipate objections forces the designer of a thought experiment to build in a battery of precautions. Indeed, thought experiment is subject to the dialectical pressures associated with all experimental inquiry. This similarity of selective factors forms an environment for convergent evolution between thought experiments and ordinary experiments. Like thought experiments, ordinary experiments are crafted to provide background evidence for the premises of an appeal to authority:

1. Psychological premise: Lazzaro Spallanzani believes that salamanders regenerate limbs.
2. Reference class identification: Spallanzi is a competent biologist who arrived at his belief by monitoring randomly amputated salamanders.

3. Accuracy claim: Most of the biological beliefs of competent biologists are true when based on a randomized study of a salient effect.

4. Salamanders regenerate limbs.

Spallanzani's experiment is crafted to provide evidence for all three premises. By contriving his experiences, the experimentalist makes himself an authority and provides a recipe for making others into authorities. Ordinary experiments must be designed to exhibit the relationship between the independent and dependent variables clearly. Thus, the procedure should be simple to follow, so that it can be repeated by others. The effect should be made as noticeable as possible in order to minimize the burden on one's perceptiveness, skill, and luck. Extraneous variables must be minimized so that your results cannot be traced to confounding factors. Experimentalists have responded to these problems with a variety of solutions that have been standardized into common practice. Thought experimenters have faced the same sorts of problems and so have developed similar sorts of solutions. So we should expect experiment and thought experiment to have similar structures, just as we expect similarity between the wing of a bird and the wing of an insect.

Of course, these safeguards do not guarantee that thought experiment will be reliable. Evidentiary law has accumulated centuries of safeguards for eyewitness testimony; yet in the last thirty years, psychologists have demonstrated that eyewitness testimony is seriously overrated. What assurance do we have for believing that testimony based on thought experiment is as reliable as touted?

3. *Supporting the accuracy premise.*   For all we *know*, thought experiment could be significantly overrated or underrated. Suggestive evidence points in both directions. On the one hand, historians present thought experiments as significant contributors to scientific progress; for they readily credit thought experiments with improving the climate of opinion. Many philosophers reverse the commonsense picture of scientific studies and present them as islands of experience amid a sea of theory. If the theory exhalters are right, thought experiment is entitled to a lot of credit. On the other hand, there is a gloomy trend in the psychology of reasoning that suggests that human beings are surprisingly susceptible to fallacies. Even professional statisticians slip into primitive styles of thinking when the problems are not packaged in a way that triggers their formal training. An enterprise that prides itself on its power to summon "intuitions" may thus encourage just the irrational modes of thought that educators labor years to correct.

Critics of thought experiment will remark that the appeal to authority is traditionally listed as a fallacy in logic textbooks. However, recent textbooks carefully distinguish between legitimate and illegitimate appeals to authority. Bad appeals confuse popularity with authority, equivocate between epistemic and nonepistemic authority, misrelativize expertise (an authority in one field need not be an authority in another), overlook degrees of expertise, and rely on unrepresentative samples of expert opinion. When I speak of thought experiments as supporting appeals to authority, of course, I intend the wholesome sort.

Normally, epistemic authority is conferred by the individual's training, privileged perceptual capacities or opportunities, or his record of successful judgment. Although these factors often play a small role in the appeals driven by thought experiments, the major factors are those associated with careful judgment. The question itself is intended to prime the respondent, to make him a clear thinker on one aspect of the issue (just as juries are shepherded into better judgment). Since the vast majority of human beings are capable of being coached into this sort of lucidity, the bulk of humanity can be quickly brought up to speed as competent thought experimenters for many issues. Hence, the mechanism underlying thought experiment must be nearly universal to human beings.

Mach correctly identified this mechanism as evolution. Of course, only organisms with beliefs face any adaptive pressure to have true beliefs and to avoid false ones. Darwinian forces have not made the asparagus wise. But if Mother Nature is going to pay for beliefs, these internal representations must contribute to the four Fs: feeding, fighting, fleeing, and reproduction. The Strategy of Falsehood outperforms the Strategy of Truth in some conceivable environments. An organism with self-destructive desires would be helpfully handicapped by stupidity. A morose creature might function best when susceptible to optimistic delusions. An animal only capable of reasoning in accordance with universal generalizations would profit by overlooking rare exceptions. But notice that the Strategy of Falsehood only works when compensating for a flaw. Although Mother Nature occasionally adds such epicycles, she generally prefers to get it right directly.

She also keeps us informed on a need-to-know basis. Knowledge consumes energy; hence, channels of useless information dry up. Witness the degenerating eyes of cave species, the vestigial status of human ear muscles, and our diminishing sense of smell. Since most truths are useless to human beings, there is a presumption of ignorance. This presumption can be overridden when knowledge of the useless fact can be explained as a by-product of an epistemic capacity that does earn its keep. But notice that the burden of proof goes against the attribution of knowledge.

Happily, the evolutionary story suggests that we stick to what we know. Ancestors who strayed to other topics wasted energy and were beached at the edge of the gene pool. There is selective pressure against traits that fail to help an organism, as well as those that harm it. Witness the flightlessness of island birds and the lack of pigment among deep sea fish. So beliefs that fail to help human beings should disappear; the corresponding questions should fail even to arise. The pleasant presumption is that we only bother to form beliefs about issues that are within our realm of competence. Thus, a thought experiment that succeeds in eliciting an opinion has also succeeded in eliciting an opinion that is apt to be correct. Hence, evolutionary theory gives us a general basis for according epistemic respect to the results of thought experiment.

Of course, this general basis is only presumptive and only a matter of degree. Here and there we find pockets of incompetence. Useless and even harmful traits linger because of their linkage with helpful traits. Reliable belief-forming mechanisms might be unavailable or too costly. Or they might start out reliable but then become outdated or supplanted by a cheap substitute. A little imagination reveals lots of ways for evolution to generate irrationalities. It also reveals lots of ways for a free market to generate wasteful business practices. But that hardly constitutes grounds for agnosticism about the connection between competition and efficiency—or rationality. Most of these pessimistic possibilities are parasitical on a general level of reliable performance. One cannot be all wart and no face. Moreover, the pessimistic possibilities are feeble; their plausibility varies inversely with the amount of biological detail we add about human beings.

The mechanisms responsible for dips in our general level of performance may also lead to jumps above this level. The ability to learn calculus might tag along with the ability to learn addition. George C. Williams has suggested that "advanced mental qualities might possibly be an incidental effect of selection for the ability to understand and remember verbal instructions early in life."[11] Highly directable children are less accident-prone. The subabilities constituting this trait can be diverted to intellectual tasks that lack a payoff in fitness.

It is natural to wonder how a practical process like evolution can give rise to abilities to do theoretical work. Nevertheless, the curiosity rests on an overdrawn contrast between practical and theoretical abilities.[12] The costs of encoding and applying information create a drive for simplicity and hence generality. Evolution is not an entirely backward-looking process. Since the future will not be exactly like the past, a well-adapted animal must be capable of coping with unprecedented conditions. Hence, rather than tightly tailoring cognition to the actual environment, Mother Nature loosely suits us to a range of possible environments. She passes over highly specialized rules of inference in favor of a diversified portfolio of rough-and-ready rules. The redundancy makes for acceptable overall performance over a wide (but still limited) range of situations. The downside is diminished accuracy within some particular situations, along with a wider but milder susceptibility to illusion and confusion.

Happily, we are self-corrective creatures who can demote a belief-forming illusion into a mere impression-forming illusion. Pilots learn to let their instruments override their sense of up and down. Physicists rightly disregard their commonsense reactions to thought experiments about fast projectiles in favor of tutored intuitions. Thus, we find thought experimenters reporting their cogitations in the language of strata. First reactions are described, along with considered judgments. Ethicists are familiar with this layered response from R. M. Hare's theory that moral thinking is conducted at both an intuitive and a critical level.[13] Unger's dualism is between dominant and recessive responses.

Evolution vindicates methodological conservatism—the view that the very fact that you believe p is reason for continuing to believe p. The degree of self-justification may be low, but it is high enough to give thought experiment its initial, general momentum.

II. Positive Theories of Armchair Inquiry

The illegitimate part of our wonder about thought experiment boggled at its *possibility*. It made us want to repudiate thought experiment or to relish it as an epistemological miracle. The legitimate portion of our wonder is directed at its mechanism. Having reason to regard a source as reliable is compatible with curiosity about its inner workings. A curious boy knows that calculators are reliable because of his personal success with the ones he has tried and their general success in the marketplace. But he still wants to know how they work. Thought experiments can be legitimately studied in the same spirit.

So how does thought experiment improve the epistemic state of the thinker without the addition of new information? Note that answers to this question may extend to regular experiments: the fact that a method enlightens by bestowing fresh information does not preclude the possibility that it simultaneously enlightens in other ways. Philosophers and psychologists have proposed four positive models of armchair inquiry. They all have merit but are sketchy. Hence, these skeletal accounts will be enfleshed with applications to particular thought experiments.

A. The Recollection Model

Mach pictured thought experiment as a device for converting kaleidoscopic intuitions into precise, self-conscious doctrine. The idea that we have a storehouse of implicit knowledge whose contents can be converted into explicit knowledge goes back to Plato's doctrine of recollection. Plato depicts Socrates as developing this doctrine in response to a dilemma posed by Meno. A questioner either has knowledge of the answer or has no knowledge of the answer. If he knows the answer, then his question is pointless. If he has no knowledge (as Socrates professed), asking is equally useless because the know-nothing has not the means to recognize the correct answer even if he receives it. Hence, inquiry is either redundant or doomed. Socrates ducks the dilemma by qualifying his ignorance: he concedes that there is a sense in which the inquirer has prior knowledge of the answer. He demonstrates this weak sense of knowledge with a slave boy. Since the boy was born in Meno's house, Meno knows that the boy was never taught geometry. Nevertheless, Socrates elicits an interesting geometrical theorem from the boy by artful interrogation. Socrates' questions are leading in the sense that he draws attention to relevant consequences, but they are not substantive hints.

Socrates takes the slave boy's final answer as proof that the boy knew the answer all along; the questions were only *reminding* the boy of what he already knew. According to the full-blown doctrine of recollection, the boy knew the answer in an explicit way prior to his birth but lost this firm grasp during the transition from that preexisting state to his life as a slave boy. All learning, says Socrates, is recollection of previous knowledge. Thus, the role of the teacher is not to impart new knowledge but to revive old knowledge. Even ordinary observations are demoted to memory cues. An empiricist says that bird-watching gives knowledge of birds by induction from many instances of birds. But Plato says the bird-watchers are only being prompted for knowledge they had of the form *bird*. This is mad-dog rationalism: no knowledge is counted as a posteriori. According to Plato, the senses only enable the thinker to be prodded into recovering what he already has. The situation is reminiscent of the wealthy brothers in Conan Doyle's *Sign of Four*. Their father had hidden a treasure on the grounds of the estate but died before revealing its exact location. In one sense, the brothers possessed the treasure; but in another, they did not.

Although Plato's explanation of recollection in terms of preexistence was never credible, philosophers have been steadily impressed with the analogy between remembering and analytic knowledge. For memory gives us a case in which a person makes an epistemic advance without learning new information. He draws the answer from within.

Norman Malcolm's distinction between pure and impure memory sharpens Plato's insight. According to Malcolm, I remember p if and only if I know p because I used to know p.[14] One apparent counterexample to this definition is the case of a neophyte bird-watcher who remembers seeing a lapwing last week though he only later learned how to identify lapwings. Another case is the lady who remembers that her childhood home faced west even though as a girl she merely noticed that it faced the setting sun. Malcolm replies that these are instances of *impure* memory in which later knowledge or inference is mixed with the memory proper. Regardless of whether this distinction saves Malcolm's definition, it does answer a question about thought experiment: What profit is there in merely replaying a past experience? The answer is that although there is little profit in pure recollection, the thought experimenter can prosper from impure memory. Like a cow chewing and rechewing its cud, the thought experimenter reprocesses remembered experiences in light of new interests, abilities, and more recently acquired information and concepts. He can also redistribute his mental assets over time. We can only attend to so much while perceiving. Our ability to replay old experiences lets us borrow attention from the future. The reprocessing lets us squeeze more mental nutrition from our old evidence.

Many thought experiments function as reminders. A standard physics thought experiment features a pilot who forgets to fasten his seat belt when performing a front-back loop. At the top of the loop, the pilot is upside down. Does he fall out of the plane? One is initially inclined to answer 'yes' on the grounds that gravity will pull him down. But since you figure the question would not have been asked if the answer were so straightforward, you check for forces that might counteract gravity. Many people are led to think of centrifugal force because the hypothetical stunt reminds them of carnival rides. They increase the similarity by imagining the pilot doing a left-right loop. The pilots feet would press against the floor, he would not be leaning just on his side. This comparison leads to a negative answer for the vertical loop case: centrifugal force would keep the pilot in the plane. Sometimes there is a follow-up question: does the speed of the plane matter? An affirmative answer is obtained by considering a sequence of slower and slower loops until the situation becomes equivalent to past ones in which you slowly turn a container upside down thereby dumping the contents.

Executed experiments also owe some of their success to their service as reminders. The topic of air resistance is sometimes introduced by having students drop playing cards onto a table. When the card is dropped thin edge downward, the falling card slides away from the table. When dropped with the edge parallel to the table, the card floats down onto the table. This slow descent reminds students of a parachute and so they are led to think about the relevance of the surface area of a falling object. The relationship between heat and air pressure is demonstrated by a bottle and coin. After chilling the empty bottle for an hour, place a wet coin on top, and wrap your warm hands around the bottle. Soon, the coin bounces up and down a bit. The resemblance to the dancing lid of a boiling pot of water helps you realize the air is expanding with the increased heat.

The Greeks knew that imagination can enhance memory because they used the method of loci, or memory palace, probably invented by Simonides of Keos (556–468 B.C.) a poet and teacher of rhetoric. The technique is directed at the task of remembering a long sequence of items. One proceeds by constructing an image to represent each item and placing those images in a familiar setting such as the rooms of one's home or the route to work. When you wish to recall the items, you need only take a mental stroll and note the images you have distributed in the setting. Although Plato's contempt for the world of particular things led him to discourage the use of mnemonic techniques (even note taking!), the method of loci does provide knowledge; for if you remember the sequence, you know the sequence. The method of loci is within the context of justification. It *appears* to be a technique of discovery because the process is a private, nonargumentative sequence of surreal images. But unlike a daydream or a hallucination, the images have been deliberately and self-consciously programmed by the thinker to remind him of facts systematically.

Some thought experiments make theoretical points more memorable by whetting our intellectual appetite. As can be detected from what sells in the news and entertainment industries, people like to think about dramatic, concrete, conflictual situations. A "crucial" thought experiment can satisfy this craving by deriving conflicting predictions about what would follow an upheaval in the ordinary world. A smart four-way example opens with the supposition that the sun suddenly disappears.[15] What happens to the illumination and position of the earth? On the Ptolemaic view, the earth is plunged into darkness but remains in orbit. Given a Copernican system in which the speed of light is assumed infinite, the instantly darkened earth flies off on a tangent. After the discovery that the speed of light is finite, the Copernican would say that earth remains illuminated for about eight minutes but still flies off instantly. Given the special theory of relativity, the earth enjoys eight minutes of continued orbit in addition to its eight minutes of illumination because gravity cannot operate faster than the speed of light. Comparing and contrasting the four theories in this dramatic way makes it more likely that we will think about the theories and increase the time we think about them.

Crucial thought experiments also enhance understanding by teasing apart principles. Intuitions based on rival principles tend to converge on easily anticipated cases. For example, all ethicists end up condemning cruelty to animals. Jeremy Bentham takes the direct utilitarian path and appeals to the requirement that we minimize pain. Immanuel Kant placed animals outside the moral community but still managed to condemn cruelty to animals on the grounds of self-corruption: today's dog beater is tomorrow's mother beater. Some Christians condemn inhumane treatment as disrespectful stewardship: to beat a dog is to abuse one of God's handiworks. This uniformity of result makes the sources hard to rate. Amartya Sen compares the problem to the one statisticians face when the explanatory variables from alternative models move together. To separate the sheep from the goats, we must use cases that "push us in different directions, even if such examples are somewhat less common than the ones in which either approach will recommend the same answer. It is methodologically wrong, therefore, to ignore the relevance of our intuitions regarding rather unusual examples which are brought for this reason into moral arguments."[16] The limit of Bentham's condemnation is marked by imagining that the beater knows the animals are insentient. To test Kant, we suppose the dog beater takes an anticorruption pill before engaging in his evening delights. The appeal to Christian stewardship is challenged by asking whether dog beating would be okay in a godless world or in a world in which beating refined the dog in the way a blacksmith's blows enhance a lump of iron.

Since the cognitive value of contrast is general, Sen's point is not restricted to ethics or even thought experiment. The artificial settings of laboratory experiments are also designed to pull large predictive differences out of small theoretical divergences. Since big differences stand out, they are easier to detect, appreciate, and evaluate.

Thought experiments also increase comprehension and memorability by providing cases especially suited to interpolation. Thus, economists study the pure command economy and the pure free-market economy because they are endpoints of a continuum from which more realistic cases can be inferred by edging inward. By focusing on the extremes, we need only remember two economic systems plus the method of calculating intermediate cases. The same trick is applied to create the physiologist's endomorphs, mesomorphs, and ectomorphs, as well as the physicist's perfect conductors and insulators. Interpolation is not restricted to *imaginary* extremes; the periodic table and the pursuit of pure samples illustrate the same cost-cutting measure.

Thought experiment is not dedicated to remembering long sequences of facts. But it may be analogous to the method of loci in that the technique of thought experiment uses imagination to remind us of facts systematically. Both methods exploit our excellent visual capacities. This visual emphasis of the method of loci was noticed long ago by Cicero:

> It has been sagaciously discerned by Simonides or else discovered by some other person, that the most complete pictures are formed in our minds of the things that have been conveyed to them and imprinted on them by the senses, but that the keenest of all our senses is the sense of sight, and that consequently perceptions received by the ears or by reflection can be most easily retained in the mind if they are also conveyed to our minds by the mediation of the eyes.[17]

It is no accident that we almost always *visualize* the scenario rather than "auralize," "tactualize," or "odoriferize" it. A large portion of the human brain is dedicated to solving the sophisticated visual problems that were crucial to our preverbal ancestors: "Because this wisdom is embodied in a perceptual system that antedates, by far, the emergence of language and mathematics, imagination is more akin to visualizing than to talking or to calculating to oneself."[18] Going visual lets us play to our strengths even though we may have trouble articulating its results. Another similarity between thought experiment and the method of loci is the importance of making the main characters few but salient, the background familiar and unassuming, and the conditions of observation as normal as possible.

It is one thing to have the data, another to draw interesting conclusions from the data. So we pay dearly for the organization of information. Librarians, statisticians, and accountants draw salaries for arranging facts in a way that increases access to shared information and which facilitates inference from the data once accessed. (Equally telling is the fact that we pay dearly for the *dis*organization of data: secret codes, disinformation campaigns, paper shredders.) The brain itself has metalevel nervous networks. Instead of gathering external information, they monitor the activity of other nerves and act as filters and amplifiers. The method of loci shows that conscious devices are also used to facilitate the storage and retrieval of information. Thought experiments may be another mental device for recovering the right datum.

B. The Transformation Model

Linguistic philosophers portray the a priori refinement of implicit knowledge as proceeding from knowing *how* to knowing *that*. The idea is that we all have knowledge of how to speak the language in which we are philosophizing. Since philosophical theses concern meaning and meaning is governed by the rules of the language, we can settle philosophical questions by codifying our mastery of linguistic rules.

By asking 'What would you say if p were the case?', I elicit a particularly telling sample of speech. The formulation of the question misleadingly suggests that I am asking for a *prediction*. If prediction were the point, we would consider factors that might lead one to misspeak (stress, inattention, misunderstanding) or exhibit idiosyncrasies. Linguists who try to predict speech errors do ask 'What would you say?' this way. But philosophers read the question prescriptively, as 'What *ought* you to say'. (Prescription, rather than prediction, is also the point behind the ethical hypothetical 'What would you do if x?'.) The procedure is to give a sample of *correct* speech ("under the breath") in response to a make-believe situation. When executed by a naive informant, the procedure is apt to have little impact on his beliefs. But when the informant is a philosopher or linguist, the procedure is rationally persuasive. By monitoring your own performance, you become living testimony for the thesis at issue. Consider Harry Frankfurt's refutation of 'A person is responsible only if he could have done otherwise'.[19] He has us suppose that a scientist has wired up a man with a fail-safe device. The device will cause the man to do a bad thing if he does not do it on his own. As it turns out, the device is not activated because the man does the wicked deed of his own accord. Is the man responsible? Since we are inclined to describe him as responsible for the deed, our mastery of English gives us evidence against the entailment rule. Knowledge of how to use 'responsible' transforms into knowledge that responsibility is compatible with the inability to do otherwise.

Are all appeals to ordinary language thought experiments? Clearly not. Sometimes the linguistic informant is not construing the situation as hypothetical. This occurs whenever the informant's speech is drawn from the field (such as Finney's legal testimony cited by Austin in "A Plea for Excuses"), the informant is asked to describe an actual situation (as when a linguist presents colored marbles and asks for color identifications), or when the informant is deceived into thinking the situation is actual. Another exception is the cooperative but naive informant who recognizes the hypothetical nature of the situation but does not understand the point of talking about it.

What about the converse? Are all thought experiments appeals to ordinary language? The analytic philosopher's fondness for semantic ascent will attract him to this reduction. However, only a proper subset of thought experiments are attempts to stimulate careful descriptions of imaginary scenes. One class of counterexamples are thought experiments that serve as self-conscious preference-revealing devices. A woman trying to settle on whom she really loves might explore her preference structure with a string of hypothetical questions: If I had a love potion, to whom would I give it? If Renaldo were poor and Pedro were rich, would I then prefer Pedro's company? Would I stand by Renaldo if my family and friends turned on him? Her answers reveal something about her particular desires, rather than her language. Often, the point of revealing the thought experimenter's preference is to show the *generality* of the preference:

> People experience losses and gains with different levels of intensity. Specifically, "losses loom larger than gains." To illustrate, imagine that you have just been informed that one of your investments has made you $10,000 and consider how happy you would feel about the event. Next imagine that you have just learned that you have lost $10,000 and consider how unhappy you would feel about this. Now try to compare the intensities of your feelings in both cases. Is a gain of $10,000 experienced more intensely than a loss of $10,000 or vice versa? For most people the displeasure of a loss is experienced more intensely than the pleasure of an objectively equivalent gain.[20]

When you learn you are more concerned about losses than gains, you acquire a premise for the conclusion that most people care more about losses.

A related class of nonlinguistic thought experiments (ones that merely reveal beliefs) have been discussed. Like the desire revealers, they are sometimes used to provide a lemma for a conclusion about a whole population. This is not quite the same as the method of empathy. When sociologists and historians project themselves into the positions of other agents, they are trying to learn their own *hypothetical* beliefs and desires in order to infer the real agent's beliefs and desires. Interestingly, the method is also used to reason from one's own hypothetical reaction to the fictional beliefs and desires of characters in novels.

A third class of counterexamples to 'All thought experiments are appeals to ordinary language' involves hypothetical experiences. This is how Frank Jackson attacks *physicalism*, the view that the actual world is entirely physical. Suppose that "Mary is confined to black-and-white room, is educated through black-and-white books and through lectures relayed on black-and-white television."[21] She learns all the physical facts about the world in this way. When Mary is released from the room, what will her reaction be? Ho, hum? No, she will learn something exciting, what it is like to see in color. So Jackson concludes physicalism leaves something out.

Lastly, a large group of nonlinguistic thought experiments investigate nonlogical necessities. Most scientific thought experiments concern what is possible relative to natural laws (physical, chemical, biological, etc.). Thus, the transformation model does not account for the bulk of these paradigm cases. Ironically, Mach can handle synthetic but not analytic thought experiments, the conventionalist can handle analytic but not synthetic ones.

Some may think the conventionalist's silence is tolerable because he only wishes to explain *philosophical* thought experiments.[22] But we metaphilosophical gradualists deny the wisdom of this issue separation. Philosophy is essentially protoscientific. It speaks to the miscellany of questions that cannot be addressed in standardized ways. One of the philosopher's jobs is to develop approaches that have some prospect of becoming a research policy for these homeless questions. When successful in a small way, the question is pushed into another, better-established field. For example, 'What is the origin of the universe?' has been recently delegated to astrophysics. Of course, there is pull, as well as push. The astrophysicsts were not twiddling their thumbs while the metaphysicians massaged the concepts of space and time: the astrophysicists' development of the big bang theory has unified cosmology in the way evolutionary theory has unified biology. Ambitious outsiders always stalk the perimeter of philosophy in the hope of picking off strays such as 'What is the fate of the universe?' (Lord Kelvin) and 'Why does morality exist?' (E. O. Wilson). When successful in a grand way, philosophy sires new disciplines. Astronomy became independent thousands of years ago, physics a few hundred years ago, and psychology at the turn of the century. Philosophy's progeny will themselves have descendents (biochemistry, psychophysics); but the line of descent can only be detected by intellectual history. Philosophy's relationship with the sciences will be further obscured by the terminological fact that its successes no longer count as philosophy. Hence, from an outside, ahistorical perspective, philosophy looks like two thousand years of fruitless debate. But it's only nonscience in the way that news is nonhistory; the difference between philosophy and science lies in our *relationship* to the subject matter, not in the topics themselves. Just as news is constituted by the external requirement of being recently known, philosophy is under the external constraint of being methodologically unsettled. Hence, the conceptual disarray that is so symptomatic of philosophy does not confine its thought experiments to the job of testing analytic propositions.

C. The Homuncular Model

There is a resemblance between the conventionalist's account and Daniel Dennett's more general discussion of cognitive autostimulation. If one pictures the mind as a thing without parts (as Descartes did), then it is hard to make sense of the common practice of asking oneself questions. The practice would be as futile as "paying oneself a tip for making oneself a drink."[23] But if we follow Dennett in viewing the agent as an imperfectly coordinated complex of cognitive systems, a crew of *homunculi* (little men), then self-questioning can be seen as a valuable communication device: "Crudely put, pushing some information through one's ears and auditory system may stimulate just the sorts of connections one is seeking, may trip just the right associative mechanisms, tease just the right mental morsel to the tip of one's tongue. One can then say it, hear oneself say it, and thus get the answer one was hoping for."[24] There are many techniques of cognitive autostimulation: talking to yourself, drawing pictures, singing to yourself, and so on. Observation of your own description belongs to the same list. Thus, many of the thought experiments smoothly explained by the transformation model are also smoothly explained as autostimulation. However, unlike conventionalism, "homuncularism" handles synthetic propositions.

Another difference between Dennett and the conventionalist is that Dennett pictures the stored knowledge as already propositional. Each part has its store of known propositions, just as different law enforcement agencies have their separate stocks of information. Dennett presents the problem as that of passing pieces of knowledge from one part of yourself to another. The conventionalist pictures the stored knowledge as nonpropositional, as ability knowledge.

The homuncular theory of armchair inquiry explains the possibility of newsless edification by modeling the intrapersonal case on interpersonal information processing. A piece of information that is useless to me may become useful when passed to someone who is better able to digest it or who already has information that can be combined with the new item. Bird-watchers contribute their individual bird counts to a central source that estimates the bird population by pooling reports. Our inefficiency in collating individual reports explains how it can be that "the left hand does not know what the right hand is doing" and how we can fail to "put two and two together." Medievals knew about the wheel, knew about the hand barrow but never made a wheelbarrow. My failure to unite the facts is analogous to a communication breakdown within a bureaucracy. Some thought experiments help by facilitating communication between the self-involved homunculi. For instance, people know that there are fewer than a million hairs on a human head and know there are more than a million New Yorkers but fail to realize that at least two New Yorkers have the same number of hairs on their heads. We can get them to pull their facts together by supposing that the New Yorkers are assigned rooms in accordance with the number of their hairs. Everyone with one hair goes to room 1, everyone with two hairs goes to room 2, and so on. Must two New Yorkers share a room? Since the answer is *yes*, people realize one of the consequences of their two bits of knowledge is that two New Yorkers have the same number of hairs on their heads.

Postulating homunculi helps to explain why people can have strange combinations of abilities and disabilities. Aphasiacs, alexiacs, and agnosiacs are puzzling because they are able to perform certain tasks without being able to perform others that seem closely related (like reading and writing, matching colors and naming them, recognizing letters and recognizing words). Once we picture the mind as much more loose and separate than common sense suggests, we can see the point of otherwise puzzling mental tricks. Psychologists have demonstrated that recall tasks are surprisingly harder than recognition tasks. Thus, mnemonic techniques such as acronymic cues (ROYGBIV for the colors of the spectrum) and the method of loci convert recall problems into recognition problems. Short-answer questions can be converted into multiple-choice questions by visualizing the alternatives. This self-prompting is evident in thought experiments having a small number of possible answers.

The homuncular model suggests that the imagination will have a redistributional role. We will play to our strengths by repackaging questions so that the problems can be contracted out to the most talented homunculi. For instance, we take advantage of our excellent visual judgment when we memorize star positions by picturing them as dots outlining familiar shapes. Another method of artful reassignment is to concretize the situation. This explains the popularity of theoretical fictions. When an astronomer is trying to calculate the amount of energy radiated by the sun, he imagines it contained by a spherical shell and then calculates the energy received by the shell. A hydrologist teaching the concept of discharge has his students picture a net strung across the river. Since the net's hole size is a square meter, the water flowing through hole in one second equals one meter multiplied by the length of the "block" of water. The river's per second rate of discharge equals the sum of these blocks.

The conceptual clarity that comes with operationalizing explains how many executed experiments manage to be worthwhile even though they are empirically trivial. In J. S. Sachs's classic recognition experiment, each subject was asked to recall a sentence that he had read in a passage eighty syllables before (20–25 seconds).[25] For the passage on Galileo's telescope, the actual sentence read was sentence 1:

1. He sent a letter about it to Galileo, the great Italian scientist.
2. He sent Galileo, the great Italian scientist, a letter about it.
3. A letter about it was sent to Galileo, the great Italian scientist.
4. Galileo, the great Italian scientist, sent him a letter about it.

Subjects frequently misidentified the sentence as one of the synonymous sentences 2 or 3 but rarely as sentence 4. This demonstrates that the subjects were remembering the meaning without remembering the grammatical form. So barring some kooky kind of amnesia, the meaning is not stored as an *English* sentence. The evidence suggests that the meaning is stored as a sentence in a language of thought and then translated into English. For our purposes, the important feature of Sachs's experiment is that its empirical content does not go beyond the commonsense observation that we remember the gist of what was said rather than a particular sentence. The outcome of the experiment surprised no one, yet the experiment confers two cognitive benefits: (1) it draws a surprising connection between "what we already knew" and the language of thought hypothesis; and (2) it repackages this common knowledge in a way that makes its empirical consequences shine. Usually, we put a claim into a testable format in order to ascertain its truth value. But sometimes the goal is to cultivate the claim's heuristic value and to display its role within a larger body of theory. In other words, verification-oriented reformulation opens lines of inquiry and links pieces of knowledge with other pieces of knowledge.

Empirical grounding is a semantic, rather than an epistemological, concept. Don't confuse it with confirmation. Grounding can be conducted without any new information at all. It is just a matter of linking a concept with tests. Scientists treasure testability and so are fond of redefining familiar terms in terms of easily verified operations. For example, *hungry rat* may be defined as a rat that has been deprived of food for twenty-three hours. One advantage of tying the concept to a test is increased coordination among speakers. The test can serve this end without being actually performed. Indeed, the cheapest way of coordinating usage is to specify a sequence of operations confined to thought alone. Hence, we should expect to find a number of operational definitions conducted by operations in thought. The following definition of 'electrical intensity' is an example: "Let us suppose that we introduce an infinitesimal charge on an infinitely small conductor. . . . The electric intensity at any point is given, in magnitude and direction, by the force per unit charge which would act on a charge particle being supposed so small that the distribution of electricity on the conductors in the field is not affected by its presence."[26] Hard-core operationalists may demand that the operations be *public* and so deny that thought experiments can provide operational definitions. But we need not settle the terminological point to see the resemblance between defining in terms of concrete operations and defining in terms of abstract ones. In both cases, usage is coordinated by reducing it to a checklist of deeds. Recourse to abstract operations allows us to define even "imaginary" scientific entities clearly, such as the equator, lines of longitude, and centers of gravity:

"Let us imagine that the stars are located on the inside of a giant *celestial sphere* centered on the earth, which is in fact the impression the night sky gives. Although the constellations are entirely the product of the human imagination—the stars in a particular constellation have no relationship to one another except the accident of being in the same part of the sky as seen from the earth—they provide a handy way to identify regions of the celestial sphere."[27] Another advantage of defining by abstract operations is that we can operationalize normative notions. Perhaps the Golden Rule should be considered as an operational definition of *morally right*.

In addition to clarifying old concepts, thought experiments introduce new ones: *gettier, reversibility, social contract, state of nature, two-boxer*. Of course, any institution (law enforcement, baseball, house cleaning) teethes in new terminology. The special significance of the vocabulary introduced by thought experiment lies in its preoccupation with conceptually challenging situations. Thought experimenters do not passively wait for a new linguistic difficulty to crop up. They actively seek out conceptual puzzles and self-consciously develop more efficient conventions for describing the recalcitrant phenomena.

Some thought experiments reveal hidden disagreements within the internal committee. I have shown how poor communication between homunculi leads to incompleteness: one part of me knows that Akeel teaches at Columbia, and another part knows that David teaches there; but since the parts fail to share information, I fail to realize that Akeel and David are colleagues. Poor communication hides inconsistency by insulating the conflicting beliefs from each other. Thought experiments can bring the conflict to a head by removing the partition between the diverging homunculi. Consider the belief that all harms must make a discriminable difference to their victims. For example, if I waste water during a communitywide shortage, then it seems that the *consequence* of my act is not bad, because no one can tell the difference. Jonathan Glover uses a colorful thought experiment to show that this belief conflicts with our belief that little differences can add up to a big difference:

> Suppose a village contains 100 unarmed tribesmen eating their lunch. 100 hungry armed bandits descend on the village and each bandit at gun-point takes one tribesman's lunch and eats it. The bandits then go off, each one having done a discriminable amount of harm to a single tribesman. Next week, the bandits are tempted to do the same thing again, but are troubled by new-found doubts about the morality of such a raid. Their doubts are put to rest by one of their number who does not believe in the principle of divisibility. They then raid the village, tie up the tribesmen, and look at their lunch. As expected, each bowl of food contains 100 baked beans. The pleasure derived from one baked bean is below the discrimination threshold. Instead of each bandit eating a single plateful as last week, each takes one bean from each plate. They leave after eating all the beans, pleased to have done no harm, as each has done no more than sub-threshold harm to each person.[28]

The key threat to homuncular theories is circularity. If the homunculi are as sophisticated as the agents they were invoked to explain, they explain nothing. Homunculi that are as sophisticated as we are as mysterious as we and so our puzzlement is merely redirected to how the homunculi are able to improve epistemically without new information. If we postulate subhomunculi to explain the abilities of these homunculi, we face an infinite regress. Dennett finesses this problem by requiring the homunculi to be less sophisticated than the agent they are invoked to explain. Thus, my homunculi will be more primitive than I and their subhomunculi would be even more primitive than they. Hence, instead of facing an infinite regress of equally sophisticated homunculi and subhomunculi, we face a finite sequence of progressively more stupid homunculi. The sequence eventually ends because some homunculi are primitive enough to be explained without subhomunculi—say, merely in terms of feedback mechanisms like a thermostat. Of course, it is one thing to promise this cascade into insentience, another to deliver it. Too often, homuncularists act as if they have discharged their obligation by taking us only a few steps down. We still await a complete homuncular reduction of a psychologically interesting phenomenon.

D. The Rearrangement Model

This model is inspired by situations in which the information at hand is made more digestible by changing its form. We have the information in the sense of possessing the data but lack it in the sense of not being able to draw needed inferences from it. The information is ungainly, like a plank grasped at an extreme end instead of at its center of gravity.

1. *Shortcuts.*   There are different kinds of awkwardness. First, your information can require many steps to reach. Think of a road atlas that requires you to flip to different pages as you cross over state lines. Even information that is in order can be unwieldly if it restricts your order of access. One of the advantages of watching a movie on a VCR rather than at a theatre is that you are free to fast-forward and rewind.

Nowadays, there are televisions that let us view two shows on the same screen. This juxtaposition frees us from reliance on memory so that we can make more accurate comparisons. James Rachels employs a parallel presentation in his critique of the American Medical Association's position on euthanasia.[29] Association policy forbids active euthanasia ("mercy killing") but permits passive euthanasia. Rachels objects on the grounds that there is no morally relevant difference between killing someone and letting him die. He tries to persuade us by placing a pair of hypothetical dirty deeds side by side. The first involves Smith, a man who will come into a large inheritance if his six-year-old cousin dies. One evening, while the boy is taking a bath, Smith sneaks in and drowns him. The second scenario involves Jones, who also has an inconvenient six-year-old cousin. Like Smith, Jones decides to kill the bathing boy. However, when Jones enters the bathroom, he sees the boy slip, hit his head, and land with his face in the water. The delighted Jones stands over the boy, ready to push him under if he recovers. But the boy drowns on his own. Smith killed his cousin; Jones merely let his cousin die. But having carefully controlled all the extraneous variables, we see the distinction does not make a moral difference. Many ordinary experiments are also designed to facilitate direct comparison. To demonstrate that water pressure increases with depth, make three holes down the length of a can filled with water. Just with a glance we can see that the length of each spout grows with the depth of the hole. The juxtaposition of the three effects lets us skip measurement of their respective lengths.

Our ability to blend scenes together lets us derive necessities from other necessities. Consider a problem frequently discussed by Gestalt psychologists. A monk begins a walk up a mountain Monday morning and reaches the summit in the late afternoon. He spends the night meditating. Tuesday morning he begins the walk down the same path and reaches the foot of the mountain in the early afternoon. Is there any place he occupied at the same time on each day? Although most people initially answer 'Probably not', a mental picture quickly shows that the correct answer is 'Necessarily so'. Superimpose the scene of Tuesday journey on the Monday journey. Since the Monday monk going up the mountain must meet the Tuesday monk going down, there must be a place that was occupied at the same time of day.

Shortcuts reduce either the number or length of the steps. Sequencing the data reduces the average number of steps needed to find the datum. Thus, files are alphabetized, grades presented highest to lowest, and account balances presented in temporal order. As an application to thought experiments, consider worst-case generalizations. Here, one shows that a property holds for the case least favorable to your thesis and so conclude that it must hold for the more favorable cases. Darwin uses this method to prove his generalization of the Malthusian principle: every species reproduces faster than the resources needed to sustain its descendents. Darwin takes the slowest breeder, the elephant, and supposes that the descendents of a single pair multiply unchecked. In five hundred years, there would be fifteen million of them!

Difficult questions can sometimes be answered by addressing other questions. One well-known strategy is to begin with a drastic simplification and then introduce complexities as part of a gradual return to the original question. For instance, Newton developed an account of planetary systems by first studying the behavior of a pointlike sun and a single, pointlike planet. Then he added more point–planets, changed the points to mass–balls, then had these balls spin and wobble and exert influence on each other. After this degree of de-idealization, he had a model that was worth fine-tuning to empirical data.

2. *Chunking.*   Meditate upon the difference between how two prisoners count their days on their cell walls (Figure 4.1). In both cases, the information is in front of your nose, but you can more easily tell how long the neat prisoner has been in jail because his notation lets us subdivide the calculation into a small number of uniform chunks. Many mathematical explanations consists of redistributing the same material into uniform chunks. According to legend, Gauss's talent was discovered by a teacher who ordered his pupils to add the first hundred integers. The teacher had hardly begun reading his newspaper

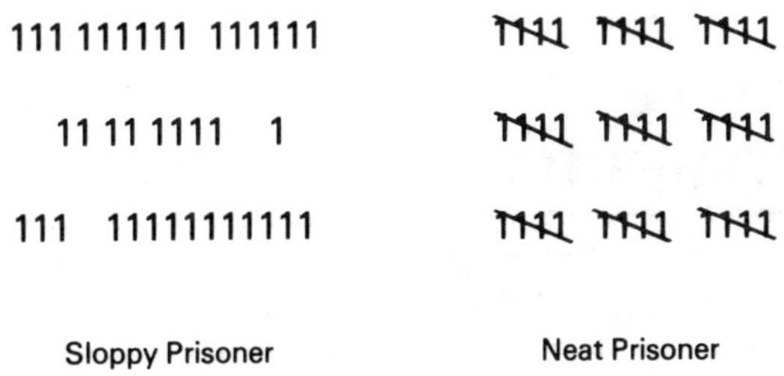

111 111111 111111   TH1 TH1 TH1

11 11 1111   1   TH1 TH1 TH1

111   11111111111   TH1 TH1 TH1

Sloppy Prisoner   Neat Prisoner

Figure 4.1 Sloppy Prisoner and Neat Prisoner

when young Gauss presented the correct answer. Instead of laboriously adding a hundred distinct numbers, Gauss rearranged the problem into a task of adding the same number one hundred times. This constant number is 50.5 since it is the arithmetic mean of the values. In general, the sum of first n numbers is $n(n + 1)/2$:

$$
\begin{array}{l}
1 + \quad 2 \quad + \quad 3 \quad + \ldots + n = S \\
n + (n - 1) + (n - 2) + \ldots + 1 = S' = S \\
\hline
(n + 1) + (n + 1) + (n + 1) + \ldots + (n + 1) = n(n + 1)
\end{array}
$$

Rearranging the problem this way makes the problem easier, just as it is easier for a child to eat his steak after his mother has sliced it for him.

3. *Angle of entry.*   Predatory birds and snakes maneuver so that their prey can be swallowed headfirst. Consumers of information display similar preferences. We can only read mirror-image writing with difficulty, hence motorists are often baffled by 'AMBULANCE' as it appears on the front of ambulances but recognize it immediately when reading it in their rearview mirror. A newspaper is easier to read when it is right side up.[30] Dramatic differences in recognition also occur when we invert photographs, geometrical figures, and maps (Figure 4.2). Children between two and five often appear indifferent to how a picture is oriented. When you see a child engaged in a book that he is holding upside down, it is natural to infer that orientation is irrelevant to children. However, recent experiments demonstrate that children have at least as much difficulty recognizing disoriented figures as adults do. The child's indifference may be simply due to the fact that it has not noticed how certain orientations improve recognition.

Occasionally, information is repackaged to *block* some inferences. Once we recognize an object, we stop inspection of the surrounding features. When one of these features is important, recognition of the foreground object will lead us to overlook an important fact. Hence, slowing recognition can bring these neglected features to our attention.

The repackaging can also get around intellectual defenses that protect cherished beliefs. Philosophy teachers sometimes use a variant of John Wisdom's

Figure 4.2

parable of the gardener this way.[31] Suppose Theodore and Atwell come upon an old garden and are surprised to see that some of the original flowers flourish among the weeds. Theodore infers that there must be a gardener who tends the flowers while no one is looking. Atwell disagrees. So they camp out and carefully watch for a secret gardener. After none is detected, Theodore concludes that the gardener must be invisible. So they erect a glass dome around the garden. The flowers still flourish but no one breaks the glass. Theodore concludes that the gardener is incorporeal. More precautions just lead Theodore to infer that the gardener is completely indetectible to scientific testing. But Theodore still claims to know that "The gardener cares" and that "The gardener is on our side." Is Theodore rational? Students castigate Theodore on the grounds that his belief has degenerated into an unverifiable dogma. Then the teacher queries: "What is the difference between Theodore's belief and a theist's?" Many of the students who confidently dismissed Theodore are themselves theists and so are startled by the similarity—more startled than they would have been had they been filtering the analogy through their intellectual defenses.

Many problem solving strategies change the perspective from which information is consumed. This is the lesson to be learnt from riddles such as "If doubling a dollar ten times yields a fortune of $1,024, how many times must the dollar be doubled to reach $512." The natural solution begins at the beginning, but the quick solution is to work backward.

In addition to making old information more informative, imaginary perspectives lubricate the entry of fresh information. Thought tours do this by supposing that the reader is receiving the information as part of a guided tour of the scene. Thus, a historian will walk us through a day in ancient Rome, pointing out important architecture, typical activities and garb, and mentioning issues of the day. The historian is not merely rearranging familiar information; he is feeding us new facts. Scientists take us on similar tours to the sun and the future and through the human body.

4. *Deformation.*   In addition to varying perspectives, we can refigure jumbled things into simpler forms. The irregular shapes of organisms obscure the significance of their surface area-to-mass ratio. W. K. Clifford unwrinkles the relationship by making the organism a cube.[32] The cube absorbs nutrition through its entire surface of six sides. It needs 2 square inches of surface area to sustain one cubic inch of mass. The cube's length begins at 1 inch. So it has 6 square inches of surface area and 1 cubic inch of mass. (In general, the cube's length = n, area = $6 \times n^2$, and volume = n^3) So the little cube only needs 2 sides to sustain itself and 4 sides to feed new growth to a length of 2 inches. But now the 2-inch-long cube has an area of 24 inches and a volume of 8 inches, so 4 of its sides are needed to merely sustain its life. Since it only has 2 extra sides for growth, its size increases more slowly. At a length of 3 inches, the cube has $6 \times 3^2 = 54$ inches of surface area and $3^3 = 27$ cubic inches of mass. Now all 6 sides of the cube must be used to sustain the animal. The creature must stop growing. Thus, the fact that mass is cubed while surface area is squared imposes a limit on growth. Clifford's cube can also be used to explain why small animals must eat more (because they lose heat faster), why small animals survive falls, and why the largest creatures are aquatic.

Since much of explanation is a matter of reducing the strange to the familiar, a thought experiment can illuminate by systematically varying a plain phenomenon into one equivalent to the vexatious one. Astronomers use this rearrangement in the course of explaining why artificial satellites travel so much faster than the moon. (Satellites take about an hour and a half to circle the earth, the moon needs almost a month.) First imagine a stone tied to string which you must keep suspended by whirling it around. If the string is shortened, the stone must be revolved more rapidly to keep it from falling to the ground. But suppose you continue to lengthen the string until the stone is at the same height as the satellite. Now it needs to make far fewer revolutions. If you let out even more line, the stone eventually reaches the far slower speed of the moon. Just as the pull of the string keeps the stone from flying off, the force of gravity confines the satellite to an orbit.

Comprehension is also assisted by exaggerating quantity. A flight from New York to San Francisco takes longer than the return trip because the North American continent has prevailing westerly winds. Do the passengers save as much time returning as they lost going? To see that they gain less than they lost, suppose the wind is moving at 499 miles per hour while the plane only travels 500 miles per hour. Since the plane only nets 1 mile per hour, the trip to San Francisco will take a few months. The return flight will net 999 miles per hour and so take only a few hours. Nevertheless, the saving is swamped by the loss. The positive and negative effects of the wind look like they cancel out because we fail to consider how long the force works on the plane. Equal forces cancel out only when applied equally. The outlandish scenario has the advantage of being a flaming counterexample. Less extreme cases would not highlight the temporal element with the same intensity.

III. The Cleansing Model

The *cleansing model* is inspired by incidents in which you recognize your own irrationality and then change your beliefs to remove the flaw. A friend of mine was a sensitive boy who became worried about the ants that he inadvertently crushed while walking about. His father assured him that he should not worry about killing ants because "there are millions of them." This satisfied my friend for a while. But his new equinamity was terminated by a thought about the growing human population. This illustrates the familiar situation where an inconsistency takes root, is detected, and is then weeded out.

Whereas the previous models cast the epistemic improvement as a matter of adding positive features, the cleansing model concentrates on subtracting negative features. These negative features are intellectual vices that diminish your efficiency at tasks such as argument, explanation, inquiry, prediction, planning, problem solving, and teaching. To be rational is to be free of the vices that encumber your performance of these cognitive tasks. This picture of rationality may seem too downbeat to explain why rationality is held in high esteem. But compare it with other absolute concepts that pick out valued conditions: health (absence of diseases), peace (absence of violence), liberty (absence of coercion and obstacles).[33] For a more specific improvement of the image of rationality, we turn to the recalibration of the imagination.

A. How Thought Experiment Corrects Imbalances

A thinker with a balanced imagination can imagine all and only genuine possibilities. The sceptics of chapter 2 provide ample evidence that no human being has a perfectly balanced imagination. However, there are remedies for these excesses and blind spots.

One solution is formalization. After centuries of failed attempts to square the circle and double the cube, geometricians became suspicious of the intuition that the tasks were feasible. So extra care was given to definitions and inference rules. As proof procedure grew in rigor, mathematicians could completely gag their imaginations by *disinterpreting* the question. Instead of treating the symbols expressing the question as representing things, they are regarded as meaningless marks that license the writing of other meaningless marks in accordance with special rules. Proof becomes a syntactic game in which the player tries to derive one set of squiggles from another. The meaning behind the symbols is revived only after the game. Among the fruits of this vigorous rigorism was a chain of impressive impossibility proofs.

For many metaphysicians, these surprising proofs became emblematic of the illusions fostered by a runaway imagination. In the eighteenth and nineteenth centuries, the idealists were abuzz refuting the reality of space, time, and matter on the grounds that they were self-contradictory or defending the necessity of Euclidean geometry, determinism, and whatnot on the grounds that their alternatives was inconceivable. As Bertrand Russell emphasized, this gave logic a restrictive reputation that only tells half the story. Revolutionary advances in logic and mathematics put twentieth-century figures in a position to probe the old impossibility proofs of the metaphysicians and determine that the concepts "are not in fact self-contradictory, but only contradictory of certain rather obstinate mental prejudices."[34] They have gone on to offer possibility proofs of non-Euclidean geometry, infinite collections, and peculiar kinds of space and time.

> Thus, while our knowledge of what is has become less than it was formerly supposed to be, our knowledge of what may be is enormously increased. Instead of being shut in within narrow walls, of which every nook and cranny could be explored, we find ourselves in an open world of free possibilities, where much remains unknown because there is so much to know. . . . Logic, instead of being, as formerly, the bar to possibilities, has become the great liberator of the imagination, presenting innumerable alternatives which are closed to unreflective common sense, and leaving to experience the task of deciding, where decision is possible, between the many worlds which logic offers for our choice.[35]

A famous illustration of the benefits of this widened choice is Einstein's use of Riemanian geometry in his general theory of relativity. The value of stockpiling mind-broadening curiosities is widely appreciated. Funding for theoretical mathematics is often justified on the grounds that today's mathematical plaything is tomorrow's engineering tool.

In Ronald Giere's terminology, these abstract models, along with the experimental and instrumentation skills needed to work with them, are "cognitive resources". A mixture of these resources, observation, and luck determine which scientists are positioned to solve problems at the intellectual frontier. Giere points out that *experiences* can provide cognitive resources because they provide a stock of metaphors. His example is Alfred Wegener's visit to Greenland. Wegener was among the many people who were impressed by the complementary shapes of Africa and South America. The congruence of their Atlantic coastlines suggests that the two continents were once joined. However, turn-of-the-century geologists were not sympathetic to the hypothesis of continental drift. Some of their antipathy was a matter of policy; the geologists were uniformitarians (believing that geological features had gradual rather than sudden, catastrophic causes). Their specific objection was they could not conceive of a force powerful enough to move continents. Notice that their complaint was not merely that there was no known force. Absence of a known causal mechanism did not prevent geologists from accepting the existence of ice ages and geomagnetism. Wegener, on the other hand, was able to imagine such a force because he had seen Greenland's glaciers and ice floes: "Millions of years ago the South American continental plateau lay directly adjoining the African plateau, even forming with it one large connected mass. The first split in Cretaceous time into two parts, which then, *like floating icebergs, drifted farther and farther apart.*"[36] Wegener's Greenland experience was mind-broadening: it enabled him to fluently conceive of more possibilities. This made him less likely to misexclude continental drift on grounds of physical impossibility.

Since imagined experiences can overcome imagination blind spots in the same way as actual experiences, thought experiments are also purveyors of cognitive resources. Recall how Darwin and Horowitz used the technique of hypothetical histories to soften up apparent incompatibilities between natural selection and organic complexity. By multiplying models, biologists hope to limber up the imagination so that a realistic explanation can be more readily recognized.

Imagination dilation is always in demand because research programs heavily constrain explanation. Behaviorism confines psychological explanations to actual and hypothetical bodily states. Classical economics forbids appeals to altruism, weakness of will, and demand-independent value. Cartesian physics rules out empty space, action at a distance, and material distinctions other than shape and motion. These constraints often appear to preclude the explanation of well-accepted phenomena such as pumps, magnetism, and heat. The adherents of the research program must then either deny the obvious or debunk the apparent incompatibility. The debunking strategy is initiated with thought experiments that thread us through the maze of constraints. As Richard Feynman observes, science requires a special kind of imagination—"imagination in a terrible straitjacket."

Since thought experiment can correct underestimates and overestimates of what is possible, it is a partially self-correcting method. Although an imbalanced imagination sires fallacious hypotheticals, other thought experiments expose some of the errors and thereby rectify imbalances.

B. Theoretical and Practical Irrationality

Absolute concepts are simple at one level because they are simply defined as universal negative judgments: G is the absence of Fs. However, agreement and simplicity at this first level of definition is compatible with complexity and controversiality at the next level (when the definiens are themselves defined). Definers of 'health' agree that it is the absence of disease but give conflicting definitions of 'disease'. Subjectivists let 'disease' encompass any unwanted bodily condition. Conventionalists define 'disease' as an excusing condition. Teleologists analyze 'disease' as designating a failure of a natural function. Parallel disagreement over definiens may ensue if we agree that 'rational' means the absence of irrationalities. But the value of the absolutist definition would survive this eventual breakdown of consensus. Speakers typically agree on most particular Fs even though they sharply disagree over how 'F' is to be defined. Rough and ready categories of irrationality (bias, circularity, inconsistency) uncover irrationalities with success comparable to that of the layman's categories of disease (gout, sniffles, fever). Of course, folk terminology is not directly accepted by theorists. But it does give them a running start. So although there will be hard cases in which we are hamstrung by disagreement over 'irrational', there will be widespread, rough agreement.

Consequently, a special theory of irrationality is not needed to sell the absolute concept of rationality. We are free to plug in different theories of irrationality. Many of the insights originating with the positive theories of rationality can be salvaged within the substantive accounts of irrationality.

Nevertheless, we should take advantage of the standard distinction between "theoretical" and "practical" irrationalities to note that thought experiments only aim at the theoretical ones. Theoretical rationality concerns belief formation, practical rationality concerns choice. Since the weak-willed smoker has the same beliefs and desires as his iron-willed counterpart, the irrationality does not occur at the level of belief or values. The smoker's irrationality consists in his inability to translate beliefs and desires fluently into action.

Some of the techniques for reducing practical irrationalities involve imagination. For instance, desensitization therapy treats the arachnophobic by having him visualize a spider on a distant wall. Once the patient tolerates this, closer visualizations follow until finally he can calmly imagine the spider on his shoulder. This completes the first phase of the clinical slippery slope that ends with the patient at ease with real spiders. Although the visualizations make the patient more rational, they are not thought experiments. For the mental exercise does not purport to prove anything or make anything questionable. After all, desensitization is not an *insight* therapy; I can also be desensitized into losing rational fears.

Only a theory chauvinist would take thought experiment's indifference to practical irrationalities as a sign that they are not as serious as theoretical irrationalities. Rationality is often better served by mentally undressing your audience than by conducting a thought experiment. Certain mental practices *systematically* increase practical rationality. The tendency to myopically discount the future is countered by vividly contrasting your small short-term sacrifice with your big, long-term predicament. Egocentricism is twisted to pieces with role reversals. The evolutionary pressures that underwrite our thought patterns are not restricted to reasoning. Reproductive success is also influenced by paths of attention, by what we find emotionally engaging, by what we take *seriously*.[37] Since practical rationality pays in the biologically circuitous way that theoretical rationality pays, we should expect imperfections. But resist the idea that our emotionality is separate or even opposed to rationality. Even chains of "free association" are structured. The amplitude and frequency of our mood swings are as much an adaptation as the swing of a walker's arms. Fantasies, fascinations, and focusings orient us toward greater *practical* rationality.

We have indirect control over the forces that encode our beliefs and desires into behavior. By changing our way of life, we affect our susceptibility to excesses and ruts. Since people are more apt to change when life goes badly, they channel into habits and social settings that promote practical rationality. Given our penchant for belief–desire psychology, these changes are frequently mischaracterized in terms of theoretical rationality. Thus, a man who becomes well adjusted because of a formative experience will try to articulate what he learned. If he recognizes his failure to specify the big lesson, he will conclude that his insight is ineffable. This mysticism is common among admirers of fiction. Grateful readers cannot express what they learned as a plausible and interesting *proposition*. Hence, they infer that the aperçu is inexpressible. However, novels can enhance our rationality nonpropositionally. Stories freshen our sympathies by fitting us into another person's shoes. They soften inhibitions with hypothetical performances. Stories can also *sensitize* us to things that go underappreciated (such as sight or security). Depiction of hardships lowers our standards and makes our assets stand out. Utopian novels recalibrate us in the reverse direction by inviting comparisons that highlight our deficiencies. Some novels merely equip us for insights. For example, George Orwell's *1984* provides a vocabulary list (*Big Brother, thought police, double-think*) that facilitates recognition of totalitarian tendencies.

The fact that a thought experiment aims at an improvement at the level of theoretical rationality does not prevent it from also boosting our practical rationality. Many of Plato's thought experiments are intended to rouse us out of intellectual lethergy. In his allegory of the cave, a row of men are chained so that they can only look at a wall within a cave. Behind them is a fire and individuals who carry figurines back and forth. Since the shadows cast by these figurines are all the prisoners ever see, they regard shadows as the only reality. Now suppose a prisoner is unchained and turns to see the source of the shadows. Let this shock be followed by ejection from the cave into the sunshine. The prisoner would be dazzled and painfully confused. He would then adjust to what he now realizes to be true reality. Eventually the liberated prisoner would pity his friends in the cave. So he gropes his way back. The enlightened man tells his cavemates that they live in a world of illusion. But they scoff at his otherworldliness and take his lost ability to read the shadows as evidence that he is daft.

The allegory of the cave is a thought experiment because it is a possibility proof. It demonstrates that Plato's theory of forms could be right even though it profoundly contradicts common sense. However, the allegory is also intended to decaptivate us from the glitter of the sensual world and reroute attention to abstractions such as triangularity, manhood, and justice. The *principal* aim of a thought experiment must be the theoretical one of answering or raising the experimental question. Hence, you will be disinclined to regard Plato's allegory as a thought experiment to the extent that you think it is mainly intended to rally us out of a failure of nerve. Of course, withholding the status of thought experiment would not necessarily be a demotion of the allegory to cerebral cheerleading. As Wittgenstein observed, thinking is like swimming underwater. We must struggle against our natural buoyancy to reach down to the depth of a problem. Thus, the devices of practical rationality have a legitimate role in promoting scholarly virtue.

Nevertheless, I raise the issue of practical rationality mainly to earn the right to ignore it. The thought experiments of Plato or Galileo or Nietzsche are designed to do more than inform: their hypotheticals have an important and wrongly despised rhetorical dimension. However, this dimension is nothing special to thought experiment. We can "add on" a practical subgoal to any instrument of persuasion. In particular, regular experiments are also fielded to improve practical rationality. For instance, surgical instructors drive home the dangers of casual hygiene by growing bacteria cultures from their students' freshly washed hands.

IV. An Eclectic View of the Mechanics of Thought Experiment

None of the five models of armchair inquiry precludes any of the others. So we are free to pick and mix. There are thought experiments conforming to each—and some that fit all the models simultaneously. Some thought experiments act as reminders or organizers; they serve as the catalyst by which dispositional knowledge manifests itself as occurrent knowledge. Other thought experiments take knowing *how* into knowing *that*; an important special case is the transformation of linguistic ability into linguistic doctrine. Our (albeit limited) understanding of how the brain works suggests that some tasks are completed in an homuncular fashion; there is sometimes a transition from one part knowing that p to another part knowing that p—or, at least, a transition that may be fruitfully considered that way. Rearrangement certainly plays a role in thought experiment: witness the popularity of bizarre perspectives on problems. Last, many thought experiments fit the fifth model, especially as inconsistency revealers. Thought experiments such as Galileo's refutation of Aristotle's theory of motion (involving the joined falling objects) bear a striking resemblance to reductio ad absurdum. Indeed, some logicians are tempted to view all thought experiments as enthymematic reductios.

Although all of the models have applicability, only one has the prospect of immediate elaboration. We have a very limited understanding of how the mind works. This psychological obscurity fogs in the positive models. We can say that thought experiments function as reminders, transformers, autostimulators, and rearrangers, but we cannot go much beyond that. Future progress may enable us to go further. But for now, the positive models only provide vague sketches of how thinkers improve without new information.

The wealth of logic contrasts sharply with the poverty of psychology. A century ago, a group of philosophers and mathematicians (Frege, Venn, Boole, Peirce, Whitehead, Russell, etc.) set out to develop a logic that would formalize the notion of *proof*. They succeeded. We now have a concept of demonstration adequate to the needs of mathematics. Contemporary logicians would like to do for natural language what their predecessors did for math. Thus logic continues, expanding and progressing. Given the wealth of logic, the model best suited for *pursuit* is the cleansing model. I advocate it in the same spirit as a soil expert who advises a farmer to cultivate one plot of land rather than the other. I am not saying that the other models are barren, only that they are less fertile and that the development of the cleansing model is the logical stepping-stone to their successful development.

The cleansing model is reassuringly topic-neutral. Other models of arm-chair inquiry only seem suited for certain sorts of thought experiments. The recollection model cannot handle thought experiments that are absolutely a priori (logic and math). Nonlinguistic thought experiments are a problem for the transformation model. The homuncular model is iffy all over; we do not know where it will ultimately touch base or where we may be left with dangling homunculi. And the rearrangement model works well with thought experiments that feature a heuristic twist but fuzzes out over the others. In contrast, the pervasiveness of irrationalities assures us that the cleansing model will bear on every field from axiology to zymology.

Hence, if we go with the cleansing model, we enjoy the benefits of logic with a low risk of irrelevance. The model tells us that thought experiments improve our thinking by exposing irrationalities—by prompting us into greater rationality. Our understanding of the nature of theoretical irrationalities compares quite favorably with our understanding of psychological notions. Of course, our understanding of the logical concepts is not perfect. But notice that the path for improvement is clear.

My plan is to be even more selective. I shall lock onto a single irrationality—inconsistency. There are five reasons for this strategy. First, logic has made it the best-understood irrationality. Second, it is a highly general irrationality. Some philosophers even think that all irrationalities are reducible to inconsistency. They say the cognitive wrongness of bias boils down to the inconsistent application of standards; circularity is condemned because it leads to inconsistent probability assignments based on the same evidence; and so on. 'Reduction' is a tricky term but I shall eventually show a solid sense in which this reductionist thesis holds for the irrationalities cleansed by thought experiment. My third reason for focusing on inconsistency is that it's the most widely acknowledged irrationality. Inconsistency is not a borderline case like absent-mindedness or a controversial instance such as selfishness. Fourth, inconsistency is the most severe irrationality: when you discover an inconsistency, you have the strongest motive to change your beliefs—indeed, the process of revision starts automatically. Last, since other studies of thought experiment have also focused on inconsistency, the strategy lets us build on this past work in a straightforward fashion.

5

Kuhntradictions

One bright day in the middle of the night
Two dead boys got up to fight.
Back to back they faced each other,
Drew their swords and shot each other.
A deaf policeman heard the noise
And came and killed those two dead boys.
Anonymous

This chapter focuses on Thomas Kuhn's account of thought experiments. It begins with what Kuhn takes to be the smart talk about thought experiments. I then detail Kuhn's amendments to this view and raise objections. Most of the opposition is directed against the notion of local incoherence. Finally, I reconstruct Kuhn's error in order to salvage the considerable insight that it contains.

I. Kuhn on the Received Opinion

Philosophers claim no monopoly on conceptual analysis, just generic expertise. Usually, scientists solve their own conceptual problems. The adequacy of a conceptual analysis depends heavily on how it reverberates through a network of theories. Scientists have intimate familiarity with this web of belief and so have a much better feel for what will smooth out wrinkles and what will snarl the lines. A philosopher without scientific training might be of some help just as an astronomically naive mathematician might be of some help in the prediction of eclipses. But the fact that a problem has a mathematical aspect does not mean that mathematicians are the ones in the best position to solve it or even that they would be of much use. The same applies when scientific problems have philosophical aspects. Understandably, it was the physicist Newton who analyzed weight into mass and force. It was the chemist Cannizzaro who drew the crucial distinction between atomic weight, molecular weight, and equivalent weight. And it was the biologist W. Johannsen who distinguished genotype and phenotype. So we should expect to find the scientist appealing to thought experiments in his more theoretical moments.

Thomas Kuhn characterizes the received view about scientific thought experiments as springing from a demand for consistency:

> On this analysis, the function of the thought experiment is to assist in the elimination of prior confusion by forcing the scientist to recognize contradictions that had been inherent in his way thinking from the start. Unlike the discovery of new knowledge, the elimination of existing confusion does not seem to demand additional empirical data. Nor need the imagined situation be one that actually exists in nature. On the contrary, the thought experiment whose sole aim is to eliminate confusion is subject to only one condition of verisimilitude. The imagined situation must be one to which the scientist can apply his concepts in the way he has normally employed them before.[1]

So according to the cognoscenti, scientific thought experiments are relevant to the scientist's conceptual apparatus, rather than to nature.

II. Misfits

Kuhn finds merit in the mainstream view but thinks that philosophers oversimplify the nature of the inconsistency. He believes there is a large family of thought experiments that pivot on a nonstandard sort of inconsistency. According to Kuhn, philosophers overlook this peculiar sort of contradiction because of a misguided loyalty to the analytic/synthetic distinction. He begins his account in case study style.

Piaget reports that children apply *faster* to the object that reaches the goal first. This goal-reaching criterion is applied even when the losing object was handicapped by a later starting time or a more circuitous route. The children only display an indirect sensitivity to these handicaps through their use of a second criterion of "perceptual blurriness" which applies to the object displaying much more hectic behavior. Piaget put his children into a state of conceptual confusion by exaggerating the race handicap so that the object finishing second was much more active. In particular, he had the children witness three races between two toy cars that I shall call A and B (Figure 5.1). In race 1 the goal-reaching criterion selects A as the winner because A reaches the goal before B. In race 2 the two cars reach the goal at the same time even though B started much later; hence, B wins by virtue of the blurriness condition. Race 3 is reminiscent of an old television commercial depicting a race between a Volkswagen Beetle and a dragster. Both criteria are activated because A reaches the goal only slightly before B which had a much later start. Hence, the goal-reaching criterion counts A as faster, while blurriness counts B as faster. After a period of ambivalence, many of the children revised their concept in a way that more closely conforms to adult usage. According to Kuhn, Piaget's puzzle situation improved the children's conceptual apparatus by revealing a hidden contradiction and serves as a model for understanding thought experiments.

Kuhn's second illustration is one of Galileo's thought experiments. The scenario features two balls dropped from C, one traveling down the inclined

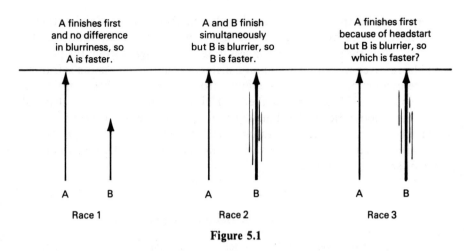

Figure 5.1

plane CA and the other down the vertical CB (Figure 5.2). Galileo's character, Salviati, stipulates that the planes are frictionless; hence, all agree that both bodies will have reached the same speed when they reach bottom, that is, the speed necessary to carry them back to the top. (To see why, think of how a ball would bounce in a vacuum—back up to the original height, no higher or lower.) When Salviati asks which ball is faster, the participants initially pick ball 2. Although this answer is supported by both the goal-reaching and perceptual blurriness criteria, it violates a necessary condition that had been introduced by the medievals: the mean speed of the faster object must be greater. After all, if the average speed of the faster object were not greater, it could not gain on a slower object in the long run. Since both objects start at rest and finish with the same momentum, they must have equal mean speeds.

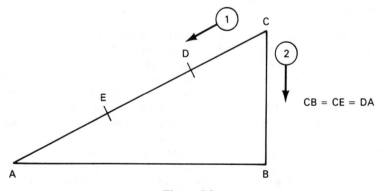

Figure 5.2

Hence, the mean speed theorem implies that neither is faster. Salviati tries to break the deadlock by noting that the faster object is normally defined as the one that covers the same distance in less time and so suggests that the problem could be resolved by finding a common distance. But this suggestion is fruitless because conflicting answers are obtained from comparisons between CB and segments of CA. If CD is chosen, then ball 2 is faster. If EA is chosen, then ball 1 is faster. Choose the intermediate segment, DE, and neither is faster. Galileo ultimately concludes that the solution is to bifurcate the notion of *faster* into two senses, one for instantaneous rates of motion and the other for completion times of specified motions.

III. The Left Hand of Logic

Galileo's demonstration would have been unpersuasive if he had instead directed the participants to imagine the object disappear at C and reappear at A; for the sixteenth-century scientists could dismiss the situation as outside the intended scope of *faster*. With the advent of twentieth-century quantum jumps, the thought experiment is not too farfetched: "If a thought experiment is to be effective, it must, as we have already seen, present a normal situation, that is, a situation which the man who analyzes the experiment feels well equipped by prior experience to handle. Nothing about the imagined situation may be entirely unfamiliar or strange."[2] Just as Aristotle demanded verisimilitude in tragedy, Kuhn demands it in thought experiment. Hence, Kuhn embraces a relativity about the *effectiveness* of thought experiments. We must ask, "Effective with respect to whom?" This is a prelude to a more radical relativity that emerges from his reflections on the contrast between two types of inconsistency.

Kuhn's second point is that thought experiments jostle out a concept's "contradictions." These are not contradictions in the sense used by logicians. Kuhn contrasts the logician's standard example of a contradictory concept, *square circle* (which is obviously inconceivable, inconsistent, and empty) with concepts such as *faster* (which are only covertly inconsistent and apply to many actual objects):

> The child's concept of faster is repeatedly exemplified in our own world; contradiction arises only when the child is confronted with that relatively rare sort of motion in which the perceptually *more* blurred object *lags* in reaching the goal. Similarly, Aristotle's concept of speed, with its two simultaneous criteria, can be applied without difficulty to most of the motions we see about us. Problems arise only for that class of motions, again rather rare, in which the criterion of instantaneous velocity and the criterion of average velocity lead to contradictory responses in qualitative applications. In both cases the concepts are contradictory only in the sense that the individual who employs them *runs the risk* of self-contradiction. He may, that is, find himself in a situation where he can be forced to give incompatible answers to one and the same question.[3]

Thought experiments uncover hidden inconsistencies by calling our attention to imaginary scenarios that activate a clash in the criteria governing the concept. Rather than being fatally flawed, the concept usually works well. Indeed, Kuhn maintains that the acceptability of the concept often depends on contingent facts.

Rather than marking confusion, the covert inconsistencies mark a commitment to a type of universe. The confusion raised by thought experiments is *relational*—a sort of personality conflict between an otherwise innocent concept and an otherwise affable universe. Because we learn our concepts and our facts together, the thought experiment teaches us about both by exposing the mismatch. This is the gist of Kuhn's response to scepticism about the informativeness of thought experiments:

> The root of the difficulty is our assumption that, because they rely exclusively upon well-known data, thought experiments can teach nothing about the world. Though the contemporary epistemological vocabulary supplies no truly useful locutions, I want now to argue that from thought experiments most people learn about their concepts and the world together. In learning about the concept of speed Galileo's readers also learn something about how bodies move. What happens to them is very similar to what happens to a man, like Lavoisier, who must assimilate the result of a new unexpected experimental discovery.[4]

Kuhn illustrates his thesis with a thought experiment of his own.[5] Picture a world in which all motions are of uniform speed. This world would never trouble Aristotle's definition of motion: "The quicker of two things traverses a greater magnitude in an equal time, an equal magnitude in less time and a greater magnitude in less time." (*Physics* 2:232a28–31); for when each object moves at a steady rate, instantaneous and average speeds never diverge. In this world we find a scientist using Aristotle's definition. Is he vulnerable to Galileo's thought experiment in the way Aristotle was? Kuhn denies that the imaginary scientist is confused:

> Nothing could go wrong with his science or logic because of his application of the concept. Instead, given our own broader experience and our correspondingly richer conceptual apparatus, we would likely say that, consciously or unconsciously, he had embodied in his concept of speed his expectation that only uniform motions would occur in his world. We would, that is, conclude that his concept functioned in part as a law of nature, a law that was regularly satisfied in his world but that would only occasionally be satisfied in our own.[6]

Kuhn views the scientist's conceptual choices as forging the limits of what is possible. This picture does not make sense given the traditional analytic/synthetic distinction; for the distinction tells us that analytic propositions determine what is possible, while the synthetic propositions "fill in logical space" by telling us what is actually the case. Structure is sharply distinguished from content.

This view is haunted by the problem of explaining how analytic propositions are informative. Synthetic propositions are informative because they eliminate possibilities. You start out not knowing which of a large number of possible worlds you inhabit. The synthetic proposition narrows down this range of possible worlds. Analytic propositions provide no knowledge about *this* world; they concern all possible worlds indifferently. Hence, there is a mystery as to how analytic propositions can apply to reality. The problem arises in specialized form for any class of propositions regarded as analytic. Thus, we have the problem of applied mathematics, applied logic, and the paradox of analysis (for philosophical propositions). If we regard thought experiments as yielding only analytic propositions, we get the problem of informativeness raised in chapter 2. If we regard them as yielding synthetic a posteriori propositions, then we avoid this problem in the fashion of Mach, described in chapter 3. Kuhn hawks a third alternative: reject the analytic/synthetic distinction undergirding the issue.

The analytic/synthetic distinction stood without serious challenge until the publication of W. V. Quine's "Two Dogmas of Empiricism" in 1951. Quine showed that the distinction could only be defined in terms of intensional notions (synonymy, necessity, belief). An adequate definition, he controversially required, must be extensional in order to have sufficient clarity. He therefore rejected the analytic/synthetic distinction and suggested that the notion of *revisability* be used instead. Analytic propositions correlate with revision-resistant ones, while more revisable propositions correlate with synthetic propositions. Hence, instead of a dichotomy, we have degrees.

Kuhn's suggestion is that thought experiments concern propositions that straddle the old analytic/synthetic dichotomy. The physicist's propositions stating the nature of speed are not clearly analytic because they do not hold in every possible world. But they are nonetheless conceptual propositions that reflect our resolution to use terminology in a certain way. Hence, they are not clearly synthetic. Propositions stating the nature of our scientific concepts are just the sort of hard cases that prompted Quine's dissatisfaction with the distinction. What we can say is that the propositions are central to our belief system and thus have a low degree of revisability. And it is to the credit of thought experiments that they can influence these important tenets.

IV. Truth or Dare?

Quine and Kuhn make strange bedfellows. Quine is a logical conservative. He takes first-order predicate logic to be a paradigm of mature, central theory. Hence, Quine is unwilling to modify logic to accommodate the difficulties of peripheral theories—such as those comprising the history or philosophy of science. If Kuhn's theory conflicts with first-order predicate logic, then so much the worse for Kuhn's theory.

A. The Incoherence of Incoherent Concepts

There are other reservations about the nature of incoherent concepts. One group of philosophers rejects the possibility of inconsistent concepts on the grounds that language is not the sort of thing that can be inconsistent. Claims made by means of language may be logically inconsistent but not the language itself because language does not *assert* anything. Concepts are just tools for asserting. No doubt, most languages are inconsistent in the nonlogical sense of being *irregular*. Witness the arbitrary pluralization of English. If you have more than one mouse, you have mice. But houses are not *hice*, nor blouses *blice*.

Granted, we speak of expressions such as 'round square' being inconsistent. But that's metonymy. When we say Moscow is angry, we are applying 'angry' to the city because it is associated with angry officials headquartered in Moscow. People are the primary bearers of anger, not cities. Metonymy is also at work in talk of happy homes and honorable names. Likewise, propositions are the primary bearers of inconsistency. Things associated with inconsistent propositions can be called inconsistent by extension: beliefs, emotions, and punishment, for example. But this extended usage should not be taken in the same way as the primary usage. We say that 'round square' is inconsistent because statements of the form 'x is a round square' express inconsistent propositions. But this does not mean that the meaning of 'round square' is inconsistent. The definition 'x is a round square just in case x is round and x is square' is a consistent report of how the term is used. From this report we can deduce that other propositions will inevitably express inconsistent propositions, but this does not imply that the report itself is inconsistent. Blaming the definition for correctly reporting the incoherency is like slaying the messenger for reporting defeat.

Some defenders of the notion of inconsistent concepts agree that there can be consistent reports of inconsistent terms (as there can be sober studies of alcoholism) and even agree that language does not assert anything. They say language could be inconsistent in the sense of having inconsistent rules. Their reasoning is that institutions such as clubs, governments, and religion sometimes formulate rules that are impossible for some individuals to completely obey. For example, the Old Testament both requires and forbids marriage to your brother's widow. There is no problem as long as your married brother remains alive; but once he dies, you are in a dilemma for which you can hardly be faulted. The rules are inconsistent.

Language puts its users in the same sort of dilemma. This can be seen at the level of schoolbook grammar. One spelling rule tells us to begin sentences by capitalizing the first letter. Another rule tells us to spell names as their bearers prefer. But what if the person is named van Snoot. Do I write 'Van Snoot snores' or 'van Snoot snores'? Many people avoid the dilemma by giving up their intention to begin with the name and instead write 'One snorer is van Snoot'. Some commentators on Grelling's paradox cite 'heterological' as a convincing instance of an inconsistent concept. Although the term was introduced to make a philosophical point, 'heterological' has gained autonomous currency. This is not surprising because the word is readily learned and works smoothly in almost all cases. Once people learn that 'autological' means 'describing itself' and 'heterological' means "not describing itself," they reliably sort words into each category. Since 'word' is a word, it is autological. And since 'rock' is not a rock, it is heterological. We can classify other words on this basis:

Autological words: polysyllabic, intelligible, English, noun, unhyphenated
Heterological words: monosyllabic, obscene, wet, preposition, alive

So far, 'heterological' seems as unproblematic as 'onomatapoeic' which is defined as 'similar in sound to that which it denotes'. But Grelling's paradox rears its ugly head when we ask, "Is the word 'heterological' heterological?" In other words, to which list does 'heterological' belong? If 'heterological' is heterological, then it describes itself and so must go on the autological list. But if 'heterological' is autological, it does not apply to itself and so must go on the heterological list. Unlike 'round square' and like 'faster', 'heterological' can be applied to lots of cases to yield truths. Only in a few cases does it behave contradictorily. But unlike 'faster', 'heterological' is a linguistic predicate whose adequacy and verisimilitude is not relative to a possible world because it is its own problem case.

Logicians solve the paradox by first drawing attention to its structural analogy with Russell's barber paradox. Suppose there is a barber who shaves all and only those people who do not shave themselves. Does the barber shave himself? Either answer implies a contradiction. So the logicians reject the apparently modest assumption that leads to a contradiction, the assumption that such a barber can exist. If this seems high-handed to you, then think about impossibility proofs in mathematics. Euclid proved that the number of primes was infinite by deriving a contradiction from the assumption that there is a largest prime number. The same strategy dissolves riddles such as "What is darker than the blackest thing?" "Is 'heterological' heterological?" presupposes a logical impossibility. There can be no predicate that applies to all and only those predicates that do not apply to themselves. As in the case of Russell's barber, we confuse an impossible entity with a possible one that resembles it. It is possible to have a barber who, with the exception of himself, shaves all and only those who do not shave themselves. Likewise, it is possible to have a predicate that, with the exception of itself, applies to all and only those predicates that do not apply to themselves. When 'heterological' is used in contemporary natural language, we are picking out a predicate that resembles the one responsible for Grelling's paradox, but not quite the same one.

The belief that language is inconsistent is nurtured by the analogy between language and activities regulated by fallible human rule givers. However, some philosophers deny that language is a rule-governed activity. Paul Ziff insists that language only has *regularities*, not rules. Since regularities cannot be inconsistent, Ziff denies that the possibility of inconsistent language. Recently, commentators have attributed to Wittgenstein a regularity account in which there is nothing more to following a rule than facts about how the linguistic community actually behaves. Another reason for resisting the notion of inconsistent concepts is the presumption against attributing inconsistency that figures prominently in Quine's discussion of radical translation (recounted in chapter 3). If we are going to attribute inconsistent concepts, we must interpret speakers as subscribing to inconsistent rules. But this violates the principle of charity.

What, then, is the explanation of the more persuasive cases involving club membership, spelling dilemmas, and so on? I trace it to an illusion created by a confusion of priorities. Rules have priority over action in the sense that actions are guided by rules. However, this guidance priority does not imply evidential priority. Indeed, we must attribute rules to agents on the basis of their actions. Their reports of their rules constitute one body of evidence about what their rules are, but self-description is not the only evidence. We also look at how the speakers actually behave, social pressures (what they reward and penalize), what puzzles them. Linguists and anthropologists weight this behavior more heavily than their informant's self-attribution of rules because introspective reports have a long record of error. The ability to articulate the rules one is following is a special skill. Those specially trained in this type of codification (linguists, anthropologists, and philosophers) are relying on introspection. They use projections of their own behavior and their neighbors as data and then engage in the same sort of curve fitting that outsiders do. The evidential priority of behavior gives the principle of charity free reign. So we are constrained to interpret people as following consistent rules even when they report following rules that are inconsistent. The appearance of inconsistent rules is created because people tend to overestimate the reliability of reported rules.

B. Violation of Logical Conservatism

The immediate objection to Kuhn's account is that his notion of a tolerable contradiction is illogical. In standard logic, any contradiction implies everything, so contradictions are never found in correct theories. The simplest way of deriving an arbitrary proposition from a contradiction is as follows:

1. $P \& \sim P$ Assume
2. P 1, Simplification
3. P or Q 2, Addition
4. $\sim P$ 1, Simplification
5. Q 3, 4, Disjunctive Syllogism

More serpentine paths exist. The attempt to block the derivation by restricting the rules of inference is like trying to write loopholes out of the tax code. Nevertheless, Kuhn insists that only a subset of contradictions are intolerable. Which are tolerable depends on the content of our theories and on which possible world we are in. Hence, Kuhn is committed to a notion of localized incoherence.

With Kuhn's conception of contradiction comes the charming irony of science—our paradigm of rationality—bubbling over with counterexamples to the adequacy of classical logic. If the counterexamples are genuine, a big lesson would have been learned from the history of science. The lesson would be music to the ears of relevance logicians. They have long labored against the notion that contradictions cause all hell to break loose. They have engineered deviant logics to reflect the intuition that contradictions only cause a *little* hell to break loose! Consistency is no absolute desideratum on their view. It is one of degree like other desiderata for theories such as simplicity, generality, and fruitfulness.

Kuhn is outradicalized by some past instrumentalists. Whereas Kuhn regards inconsistency as bad but not catastrophic, radical instrumentalists say that some contradictions are good to have around. The background for this daring liberality is serious adherence to the distinction between scientific realism and instrumentalism. Whereas scientific realists think that science aims at a true description of reality, instrumentalists take the aim of science to be the prediction and control of nature or experience. Since reality must be consistent, realists assert, inconsistent propositions are automatic falsehoods and so are not part of the true description of reality; hence, they conclude that inconsistencies have no intrinsic scientific value. Instrumentalists are more open-minded. Their motto is, Do not ask whether the hypothesis is true, ask whether it is useful. We can see how harboring a contradiction may incommode us in various ways. But the contradiction may offer compensating benefits.

Hans Vaihinger catalogued these benefits in *The Philosophy of the "As If."* He protested against the chauvinism behind the demand for consistency. "Many of the fundamental ideas with which science operates are fictions, and the problem is not how to do away with these contradictions—that would be a futile undertaking—but to show that they are of utility and advantage to thought. It is wrong to imagine that only what is logically non-contradictory is logically fruitful. Such an attitude—if consistently adhered to—would bring us to the conclusion of Agrippa of Nettesheim, that all science is valueless."[7] Vaihinger explained our realist tendencies as a consequence of our general propensity eventually to value means as ends:

> Our conceptual world lies between the sensory and motor nerves, an infinite intermediate world, and serves merely to make the interconnection between them richer and easier, more delicate and more serviceable. Science is concerned with the elaboration of this conceptual world, and with the adjustment of this instrument to the objective relations of sequence and coexistence which make themselves perceptible. But when science goes further and makes of this instrument an end in itself, when it is no longer concerned merely with the perfecting of the instrument, it is to be regarded strictly as a luxury and a passion. But all that is noble in man has had a similar origin.[8]

When stood up against Vaihinger, Kuhn appears pale and tame. Kuhn dips his big toe into the pool of incoherence. Vaihinger belly flops.

The majority of logicians are more conservative than these characters; for the intuition that contradictions do not imply everything is an expensive one to appease. At a minimum, one must reject the inference rule of disjunctive syllogism: P or Q; not P; therefore Q. Most logicians find rejection of such a compelling rule preposterous. Recall G. E. Moore's incredulity at the idealist's claims, "Material things do not exist" and "The past is unreal." Which is more likely, Moore would ask, that common sense is mistaken on this issue or that there is a false premise leading to the strange philosophical conclusion? The history of philosophy is littered with dead principles. Common sense has a far better record. Hence, the wise man sides with common sense. A similar appeal to relative likelihood can be made against those who deny theorems of classical logic. It is not as if relevance logic were the only way to explain our inferential caution. The mere fact that we assign probabilities of less than one to our premises will prevent us from performing blind deductions. We are circumspect because we know that our premises are occasionally false. We sensibly use the weirdness of our results as a reason for double-checking our point of departure. Only once these suspicions are cleared do we continue the inquiry. Thoughts like these lead most logicians to dismiss relevance logic as a misguided attempt to reflect our inferential caution in logical, rather than epistemological, terms. By extension, Kuhn also faces a hard sell. His audience will be inclined to regard his insights as mispackaged.

For example, James Cargile has objected that Kuhn confused the definitional role of criteria with their role in expressing empirical laws.[9] As a definition 'Man is the linguistic mammal' is refuted by the logical possibility of talking dolphins. But it holds as an empirical generalization if no other mammals talk. When criteria play the role of empirical laws, they are immune to farfetched counterexamples. Imaginary counterexamples are sometimes relevant because the truth of an empirical law has to be due to more than luck. But these hypothetical scenarios can only demote the criterion to the status of accidental generalization if they are the sort of case the scientist could expect to meet. So our practice of counting people by counting heads is not impugned by the logical possibility of two-headed men. But when criteria play a *definitional* role, all possibilities are relevant. The failure to distinguish the empirical and definitional roles of criteria leads Kuhn to mix concepts with laws of nature, to view concepts as having contingent adequacy and as being insulated from farfetched counterexamples.

Perhaps Kuhn would protest that Cargile's attribution of confusion begs the question, since it is based on the very conception of definition and laws that Kuhn challenges. Whole-hearted rejection of the analytic/synthetic distinction does blur the distinction between definitions and laws. So Cargile should not simply assume that there is a sharp distinction and then infer that Kuhn has blundered. But Cargile is free to offer the sharp law/definition distinction as part of a rival explanation of the phenomena that explains away our Kuhnian intuitions. Empiricists have frequently resorted to this sort of debunking when rationalists claim to have found a synthetic a priori proposition. The empiricist's diagnosis is that a sentence such as "Light travels in a straight line" is ambiguous as between a tautology and a well-established law. If we fail to distinguish between the two readings, we will have the impression that it is an a priori proposition about the world. Empiricists who try the law/definition move on Kuhn's cases are not thereby begging the question. They are just performing a variation of a move they favor against the rationalist's candidates for the title of synthetic a priori.

Kuhn does not say that all thought experiments involve local incoherence. He only claims to be characterizing an interesting subclass. Nevertheless, it is still important to stress that Kuhn's subclass does not encompass all of the inconsistency-revealing thought experiments. For instance, topologists have shown that the definition of 'wind' precludes the possiblity of worldwide wind. But rather than reject the definition, we accept the necessary existence of windless spots on balls such as earth. The surprising impossibility theorem is regarded as a veridical paradox.

It might be suggested that this is what Kuhn means when he says that the inconsistencies of a scientist's concept mark his commitment to a type of universe. But the scientist would then be guilty of the sort of overkill for which he chides the philosopher (converting new a posteriori truths into a priori ones). Marking an empirical impossibility with a definitional exclusion exaggerates the commitment.

V. Reconstruction of Kuhn's Error

Although these objections show that Kuhn's notion of local incoherence is itself incoherent, we still need a *diagnosis* of the error. What leads Kuhn to his strange conclusion? The error path is interesting because it embodies insight. Indeed, my goal is now to resettle this insight into a logically conservative account of thought experiments.

A. The Guts of Paradox

Kuhn's two cases intrigue us because they involve *paradoxes*. A paradox is a small set of individually plausible but jointly inconsistent propositions. Since plausibility varies from individual to individual, what is a paradox to me need not be a paradox to you, just as what is a surprise to me need not be a surprise to you. Nevertheless, the set containing the following propositions is a paradox for most people—the paradox of self-deception:

1. Self-deception is possible.
2. If A deceives B into believing p, then B believes p.
3. If A deceives B into believing p, then A does not believe p.

In the case of self-deception, A = B, so A must believe p (because of (2)) and not believe p (because of (3)). Most people believe each member of this set and so have inconsistent beliefs. Since people dislike inner conflict, they try to regain consistency by rejecting a member of the paradox. So, revealing a paradox to its victim tends to induce a change of belief. Since it is rational to minimize inconsistency, the paradox has the power of rational persuasion. This provides an answer to the question how some thought experiments manage to be informative without adding new empirical information. By exposing an inconsistency between scientific beliefs, they stimulate a scientific change of mind.

This explanation of the rational persuasiveness of thought experiments is close to Kuhn's claim that thought experiments unearth hidden contradictions of a concept. But notice that I am not attributing inconsistency to a concept. Indeed, there need be no *absolutely* inconsistent member of the paradox. A member of a paradox need only be inconsistent *relative* to the conjunction of the remaining members; that is to say, the conjunction of all the members of a paradox is a contradiction, but the members themselves are not (usually) contradictions. The point can be symbolized in terms of impossibility:

p is impossible: $\sim \Diamond p$

p is impossible relative to q: $\sim \Diamond(p \ \& \ q)$

Since every proposition is inconsistent relative to some proposition, it is never informative just to say that a proposition has relative inconsistency; your audience must know relative to *what*.

Relative inconsistency is not restricted to theoretical contexts. It is common in dreams and fiction. In Conan Doyle's stories, Watson is said to have exactly one war wound; but the wound is sometimes described as being in his arm, sometimes as being in his leg. "Impossible" pictures are visual instances of relative inconsistency (Figure 5.3). As Escher himself wrote of his *Waterfall*, "If we follow the various parts of this construction one by one we are unable to discover any mistake in it. Yet it is an impossible whole because changes suddenly occur in the interpretation of distance between our eye and the object."[10]

B. Conflationary Factors

Now that we have a feel for the difference between relative and absolute inconsistency, let us consider the three factors that lead us to confuse the two.

First is inexplicitness. Only rarely do we find all the members of a paradox neatly assembled like suspects in a police lineup. Usually, the victim of a paradox does not differentiate between the members. He is apt to merge, omit, and add superfluous elements to the imbroglio—like the flustered target of sniper fire. The victim of a paradox has only a vague feel for the clash of beliefs. Since he often lacks the resources to delineate the engagement, the conflict is often left as an amorphous hostile presence.

The second factor fostering confusion between relative and absolute inconsistency is the foreground/background effect. When we attend to something,

Figure 5.3 Photograph of author holding impossible cube casting impossible shadow.

we discern many of its features. However, surrounding objects are hardly noticed. Indeed, their existence is often overlooked even though they are influencing one's observation. For example, merchandise within an attractive setting appears to be of higher quality than identical merchandise in shabby surroundings. Background also influences the perception of paradox. Typically, one member of the paradox occupies the foreground of the victim's attention while the rest quietly manipulate his thinking from the background. The result is that the conflict generated by the paradox feels as if it is emanating from the foreground proposition.

Third is loose usage of the term 'contradiction'. Contemporary logicians use 'contradiction' in a narrow way to cover only those statements that could not be true. In this narrow sense, 'contradiction' is a property of a statement, not a relation between statements. But nonlogicians tend to use it relationally, so that it is natural to talk of one person's report contradicting another's. There is also a tendency to extend 'contradict' to what contradicting statements represent. Hence, we ordinarily speak of observations contradicting theories and science contradicting religion. Hegelians and Marxists are notorious for using 'contradiction' interchangeably with 'conflict'. Nonlogicians realize that a single statement such as 'It is raining and it is not raining' can have an *inner* conflict. They tend to mark the nonrelational aspect with the reflexive term 'self-contradiction'. However, plain folks feel uncomfortable in extending 'contradiction' to a statement that cannot be divided into opposed parts. For example, many students enrolled in their first philosophy class are reluctant to call 'Some bachelors are married' a contradiction even though they agree that the statement is necessarily false. The quickest way to overcome their resistance is to show that it implies two statements that contradict each other. Hence, the ordinary usage of 'contradiction' is in some respects broader, and in others, narrower than the logical—but mostly broader.

Contemporary logicians realize that 'contradictory' has a relational use. Indeed, classical logicians used it this way in the traditional square of opposition. Although the square introduces one point of agreement, it introduces another point of disagreement. Even Aristotle diverged from everyday usage in distinguishing between contradictories and contraries. Two propositions are contradictories if and only if they must have opposite truth values. They are contraries exactly if they cannot both be true. Unlike contradictories, contraries can both be false, as with 'All sand is wet' and 'No sand is wet'. Since conflicting reports are often both false, we can see that close adherence to Aristotle would lead us to describe the opposed reports as contrary rather than contradictory.

Our trouble understanding Kuhn can be dispelled by observing how he confuses relative and absolute inconsistency because of the interplay of the three factors *inexplicitness, foreground/background effect*, and *equivocation* (between the ordinary and technical senses of 'contradiction'). Recall that Kuhn talks of inconsistent concepts that apply to things even though the concept harbors "hidden contradictions" that are tolerable in some possible worlds. This is gobbledygook if we interpret 'contradiction' as logicians do. But if we let ordinary language be our guide, we can coax some sense out of it. Kuhn is speaking of statements that contradict presumptions. Since we tend to overlook background assumptions, statements that contradict a presumption look like nonrelational contradictions, that is, self-contradictions.

Although Kuhn is interested in statements that only covertly *preclude presumptions*, the nature of presumption precluders is best studied by first concentrating on overt cases. For example, Groucho Marx's assertion, "I would never join a club that would have me as a member" is flagrantly bizarre. Many people respond to the peculiarity by calling it a contradiction. However, the statement is consistent. Groucho's quip merely precludes something we take for granted, 'I might join a club'. The point is made more suggestively by saying that Groucho's quip *contradicts* the presumption. A second example is William Archibald Spooner's "I remember your name perfectly, but I just can't think of the face." This statement strikes people as contradictory because they presume it is part of a face-to-face conversation between normal human beings with typical abilities.

A famous literary example of presumption preclusion occurs in Joseph Heller's novel *Catch-22*. Yossarian is a World War II bombardier who has requested to be taken off duty on grounds of insanity. Doc Daneeka explains that the request cannot be fulfilled. Although the doctor is obliged to ground all crazy soldiers, he can only ground soldiers who request it. The catch is that such requests demonstrate a concern for one's safety, which in turn suffices for sanity in the eyes of the regulators.

There was only one catch and that was Catch-22, which specified that a concern for one's own safety in the face of dangers that were real and immediate was the process of a rational mind. Orr was crazy and could be grounded. All he had to do was ask; and as soon as he did, he would no longer be crazy and would have to fly more missions. Orr would be crazy to fly more missions and sane if he didn't, but if he didn't want to he was sane and had to. Yossarian was moved very deeply by the absolute simplicity of this clause of Catch-22 and let out a respectful whistle.[11]

Most readers infer that the regulations are contradictory. However, the regulations could all be satisfied if there were no crazy soldiers. To see this, reformulate the rules to make their logical properties shine:

1. All insane soldiers must be grounded.
2. All soldiers who are grounded because of insanity must request to be grounded out of concern for their own safety.
3. All soldiers concerned with their own safety are sane.

In a world where all soldiers are sane, there is no problem. But if there are crazy soldiers, as there were in the world of *Catch-22*, it is impossible to satisfy the rules.

Since presumptions lie in the background, we are usually slow to notice their presence when they have been contradicted. Nevertheless, some assertions are intended to have their contradicted presumption recognized. Consider Frederick Douglass's statement: "A gentleman will not insult me, and no man not a gentleman can insult me." When we read it, we take 'Some man can insult Frederick Douglass' for granted and so are initially inclined to regard the statement as contradictory. But Douglass nevertheless intends the reader to work through the apparent contradiction and conclude that no man can insult Douglass.

The statements that interest Kuhn are *covert* presumption precluders. Unlike the overt versions just discussed, the covert cases lack immediate absurdity. They look contradictory only *after* reflection, not before analysis. At first glance, covert presumption precluders are as innocent as new-laid eggs. The point can be illustrated with a thought experiment discussed in chapter 5. Recall that Galileo refuted Aristotle's theory of motion by having us suppose that a heavy falling object is joined to a lighter one. The refutation works against the first member of the following paradox by winning agreement with the remaining four:

1. Aristotle's theory of motion is correct.
2. If so, then it is physically necessary for heavier bodies to fall faster and faster bodies to be slowed by being joined to slower ones.
3. If this were a physical necessity and a heavy faller were joined with a lighter faller, then the composite body would simultaneously fall faster and slower.

4. Nothing can be simultaneously faster and slower than another body.

5. Two falling objects of unequal weight can be joined.

Supposition 5 is salient at the beginning of Galileo's discussion and initially seems harmless. But as the Aristotelian comes to recognize the incompatibility, he experiences a sort of ambivalence about that supposition. It is as if life just spat in the face of logic. In an effort to release the tension, he will toy with the idea of rejecting other members of the paradox. Thus, Aristotelians suggest that supposition 2 can be rejected by refining the meaning of 'body'. In particular, no proper part of body is itself a body. Otherwise 'Heavier bodies fall faster' would have the even more absurd consequence that all proper parts of a body fall slower than the body itself!

C. The Phenomenology of Inconsistency

I remember feeling Kuhnian ambivalence as a five-year-old boy. Propositions 1–4 were my background beliefs about the difference between adults and children:

1. All people start small and ignorant and slowly grow bigger and more knowledgeable.

2. If so, all small people are children and all knowledgeable people are adults.

3. If the small people are children and the knowledgeable people are adults, then a small knowledgeable person would be both a child and an adult.

4. No child can be an adult.

5. Small but knowledgeable people are possible.

My background beliefs served me well until I met at the playground what I now realize was a midget. He led me to add proposition 5 to my stock of beliefs. The small size of the midget initially led me to treat him as a child. The midget said that he did not care to play with me because he was actually an adult. I was sceptical. But as our conversation continued, I found it increasingly difficult to explain all the things he knew as due to the fact that he was a short older child. After the midget left, I vacillated between thinking that (~1) he was a tiny adult or a fantastically learned child, (~4) a double being, both child and adult, and (~5) a child faking wisdom. The vacillation was structured by the possible resolutions of my inconsistency. Of course, I could only vaguely discern the alternatives. Also it was not clear to me that the difficulty was due to *my* beliefs rather than *his* existence. Indeed, I felt a strange animosity toward him, as if the midget were a metaphysical monster. He seemed a contradiction come alive, stalking the playground of Woodward Parkway Elementary School.

But I had really fallen prey to the "pathetic" fallacy. This is the fallacy of treating a psychological relation as an independently existing property. For example, if you are disgusted with something, the disgust is a relationship between you and the object. But it is tempting to oversimplify the situation by thinking that the object is itself disgusting independent of your reaction to it. Thus, we will say that a fungus is just foul and overlook our own contribution to the fact. Feeling disgusted with the fungus is erroneously pictured as *recognition* of its foulness, just as feeling the edges of a die constitutes recognition of its cubical shape. The logical form of the misrepresentation is that of oversimplifying Rxy into Rx, thereby "dropping a variable."

The pathetic fallacy is also at play with conflict attitudes such as ambivalence. We pull these attitudes inside-out. We project the feeling of inconsistency caused by the external thing onto the external thing itself. Thus, we have the eerie impression that reality has gone haywire, that there are elements within the world order operating in gay defiance of logic.

Obviously, my attempt to do justice to this childhood trauma oversophisticates the experience. Five-year-olds are philosophical innocents. Hence, my description should be taken in the same spirit as the microscopist's stained slides. We are free to add to it in the cause of clarity just as long we draw attention to this liberty. What I wish to emphasize is that the phenomenology of these conceptual challenges is just as Kuhn describes. There is the feeling that the concept was behaving just fine until it came upon the puzzle case, an enigma revealing a pocket of incoherence that might best be handled by avoiding its general territory as we might a bog. Although the psychology jiggled up by the phenomenon is fascinating, it is also thoroughly misleading. Once we notice what we are presupposing and keep our logical terminology straight, we come to see that the ontic horror is actually the distorted appearance of a smaller, natural problem. Thus, a steady application of tried and true logical principles illuminates the exit from this fun house.

D. Counteranalysis of Kuhn's Cases

Piaget's experiments show that what happened to me at the playground can happen to children at the hands of developmental psychologists. Piaget was staging the sort of conceptual confrontation that led me to the insight that 'adult' does not imply large size. The paradox for his little research subjects was as follows:

1. Cars that reach the finish line first are faster, and cars that move more hectically ("blurriness") are faster.
2. If so, then the goal reaching and blurriness criteria are each logically suffecent conditions for applying 'faster'.
3. If these criteria held and a blurry car reached the goal later than a nonblurry one, each would be faster than the other.

4. 'Faster' is asymmetrical.

5. A blurry car can reach the goal later than a nonblurry one.

Piaget reports that the race demonstrating the truth of proposition 5 befuddled the children. They were liable to switch verdicts as to which is faster, deny that either is faster, or even maintain that both are faster. There was no stable response pattern. The structure of their vacillation can be read off the five members of the paradox.

As Kuhn grants, the Piaget case is not a thought experiment, because the race between the toy cars is actual, rather than hypothetical. If Piaget had merely asked the children to visualize the race to make the same point, he would have been conducting a thought experiment. The lesson would not have been as effective, because children have little aptitude for abstract reasoning. They find vivid, concrete events much more persuasive. Adults (at least, a substantial minority of educated adults) are more flexible. They are at home with hypothetical scenarios and reason them out with a fluency rivaling their performance with actual scenes.

Kuhn's second case is a genuine thought experiment. It is Galileo's hypothetical scenario involving the two balls, one rolled down the plane, the other dropped from the same height (see Figure 5.2). The paradox for the Aristotelian is as follows:

1. Aristotle's definition of motion is correct.

2. If so, then it is necessary that "the quicker of two things traverses a greater magnitude in an equal time, an equal magnitude in a less time, and a greater magnitude in less time."

3. If these conditions held and Salviati's experiment were conducted, then ball 1 would be quicker because it traverses CB in a shorter time than it takes ball 2 to traverse CE; yet ball 2 would be quicker because it traverses DA in a shorter time than it takes ball 1 to traverse CB.

4. Nothing can be both quicker and slower.

5. Two balls can be dropped as described.

Since Galileo's audience is more sophisticated than Piaget's, he need not bother with a physical set up. In principle, his victims could escape inconsistency by denying that the imaginary adventure of the two balls could take place. However, this rejection of proposition 5 would be ad hoc. Galileo is having us visualize a situation that closely resembles ones we have experienced. To deny proposition 5, we would need to draw a principled distinction between the imaginary scenario and our recollection of the dual descents of humdrum objects. This suggests that the real function of Kuhn's verisimilitude requirement is to confer ad hoc status on any attempt to reject the hypothetical happening.

The reconstruction of Kuhn's error can be summarized as a five-step error path:

1. Victim encounters a thought experiment that reveals a situation for which the apparent rules governing a term are inconsistent relative to some background assumptions.

2. Victim presupposes those background assumptions because they are so plausible.

3. Victim then overlooks the presupposed background assumptions, thereby masking the relativity of the inconsistency, which in turn makes the conjunction of the rules look absolutely inconsistent.

4. Victim notes that rules are consistent relative to other situations and so gains the impression that the term is consistent in some situations and inconsistent in other situations.

5. Victim infers that the appearance is accurate and so concludes that the term is *locally* incoherent, that is, consistent when applied to some cases and inconsistent when applied to others.

In addition to providing a diagnosis of Kuhn's error, the reduction of thought experiments to paradoxes also supplies an answer to the question of how they manage to be informative without adding new empirical information. The explanation begins by appealing to the rational persuasiveness of deduction. Although paradoxes are not deductions, each paradox systematically suggests a battery of deductions. Since the n members of a paradox are individually plausible but jointly inconsistent, n deductions can be derived by taking the negation of one member as a conclusion and the remaining members as premises. Each of the deductions will be valid and will have compelling premises. Hence, insofar as one accepts deductions as informative, one must accept thought experiments as informative.

However, the informativeness of thought experiment is not exhausted by its ability to cycle us through a sequence of arguments. Other instruments of persuasion are also analyzable as purveyors of paradox even though their informativeness goes beyond the corresponding deductions. For instance, many normal experiments can also be explicated as paradoxes even though their most salient kind of informativeness is the empirical data deployed in support of selected members of the set.

E. Taxonomic Prospects

The paradox analysis provides an opportunity for detailing this material informativeness because it paves the way for an instructive classification of thought experiments. Writers such as Roderick Chisholm and Nicholas Rescher have noted that many philosophical positions can be classified as responses to central paradoxes.[12] For example, Chisholm is able to classify three ethical positions as rival resolutions of the following paradox:

1. We have knowledge of certain ethical facts.

2. Experience and reason do not yield such knowledge.

3. There is no source of knowledge other than experience and reason.

The sceptic denies (1), the naturalist rejects (2), while the intuitionist argues against (3).

The clarity achieved by Chisholm's classification scheme is admirable. Sadly, the history of philosophy has been conducted without the guidance of Chisholm's criteria. Although many philosophical positions can be made to match the scheme after the fact, many have grown their own way and will not fit. Furthermore, not all positions are responses to paradoxes. The metaphysical views of Schopenhauer and Whitehead seem more like speculative descriptions of the general nature of the universe. Others are more specific protoscientific theories. Thus, an attempt to apply Chisholm's classification scheme to all philosophical positions would be incomplete, unconservative, and unnatural: it would be a good idea pushed too far.

However, the idea does not face the same obstacles as a scheme for classifying thought experiments: the scope of the classification is much smaller; since there is no long tradition of commentary on the nature of thought experiments, there is no competitive terminology; and the classification system would fulfill a real need to order the chaos. So in chapter 6 I shall attempt to emulate Chisholm's clarity by applying his basic idea to the classification of thought experiments.

6

The Logical Structure of
Thought Experiments

In the study that I am making of our behavior and motives, fabulous
testimonies, provided they are possible, serve like true ones. Whether they
have happened or no, in Paris or Rome, to John or Peter, telling imparts
useful information to me. I see it and profit from it just as well in shadow as in
substance.

Montaigne

This chapter lays out a classification scheme for thought experiments. Don't
worry about whether this is the uniquely correct scheme. The adequacy of a
classification system is more a question of efficiency and suggestiveness. A
good scheme consolidates knowledge in a way that minimizes the demand on
your memory and expedites the acquisition of new knowledge by raising
helpful leading questions. For example, one of the virtues of the upcoming
taxonomy will be that it doubles as a theory of fallacy. The reason is that once
one understands how each part of a thought experiment is organized toward
the ultimate end of modal refutation, one can systematically diagnose failure.
Admittedly, this analysis plays up the logical aspects of thought experiment at
the expense of its psychological side; but the overrepresentation will be evened
out in chapter 10.

I. Attributing Thought Experiments

Whenever we describe the mental life of another person (and even ourselves),
we interpret. This may seem illegitimate; we ought to be *discovering* the
person's ideas, not *inventing* them. But you can only unearth psychological
facts if you dirty your hermeneutical shovel. Dream description poses the same
problem. If the dream is simple, as when you dream you are eating a onion,
then you can report it in good conscience—but probably won't because it's a
bore. If the dream is complicated, as when you dream of hunting predatory
onions beneath the surface of thirtieth-century Io, then your report must order
the chaos. No justice can be done to the dreamy quality of the story because

reports must impose categories familiar to both speaker and audience. The only way to avoid distortion is to abstain from reporting the dream; but then, no interesting dreams are told. Although interpretative distortion is less severe when you are articulating your own beliefs, there is the lingering feeling that you are touching them up as you go along.

These problems intensify when we describe the beliefs and reasoning of other people. Interpretative intrusion is inevitable. So grin and bear it! Since the distorted version is usually of higher logical quality than the raw original, make an upbeat comparison; touched up beliefs are like computer enhancements of dull, blurry photographs.

Our interpretative machinations surface in a muted form with regular experiments. No experiment is completely public. The reason is that the identity of an experiment is partly determined by the experimenter's intentions. So two behaviorally equivalent experiments can be distinct. Otto von Guericke proved that sound travels through water by always ringing a bell before he fed fish in a pond. Eventually, hungry fish would come as soon as the bell rang. If this format were followed in order to demonstrate classical conditioning, it would be a psychological, rather than a physical experiment. It is tempting to say that the experimenter's intentions are always decisive in determining the nature of the experiment. But this would be a version of the *intentional fallacy*, which occurs when the artist's intentions are given decisive say over the nature of the artwork. The problem of experiment attribution is magnified in the case of thought experiments because the physical component is smaller or nonexistent. There are no experimental instruments to examine, no manipulations to witness, no subjects or substances to probe. Thought experiments are done in the head.

Thought experiment attribution is profitably compared to the attribution of arguments. The interpreter must first decide *whether* an argument is being propounded. Sometimes the speaker's limited stretch of discourse makes the matter plain. Otherwise, we must ask the speaker whether he is *arguing* for a point or merely asserting something. Unhappily, most people are unfamiliar with logical terminology and so are crude informants of their own reasoning. If we decide that the speaker is indeed arguing, then his logical ignorance will hamper our efforts to specify the argument.

Arguments are typically presented as enthymemes. An enthymeme is an abbreviated formulation of the argument, that is, a formulation missing a premise or conclusion. For instance, 'Only citizens can vote, so Carsten cannot vote' omits the premise 'Carsten is not a citizen'. Since the gain in brevity is constrained by the desire to be understood, there is a limit on how short an enthymeme can be. If we clip too much, the misshapen utterance becomes ambiguous or incomprehensible. Be mindful of the sloppiness that is bred by brevity. Logic helps us avoid this vice by giving us the ability to make arguments explicit. The premises of the regimented argument can be carefully inspected for plausibility and relevance. This naked state also provides the best basis for comparison with other arguments: hidden parts become visible, and false parts disappear. Thought experiments have lacked comparable quality control. The upcoming classification system is intended to fill this regulative gap by requiring structural details.

Admittedly, the unclarity of enthymemes and ellipsis is not always a matter of remedial obscurity. Sometimes the problem is vagueness. I believe that men are taller than women but I do not believe that all men are taller than all women or merely that some men are taller than some women. I believe something in-between but cannot exactly specify which proposition. A similar vagueness dogs arguments. The trouble scholars have had in formulating Wittgenstein's private-language argument suggests that Wittgenstein may not have had any exact argument in mind. Thought experiments can be equally elusive.

Thought experiment attribution is further complicated by the principle of charity. This principle tells us to maximize the rationality of interpretees (choose interpretations that minimize the vices of bias, circularity, inconsistency, etc.). Hence, when we attribute a thought experiment, we should try to make it harmonize with other claims made by the thinker. For example, David Hume's "missing shade of blue" experiment appears to refute his principle that basic ideas must be derived from simple impressions. To know how a pineapple tastes, you must first taste it; to know what 'purple' means, you must first see it—and so on for all ideas that are not built out of other ideas. Hume begins by noting that shades of color are also simple, that they can resemble each other and can be laid out in a gradation uniting quite different colors.

> Suppose therefore a person to have enjoyed his sight for thirty years, and to have become perfectly well acquainted with colours of all kinds, excepting one particular shade of blue, for instance, which it never has been his fortune to meet with. Let all the different shades of that colour, except that single one, be plac'd before him, descending gradually from the deepest to the lightest; 'tis plain, that he will perceive a blank, where the shade is wanting, and will be sensible, that there is a greater distance in that place betwixt the contiguous colours, than in any other. Now I ask, whether 'tis possible for him, from his own imagination, to supply this deficiency, and raise up to himself the idea of that particular shade, tho' it had never been conveyed to him by his senses?[1]

This thought experiment is readily classed with others aimed at the refutation of general principles. In this case, the target is *concept empiricism*—the view that all our simple ideas are obtained from experience. Hume himself is a famous exponent of this doctrine. The mystery for historians is the cavalier manner in which he dismisses this apparently deadly counterexample. Hume blithely continues, "I believe there are few but will be of opinion that he can; and this may serve as a proof, that the simple ideas are not always derived from the correspondent impressions—tho' the instance is so particular and singular, that 'tis scarce worth our observing, and does not merit that for it alone we should alter our general maxim."[2] Perhaps we are entitled to ignore exceptional situations when designing a simple set of rules or when streamlining for pedagogical efficiency. But Hume is laying down the foundation of his philosophical system, an enterprise in which practical considerations do not justify dismissal of rare counterexamples.

II. Thought Experiments as Alethic Refuters

Although there are a number of ways to classify thought experiments, a refutational format scores the most points when judged by familiarity, specificity, and simplicity. According to this scheme, thought experiments aim at overturning statements by disproving one of their modal consequences. Modalities are operators that are applied to propositions to yield new propositions. There are deontic modalities (*permissible, forbidden*), epistemic modalities (*know, believe*), and alethic modalities (*possible, necessary*). The alethic modalites are the best-understood and more-basic modality. Hence, we won't miss anything by concentrating on them.

There are two alethic operators. Necessity is symbolized with a box, □, and works like the universal quantifier 'all'; that is, when you say that p is necessary, □p, you are saying that p is true in all possible worlds. An example is that in all possible worlds, triplets have siblings. Possibility is symbolized with a diamond, ◇, and works like the existential quantifier 'some'. When you say that p is possible, ◇p, you are saying that p holds in at least one possible world. For example, in some possible world, moths outnumber spiders. The impossibility of p is symbolized as ~◇p and means that p is not true in any possible world. For instance, in no possible world is there a bladeless ax missing a handle.

Picture thought experiments as expeditions to possible worlds. The mission is to refute a source statement that has an implication about the constituents of these worlds. Sometimes the consequence is that p fails to hold in any possible world. Hence, if we find a possible world in which p is true, we have refuted the consequence and thereby the source statement. Likewise, if the source implies that p holds in some possible world and we show that it holds in none, then the source is indirectly refuted by the refutation of ◇p.

A. Necessity Refuters

A necessity refuter is a supposition designed to refute a statement by showing that something ruled out as impossible by that statement is really possible after all. The necessity argument can be extracted from the following quintet of jointly inconsistent propositions:

   i. S *Modal source statement.* Fertile sources of modal propositions include semantic theses (definitions, synonymy claims, entailment theses), testability theses (unverifiability, unfalsifiability, indetectability), feasibility claims, law statements, disposition and intention attributions, validity verdicts, and clusters of these—theories.

ii. $S \supset \Box I$ *Modal extractor.* This proposition draws the relevant modal implication from the source statement.

iii. $(I \& C) \Box\rightarrow W$ *Counterfactual.* This proposition is read as a subjunctive conditional: if I and C were the case, then W would be the case. This proposition claims that the antecedent, which is the conjunction of the implication and the imagined situation, has a weird consequence.

iv. $\sim\Diamond W$ *Absurdity.* This proposition explains the weirdness as an impossibility.

v. $\Diamond C$ *Content possibility.* This asserts that the content of the thought experiment is a possibility.

B. The Five Responses to the Quintet

Since the above five propositions are jointly inconsistent, one cannot hold all five. This means that there are at most five consistent responses to the set.

1. *Bad source statement.* The aim of a thought experiment is to refute the source statement, just as the aim of chess to checkmate your opponent. Although our intentions normally match the *aims* of our activities, we are sometimes driven by other goals (call these non-aim goals "motives"). A chessplayer's motive may be fame, the amusement of a friend, or the demonstration of an opening. Likewise, a conductor of thought experiments may have sundry ulterior motives for deploying his hypothetical case. We must nevertheless understand his moves as means to the end of refuting a targeted statement.

a. *Definition busters.* We can illustrate the process with a thought experiment that started a storm of philosophical analysis. In 1963 Edmund Gettier published a small discussion note entitled "Is Justified True Belief Knowledge?"[3] His target was the "JTB" (justified true belief) definition of 'knowledge': A knows that p if and only if (1) A believes p, (2) A is justified in believing p, and (3) p is true. Nearly all epistemologists accepted this definition or a slight variant. Gettier's objection was that the definition is too broad. To prove the point he asked us to suppose that Smith and Jones are candidates for the same job. We are then to suppose that Smith acquires justified belief that

a. Jones is the man who will get the job, and Jones has ten coins in his pocket.

For example, Smith's evidence might be that the president of the company told Smith that Jones will get the job and that Smith has previously counted the coins in Jones's pocket. Smith notices that (a) entails

b. The man who will get the job has ten coins in his pocket.

and so comes to justifiably believe (b). Now comes the twist; despite what the president said, *Smith* is the man who will get the job and, as it happens, Smith has ten coins in his pocket. Hence, b is true, and Smith justifiably believes it. Yet Smith does not know b. Too much luck was involved. Thus, the JTB definition of knowledge is false.

Gettier's thought experiment can be recast into regimented form:

G1. The definition of knowledge is justified true belief.

G2. If knowledge is justified true belief, then necessarily, if a person has justified true belief that p, then he knows that p.

G3. If all justified true believers that p have knowledge that p and Smith is justifiably right but for the wrong reason, then Smith knows that (b) because of luck.

G4. It is impossible for anyone's knowledge to be due to luck.

G5. It is possible for Smith to be justifiably right for the wrong reason.

Almost all epistemologists agree that the Gettier cases are counterexamples to the JTB definition of knowledge and so reject the first member of the set. This was the response Gettier intended to elicit. A few commentators deny the possibility of a justified false belief and so deny the fifth proposition on the grounds that Smith was not justified in believing in (b).

i. *Refuting full definitions.* Reportive definitions, the kind dictionaries market, *describe* a pattern of usage and so are vulnerable to empirical refutation. If you can show that people do not really use 'sweat' in the way Merriam-Webster defines it, then Merriam-Webster is mistaken. Reportive definitions can be refuted by sampling actual usage found in books, newspapers, and conversations. Occasionally, these observations are supplemented with experimental evidence in the form of interviews, discourse prompts, or questionnaires about accepted usage. But definition testers also make great use of hypothetical scenarios. Purely stipulative definitions cannot be proved false because they lack a truth value. If I declare that 'flern' means the food residue left on a knife, then I am laying down a rule, rather than describing a fact. But this does not make the stipulative definition immune to criticism; it just changes the criteria of criticism. We criticize fiats, rules, and orders chiefly on grounds of inefficiency. The stipulator has some special goal. So my stipulation can be undermined by showing that it cannot achieve its goal or that its goal can be achieved in an easier or more reliable way. Stipulators are also beholden to background goals that create the potential for further efficiency critiques. Hence, definers are pragmatically committed to defending more than their definition. By bringing the definition to our attention, they suggest that it is worthy of our attention. But it is only worthy of our attention if it scores well on the standard criteria for grading definitions. Hence, the definer commits himself to the claim that the stipulation achieves a variety of objectives (or would achieve them if the definition were adopted). This metaclaim about the efficacy of the definition does have a truth value and is open to proof or refutation.

Other definitions combine a reportive element and a stipulative element. Different sorts of definition have different purposes for the stipulation. Precisifying definitions, for example, are intended to reduce vagueness. If I offer a million-dollar prize for the biggest puppy, then a reportive definition of 'puppy' (such as 'A puppy is an immature dog') would cause disputes and confusion as to what counts as a puppy. Although the definition correctly reports ordinary usage, it leaves too much open to dispute along the borderline between immature and mature dogs. So for the purposes of my big puppy contest, I should define 'puppy' as, say, 'dog younger than one year'. Administrators of government subsidies have also learned the value of "precisifying" definitions and so provide precise criteria for 'poor', 'citizen', and 'dependent'. "Precisifiers" do not have carte blanche; they are obliged to confine most of their stipulation to the gray areas. Their definition of 'poor' cannot imply that millionaires are poor or imply that penniless mothers are not poor. Moreover, their stipulation must reduce vagueness. Defining 'poor person' as 'person without enough money to achieve an adequate standard of living' fails to eliminate borderline cases. So precisifying definitions have a stipulative element and a reportive commitment, and either aspect can be assessed by thought experiment.

The stipulative element of operational definitions is intended to increase the verifiability of the term. In the case of theoretical definitions, we aim at integrating the *defiendum* with key terms of the theory and tightening its link with laws, tests, and constants. Defining 'acid' as a proton donor strengthens its relevance to established chemical and physical theory, thereby maximizing its informativeness. Defining 'sound argument' as 'a valid argument with true premises' links soundness with validity and thereby to the concept of logical form and generalizations about truth preservation. Legal definitions also aim at this integrative goal, so that users of the term are quickly directed to the relevant parts of the law.

Theoretical definitions are not restricted to science. Anywhere there is theory, there is theoretical definition. Consider theology. Here we find efforts to systematize vocabulary and eliminate conceptual problems with special definitions of 'omnipotence', 'sin', 'faith', and so on. Thought experiments abound. In attempting to define 'resurrection', Thomas Aquinas used the possibility of pure cannibals to eliminate the following definition: 'resurrection' means the reconstitution of a person from his old matter. Who would get the matter of a cannibal who ate nothing but people, the cannibal or his victims? Victims and cannibals must be resurrected for Judgment, yet there is not enough matter to go around. So Aquinas concluded that the definition was mistaken and recommended that we require at most the same *type* of matter, rather than the same *particular* matter.

ii. *Refuting partial definitions.* Thought experiments do not need sharply demarcated targets. They can be deployed against fuzzy, partial definitions and rough notions of what a word means—"folk semantics."

Some partial definitions are just broad, sufficient conditions or necessary conditions. Others come close to being full definitions, providing both necessary and sufficient conditions (which does not imply any one condition that is both necessary and sufficient). We should distinguish between $(\exists x)(Nx \ \& \ Sx)$ and $(\exists x)Nx \ \& \ (\exists x)Sx$. For example, being a mammal is a necessary condition for being a dog, and being a poodle is a sufficient condition for being a dog; but being a mammalian poodle is not a necessary and sufficient condition of being a dog.

Although most speakers are poor definers, they know enough about a term to suggest preliminary definitions and to recognize most definitional errors. So although we are rarely in a position to credit a speaker with a full and explicit definition, we are normally in a position to attribute partial and implicit definitions. All we need is a group of entailment theses concerning the same term. If we are only given necessary conditions for a term or only given sufficient conditions, we lack the premises needed to refute the definition. Since sufficient conditions only say 'The term applies under condition C', we will never deduce conflicting directions from them alone. Since necessary conditions only say "Apply the term only if condition C holds," no conflict can be deduced from them only. Conflicts can be deduced only from mixed premises.

The Ship of Theseus is a philosophical example of the clash between two sufficient conditions and a necessary condition. It dates back to Thomas Hobbes and is directed at our conception of identity. A necessary condition of 'x is the same ship as y' is that y be a single ship, not two or more. One sufficient condition for 'x is the same ship as y' is that x be spatiotemporally continuous with y. Another sufficient condition is that x be constituted from the same parts as y. Now Hobbes has us suppose that the Ship of Theseus has its parts gradually replaced (yielding ship 1) and that the old parts are stored and eventually made into another ship (ship 2). Which is the original ship?

Other thought experiments foment conflict between two necessary conditions and a sufficient condition. A necessary condition of x killing y is that y would not have died anyway. Now suppose that the night before Mr. Victim is to travel in the desert, Mr. Poisoner poisons Victim's canteen. Mr. Saboteur also wants Victim dead but is unaware of Poisoner's action. So Saboteur pokes a hole in Victim's canteen. The next day, Victim goes out into the desert and eventually dies of thirst. Who killed Victim? We cannot say that Saboteur killed him because Victim would have died even if Saboteur had not put a hole in the canteen. (Indeed, Saboteur actually lengthened Victim's life by saving him from quick death by poison.) Yet a second application of the necessary condition also excludes Poisoner: we cannot say that Poisoner killed Victim because Victim would have died regardless of whether the leaked water was poisoned. But this conflicts with the sufficient condition for killing: life-shortening. After all, Victim has been killed by someone because the activities of Poisoner and Saboteur did decrease Victim's life span.

iii. *Refuting definitional policies.* Some doctrines provide a general framework for defining a whole family of terms. For instance, behaviorism says that adequate definitions of psychological terms (*hunger, hope, grief*) are to be drawn in terms of the behavior of organisms to which those terms are applied. A more recent view is functionalism. According to this doctrine, mental vocabulary refers to functional states of the cognizers in question. A consequence of this view is that functionally equivalent systems have the same psychological attributes regardless of differences in what they are made of. Hence, functionalists say a computer would think if it shared your input/output relations.

John Searle attacks functionalism with his Chinese Room thought experiment.[4] He has us imagine that he is in a room with a manual telling him how to write Chinese symbols in response to incoming Chinese symbols. The symbols are meaningless to Searle because he does not know how to read Chinese. However, the manual is in English and only requires that he recognize the shapes of the symbols and correlate them with other shapes. Searle blindly follows the manual and produces outgoing batches of symbols in response to incoming batches. Chinese speakers outside the room think they are having a conversation with Searle. They submit questions in Chinese and Searle writes back with "answers" that seem sensible and fluent to his audience. Despite appearances, Searle does not understand Chinese. The lesson is that purely syntactic competence does not suffice for comprehension. But note that computers will only be able to "communicate" with us by the sort of figure shuffling Searle simulates in the Chinese Room. Since functionalism implies that the symbol manipulation would be enough for understanding, Searle concludes that the doctrine is false.

iv. *Scope expanders.*   Scope theses are nice examples of nondefinitional source statements. Scope propositions say that a principle only applies within a certain domain of phenomena. For example, the uncertainty principle is often assumed to operate only within the realm of elementary particles. However, there are various thought experiments designed to show how the uncertainty can creep into the behavior of middle-sized objects. Imagine two spheres each the size of a baseball.[5] The first is glued to the floor of a vacuum chamber while the second is dropped on the first. Given perfect aim and the absence of air resistance, vibrations, and so on, the ball will bounce up and down indefinitely after striking the stationary sphere's apex. But the uncertainty principle says that there is a chance that the ball will land a little bit off center. These tiny deviations will then accumulate with each bounce until the ball lands on the floor.

v. *Enriched source statements.*   Necessity statements can also be derived from equivalence, implication, and incompatibility claims. Our discussion of definition and synonymy illustrates how thought experiments are used against *semantic* equivalence claims. But since the notion of necessity can be strengthened beyond logical necessity, nonsemantic equivalence theses should be recognized: "Water is H_2O," "A straight line is the path of an undisturbed moving object," "A species is an interbreeding group of organisms." Just as logical necessity is compatibility with logical laws, physical possibility is compatibility with physical laws, biological possibility is compatibility with biological laws, and economic possibility is compatibility with economic laws.

Principles professing to state empirical laws can be refuted by thought experiment. This is done by interpreting the modal operators as governing natural necessities instead of merely logical ones. Hence, a biological principle can be refuted by showing that it entails biological impossibilities such as reverse evolution (forbidden by Dollo's law).

If we interpret the modal operators in terms of compatibility with *moral* laws, we can treat many of the famous hypothetical counterexamples of ethics. For example, some people defend their meat eating by appealing to the fact that human beings are superior to animals. Vegetarians counter that this defense is adequate only if it is necessarily the case that any superior being is entitled to eat any inferior being. But if this were so and superior aliens landed, then they would be entitled to eat human beings for just the reasons we have (taste and convenience). But that would be inhuman! So since extraterrestrial gourmets are a possibility, vegetarians urge omnivores to recant the superior-being defense.

2. *Misconnection.*   Since modal fallacies are common, expect some thought experiments to quickly derail. One modal muddle was a favorite target of medieval logicians: the confusion between a conditional being necessarily true—$\Box(p \supset q)$—and its consequent being necessarily true—$p \supset \Box q$. Sceptics commit this fallacy when they argue as follows:

1. If you know that p then you must correctly
   believe that p is true. $Kp \supset \Box(p \ \& \ Bp)$

2. It is possible for you to mistakenly believe that p. $\Diamond(\sim p \ \& \ Bp)$

3. You do not know that p. $\sim Kp$

If we interpret the argument according to the symbolism on the right, the argument is valid, but the first premise is false. The premise is true if interpreted as giving the necessity operator large scope over the whole conditional: $\Box[Kp \supset (p \ \& \ Bp)]$. But then the argument is invalid. The fallacy is also committed by some hard determinists. Their first premise is that if you perform an action, then you cannot do otherwise. They continue by saying that a man performs a free action only if he could have done otherwise and so conclude that no man is free. Here, the culprit is the ambiguity between the true but trivial conditional $\Box(Ap \supset Ap)$ ('Necessarily, if you act to bring about p, then you act to bring about p') and the false but frightful conditional $Ap \supset \Box Ap$.

"Logical" fatalists commit another modal fallacy, confusing the necessity of a disjunction, $\Box(p \ v \sim p)$, with the necessity of its disjuncts, $\Box p \ v \ \Box \sim p$. They note that it is necessarily the case that either I will surf tomorrow or not. But if it is necessarily true that I will surf tomorrow, I am not free to not surf. And if it is necessarily true that I will not surf tomorrow, then I am not free to surf.

Hence, either I am not free to surf or I am not free to not surf. Therefore I am fated one way or the other.

Sometimes the thought experiment is intended to show that a statistical "certainty" has been mistakenly identified as a physical necessity. In the nineteenth century, the basis for the second law of thermodynamics was Clausius' axiom that it is impossible for heat to pass from a colder to a warmer body unless some other change accompanies the process. In 1871 James Clerk Maxwell suggested that a creature small enough to observe and handle individual molecules could violate the law. He introduced what has become known as "Maxwell's demon":

> If we conceive a being whose faculties are so sharpened that he can follow every molecule in its course, such a being, whose attributes are still as essentially finite as our own, would be able to do what is at present impossible to us. For we have seen that the molecules in a vessel full of air at uniform temperature are moving at velocities by no means uniform. . . . Now let us suppose that such a vessel is divided into two portions, A and B, by a division in which there is a small hole, and that a being, who can see the individual molecules, opens and closes this hole, so as to allow only the swifter molecules to pass from B to A. He will thus, without expenditure of work, raise the temperature of B and lower that of A, in contradiction to the second law of thermodynamics.[6]

Although Maxwell once said he wished "to pick a hole" in the second law, he did not believe the law to be false. The chief end of his creature was "to show that the 2nd Law of Thermodynamics has only a statistical certainty."[7] More picturesque is Maxwell's famous verdict: "Moral: The 2nd law of thermodynamics has the same degree of truth as the statement that if you throw a tumblerful of water into the sea, you cannot get the same tumblerful of water out again."[8] Hence, although the thought experiment is aimed at the law, Maxwell aims to refute only the claim that the law states a physical, as opposed to a statistical, necessity.

Many economic laws have been criticized as having only statistical necessity. Consider the law stating that the quantity sold varies inversely with its price. We can show that it is not universally true with a thought experiment featuring an isolated village of peasants. The villagers can subsist on a weekly ration of one pound of rice and two pounds of beans or can subsist on the tastier and more nutritious fare of one pound of rice, one pound of beans, and one pound of fish. Rice and beans each cost $1.00 per pound, but a pound of fish costs $1.20. Since the peasants are paid $3.20 a week, they spend it all on the rice, beans, and fish diet. What happens if the price of beans rises to $1.10? (It is impossible to buy less than a pound of anything.) Bean sales double because the only adequate diet becomes the one with one pound of rice and two pounds of beans.

3. *Erroneous counterfactual.* We have already witnessed how the medieval recipe for making a vacuum by freezing a container full of water was ruined by

a false counterfactual; that is, since cooling contracts most substances, the medievals naively inferred, 'If the water froze, it would contract'.

There are subtler counterfactual errors. When Galileo realized that *changes* in motion were important, not uniform motion, he became more interested in acceleration. But there were two competing definitions: change of velocity with time and change of velocity with distance. The latter was favored by Leonardo da Vinci. Since Galileo believed that a body in free-fall has uniform acceleration, he tried to derive different empirical predictions from the definitions. This led him to an a priori argument against Leonardo's definition. If the velocity of a body is proportional to its distance, then a peculiar race could be arranged. For suppose we roll two balls down the same inclined plane but stop one of them at the halfway mark. How much longer before the second ball reaches the finish line? Galileo's answer: the second ball would arrive there instantaneously. At the halfway mark, the first ball has only half the velocity that the second has at the finish line. Since the second ball has twice the velocity, it therefore travels twice the distance and so is at the finishing line when the first ball is halfway to the finish line. Since the balls obviously would not behave this way, Galileo rejected the definition of velocity as change with distance. Although later physicists agreed with Galileo's thesis that velocity is change of distance over time, they dismissed his thought experiment as fallacious; for it rests on a confusion between terminal velocity and average velocity. Leonardo is committed to saying that the ball that has traveled twice as far has twice the velocity, but he is not committed to saying that the ball immediately acquires the twofold speed. Leonardo is only saying that the ball will eventually increase up to double the speed, not that it will average double the speed throughout its trip.

At the Solvay conference in Brussells in 1930, Einstein presented a thought experiment that was intended to refute a version of Heisenberg's uncertainty principle, which says that an energy change and its duration cannot be *exactly* measured simultaneously. The imaginary experiment features a light-tight box. The box contains a clock that is set to quickly open and close a shutter. Also contained is a gas of photons. Opening the shutter releases a single photon. By comparing the before and after weights of the box, one could determine the mass (and so the energy of the photon) with as much precision as desired. Einstein concluded that indeterministic quantum theory had to be rejected. After a sleepless night, Neils Bohr found a flaw in the hypothetical. When the photon escapes, it imparts an unknown momentum to the box. This kick causes the box to move in the gravitational field. By Einstein's own theory of general relativity, movement in a gravitational field affects the rate of the clock. Since the clock is being affected by the unknown momentum, the time it measures inherits the uncertainty. So rather than refuting Heisenberg's principle, the "clock-in-the-box" confirms it.

A popular tactic against counterfactuals is the method of counterdescription. In chapter 2, the method was illustrated with Bernard Williams's redescription of "body transfer" cases. Another neat reversal is found in David

Cole's critique of the Chinese Room.[9] Searle had deemed it plain that he would not understand Chinese even though he could transform strings of symbols expressing Chinese questions into strings expressing Chinese answers. Cole insists that Searle would understand Chinese. Cole first points out that the symbol manual and writing paper are red herrings; Searle could in principle memorize the rules and speak with the same fluency as a native. Cole's next point is that although bilinguals can usually translate, their failure to do so would not entail that they do not know the languages. They would only be abnormal bilinguals; linguistic schizophrenics who lack an ability most have. Compare them with aphasiacs who have peculiar combinations of abilities and disabilities. Thus, by redescribing the situation, Cole makes the attribution of understanding less outrageous and so can challenge Searle's assessment of the counterfactual.

Sometimes the counter-thought-experimenter only aims for a draw and can even target a counterfactual implied in an executed experiment. Both of these uses are illustrated by Mach's critique of the experiments, observations, and thought experiments deployed by Newton as evidence for absolute motion. Absolute motion is movement relative to absolute space. Relative motion is movement relative to a material object. Suspicion of absolute space led relationists to argue that all motion is relative motion. Newton argued that acceleration is absolute motion because no material object could serve as the frame of reference. (Both sides agreed that movement had to be relativized to something—they just disagreed over the *relata*. Saying 'x moved but did not move relative to something' was compared to 'x drank but did not drink something'.)

The centerpiece of Newton's case was a simple experiment that he actually performed (Figure 6.1).[10] Newton hung a bucket by a cord and turned the bucket round and round until the cord was strongly twisted. The bucket was then filled with water. This completes the first stage of the experiment. Newton began the second stage by letting go. The bucket whirled. For a while, the bucket was in motion relative to the water. But gradually the initially smooth surface of the water became concave as friction with the sides imparted a centrifugal force to the water. When the speed of the water caught up to that of the bucket, it was at rest relative to the bucket, giving us stage 3. At stage 4, Newton suddenly stopped the bucket. So once again the water was in motion relative to the bucket. Yet the situation is not the same as stage 2: the water at 4 has a curved shape. Can we completely describe the water's motion by taking the bucket as our frame of reference? No, because the water is in motion relative to the bucket at stages 2 and 4, yet the water's behavior is not equivalent: at stage 2 it is flat, and at stage 4 it is curved. Can we, instead, relativize to the laboratory, the countryside, or the earth itself? These candidates are disqualified by the pervasiveness of centrifugal effects. We need a referent that works even for huge rotating objects such as orbiting planets and galaxies. No *ordinary* material thing meets this requirement. Hence, we are forced to relativize to an extraordinary thing, namely, space itself. Newton stressed that this basic frame of reference is not just an abstraction from the

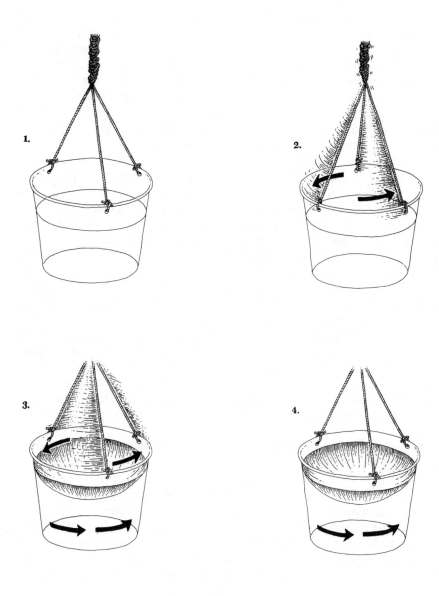

Figure 6.1

material things occupying it: "Absolute space, in its own nature, without relation to anything external, remains always similar and immoveable."[11] Newton knew that many would hope to wriggle out of commitment to this weird thing by saying that he must have overlooked an ordinary reference frame that has yet to occur to anyone. To foreclose this possibility, Newton deployed a thought experiment that stipulates all these hidden *relata* away.

Imagine two globes that are tied together with a cord and then twirled about their center of gravity (Figure 6.2). If all motion is relative, then there would be no difference between stationary observation of the rotating globes and rotating observation of stationary globes. But only in the first case would the cord become tense: "And thus we might find both the quality and the determination of this circular motion, even in an immense vacuum, where there is nothing external or sensible with which the globes could be compared."[12] Newton further supplements the case for absolute motion by appealing to the flattened poles of large rotating bodies such as the planet Jupiter.

Absolute space is unobservable and radically different from all familiar objects. Hence, Mach urged perseverance: if we cannot relativize to the bucket, then we should try relativizing to some other body or group of bodies such as the fixed stars. According to Mach, the fixed stars qualify as an inertial frame as long as we are willing to postulate a universal force whose magnitude depends on the acceleration and mass of the interacting objects and which can operate well over enormous distances; for we can then say that water rotated relative to the stars activates this force: the faster the water spins, the more the huge mass of the stars distorts the shape of the water. Sounds like a funny force? Well, most forces other than contact forces strike us as peculiar: gravity, electricity, magneticism. Certainly, Mach's force is no stranger than Newton's alternative of a force exerted by space itself. If space is changeless, how would we reconcile spatial force with Newton's third law of action and reaction? Mach's force also satisfied the positivist requirement of being verifiable. The force might be detected with a modified bucket experiment featuring a bucket with sides that were several miles thick. For the difference in local mass might then be large enough to make the water curve a bit more than with a thin bucket.

But isn't Mach's force refuted by Newton's globe-twirling thought experiment? Mach defends his use of the fixed stars by preaching intellectual humility. What basis have we for predicting what would happen to the cord connecting the globes in an immense vacuum? Mach writes, "No one is competent to predicate things about absolute space and absolute motion: they are pure things of thought, pure mental constructs, that cannot be produced in experience."[13] He rubs the point in with a rhetorical challenge. If Newton were right, we could hold the bucket steady in absolute space and rotate the heavens around the bucket without thereby deforming the water. Does the Newtonian really think he knows what would happen under such unprecedented circumstances?

Newton's appeal to the oblateness of Jupiter is equally empty. We only feel that the Jovian bulge confirms absolute space as long as we *presuppose*

Figure 6.2

absolute space. Mach applies this point to a hypothetical variant of the Jupiter observation devised by Carl Neumann. Suppose that as a planet rotates the rest of universe abruptly vanishes. Isn't it obvious that the planet would continue to be flattened at the poles? Mach answers with the charge that Neumann has "made here too free a use of intellectual experiment, the fruitfulness and value of which cannot be denied. When experimenting in thought, it is permissible to modify *unimportant* circumstances in order to bring out new features in a given case; but it is not to be antecedently assumed that the universe is without influence on the phenomenon in question."[14] What qualifies as an *important* feature of a thought experiment varies with what one is trying to demonstrate. The point holds for regular experiments as well. We are free to use a mercury thermometer to investigate freezing points but become enmeshed in circular reasoning if we use it to demonstrate that objects expand with rising temperature. The theory describing the behavior of our experimental apparatus must be independent of the theory it is used to test.[15]

4. *Pseudo-absurdity.*   When the absurdity is a contradiction, it is not open to plausible challenge. But when the consequence is surprising without being contradictory, one has the option of denying that the consequence is an impossibility. The four ways of doing this parallel the four ways a defendant can plea against an accusation. I shall lead with the boldest and work down to the mildest.

The first submove is to deny even the appearance of absurdity. This is analogous to the claim that the court has no grounds for thinking a crime has been committed. In its baldest form, the denial of apparent absurdity amounts to the countercharge that the objectors are insincere. Less feisty is the claim that there is no *legitimate* appearance of absurdity. Recall how Mach replied to Neumann's thought experiment by insisting that the consequence is strange only to those who beg the question. Alternatively, we may agree that the consequence is strange but only to one under the spell of a seductive misinterpretation.

A second option is to "bite the bullet"—agree that the consequence is strange but insist that it is a veridical rather than a falsidical paradox. This resembles the legal defense of justification. There are many famous examples of apparent absurdities that turn out to be surprising truths. For instance, the fact that the earth goes around the sun was long called the Copernican Paradox.

The analogue to the third tactic is to grant that a crime has been committed but to pin the blame on someone else (e.g., I may admit that the banjo in my attic is stolen but insist that I unwittingly "purchased" it from a thief). The defense as applied to thought experiment can be put under the slogan "Strangeness in, strangeness out". For the strangeness of the result is traced to the strangeness of the supposition. This strategy has been applied to a thought experiment inspired by an ancient tale of fatalism (a variant of which appears in Somerset Maugham's play *Sheppy*.) The story begins with a man who sees Death in Damascus. Death is startled but then warns "I am coming for you tomorrow." So the man takes the first camel out of town. The next day, Death finds his victim in Aleppo. "But I thought you'd be looking for me in Damascus!" sputters the exhausted traveler. "Not at all," replies Death; "that is why I was surprised to see you in Damascus—I knew I was to find you in Aleppo." The thought experiment modifies this story by stipulating that the traveler *knows* that he has an appointment with Death. He also knows that the date is fixed weeks beforehand and that Death's appointment book is virtually infallible. Hence, although the traveler's choice does not cause Death to be at the appointed place, his choice is excellent evidence of Death's location. For the sake of simplicity, we further suppose that the traveler must choose between one of two places, Aleppo or Damascus. Notice that the traveler can foresee that he will regret whichever choice he makes. If he chooses to go to Aleppo, Death will meet him there, and the traveler will wish he had gone to Damascus. If he chooses to stay in Damascus, Death will meet him in Damascus, and he will wish he had gone to Aleppo. But since it is irrational to do what you know you will regret, the traveler is doomed to irrationality. This strikes most people as absurd, because they believe that if you rationally ought to do x, then you can do x. Allan Gibbard and William Harper parry the attack by urging that we instead give up 'ought' implies 'can'.[16] They say the traveler is just a victim of freak circumstances that force him into an irrational choice. Hence, Gibbard and Harper contend that the absurdity has been fathered by the thought experimenter, not the decision theorist.

The analogue to the fourth defense against the charge of absurdity is to *excuse* the action on the grounds that it was the lesser evil. (This is popular among those responding to Shoemaker's fission case mentioned in chapter 1.) More specifically, this gambit appeals to the conservation of incredulity. Grant that the consequence of your position is hard to believe but emphasize that *any* position on the issue will have a counterintuitive consequence. Exhort your audience to concentrate on the *relative* counterintuitiveness of your solution rather than its absolute counterintuitiveness. There is a general rationale for this comparative theme. When faced with a paradox, one has a conflict of loyalties, therefore there must be a painful disownment of a previously cherished belief. You can redistribute the cognitive shock associated with the paradox, but you cannot eliminate it.

5. *Impossibility theorem.* According to Aristotle, terrestrial objects move only when pushed or pulled by an outside mover. The velocity of the object depends on the degree of push or pull and the resistance offered by the object and the medium. Hence, Aristotle's "equation of motion" was $v = F/R$. Critics posed a hypothetical featuring an arrow traveling in a vacuum. If the equation held for this arrow, the arrow would either be motionless because the driving force was zero or it would fly at infinite velocity because the resistance to motion was zero. But since both of these outcomes are physically impossible, Aristotle's theory of motion is mistaken. In our regimented form, the paradox for Aristotle is as follows:

1. Aristotle's theory of motion is correct.
2. If so, then velocity must equal force divided by resistance.
3. If $v = F/R$ and an arrow were in a vacuum, either $v = 0$ or $v = \infty$.
4. Neither v $= 0$ nor v $= \infty$ is possible.
5. It is possible to place an arrow in a vacuum.

Aristotelians responded that the content of the thought experiment was impossible; arrows cannot move in vacuums because empty space is impossible.

Since they were willing to accept the remaining four propositions in the quintet, the Aristotelians turned the thought experiment into a principle of sorts by regarding the four propositions as premises of an argument that concludes with the negation of the fifth as an impossibility theorem. A modern instance of this modal opportunism is George Schlesinger's handling of Newcomb's problem.[17] He argues that if there were a Predictor of human decisions, rationality would require the one-box choice and rationality would require the two-box choice. But since this is absurd, it follows that the Predictor is impossible and consequently we are free agents.

This flexible use of the set of propositions constituting the thought experiment invites a spurious charge of radical inconstancy. Recall Charles Schmitt's criticism of the medieval debate over vacuums, discussed in chapter 2. The vacuists urged that a void could be created by stretching an impregnable container so that its volume was enlarged without the addition of new content. Thus, they were using the scenario to *attack* the principle that nature abhors a vacuum. Schmitt complains that the plenists used the same scenario to *defend* this principle. The aura of arbitrariness is reduced when we distinguish between fortifying your position and disarming your adversary. A fortifying defense supplies evidence that reinforces the principle. A disarming defense undermines an attack on the principle. (It is one thing to falsify a *premise* of your adversary's objection and another to falsify the *conclusion* of the objection.) The plenists were only using the thought experiment in a disarming defense of "Nature abhors a vacuum." They wanted to undermine the vacuist's premise that the tasks of stretching and sealing are cofeasible. The plenists were willing to grant that you could build an impregnable container, and they were willing to grant that you could have a force powerful enough to stretch any container. What they denied was that you could have both simultaneously. Proving this would give the plenists an impossibility theorem that would remove an entire class of threats to their position.

Hidden impossibility results are often obtained by sweating more detail out of the supposition. This cross-examination technique is frequently used to debunk schemes for violating limits. For instance, one proposal for transcending the speed of light begins with a long rigid rod growing out into space from the equator. The longer the rod, the faster its tip whips about, because the earth is a rotating sphere. Therefore, if the rod continues to grow, the tip must eventually move faster than the speed of light. Physicists counter with close questioning. Where does the material for the rod come from? If it is being drawn from the earth, the conservation of angular momentum requires that the angular velocity of the earth–rod mass decreases in the way a skater slows down as he extends his arms. As more of the earth turns into the rod, the tip slows; once all of the earth is converted into a celestial wand, the tip will be very slow. On the other hand, if there is an extraterrestrial source for the rod's material, then the additional mass must be given extra kinetic energy to keep it moving in pace with the angular velocity of the earth. But we will run out of energy before the tip reaches the speed of light. Hence, the appearance of possibility is a product of intellectual laziness.

Modal inquests are powerful but perilous stimulants. Close interrogation tends to inflate standards of possibility. Is it possible for an iron bar to float on water?[18] That's a bar, not a needle, right? So we can rule out surface tension. Genuine H_2O? Do you mean real iron or an iron look-alike? Bear in mind that **Fe** has a specific gravity of between 7.3–7.8. And you meant *float* didn't you? To *float* means for a thing to have a lower specific gravity than its medium. Since water has a specific gravity of 1.0, it only *seems* like an iron bar can float on water. Think it through, dummy!

Notice the slide from mere logical possibility to chemical possibility. The same equivocation of standards is evident in dialogues with determinists. You say that you could have chosen to read a different book? But didn't your choice issue from a particular set of beliefs and desires? Aren't those psychological properties a product of your genetic makeup plus your upbringing? Just when did your future become unfixed, open, free? The determinist traps his interlocutor by subtly changing the question. We begin with issue of whether alternate book selection is consistent with a loose set of laws and initial conditions and wind up with the question of whether the selection was consistent with a highly specific set of laws and initial conditions. By supersaturating the context with detail, one makes the event inevitable.

Sometimes the denial of possibility amounts to a charge of meaninglessness, rather than falsehood. This was a favorite among the logical positivists. Their verification criterion of meaningfulness declared that a statement is meaningful if and only if it is either analytic or verifiable. For example, 'All vixens are female' is meaningful because we can check the rule for 'vixen' in a dictionary and then deduce the truth of the statement. 'The earth bulges at the equator' is verifiable because we can confirm or disconfirm it with a geographical survey. Contrast these claims with 'Ghosts exist' as uttered by a believer who has de-empiricalized the notion of ghosts with the qualifications that ghosts are intangible, invisible, inaudible, and in general, undetectible. Given all these meaning-leeching provisos, we scornfully conclude there is no difference between there being ghosts and there not being ghosts. The claim only appears to have content. It is actually meaningless, just as the verification criterion dictates. The positivist goes on to spurn 'The universe doubled in size last night' and 'The universe popped into existence five minutes ago' as especially deceptive pseudostatements because they more closely mimic meaningful ones. At this point, the positivist draws our attention to the general pattern of sceptical thought experiments. From the assumptions that knowledge implies certainty and certainty implies the impossibility of error, the sceptic infers that knowledge is undermined by the mere possibility of mistake. Hence, the sceptic tries to refute the claim that you know you are holding a book by appealing to the possibility that you are dreaming or that your brain is being stimulated by a deceiving neurosurgeon or whatnot. These doubts are designed to be ineliminable by empirical tests. So we seem mired in uncertainty and hence profound ignorance. But now the positivist advises us to turn this problem into the solution. Applying the verification criterion lets us eliminate these counterpossibilities as being meaningless, rather than false.

Critics of thought experiments have also harnessed a contrast principle: a term has application only if its opposite has application. The problem of evil provides the backdrop for an illustration. Some theists clamor for details about how the universe could be better than it is *all things considered*. David Hume responded to the challenge by envisioning a world in which there is no pain, only various levels of pleasure. Since putting your hand on a hot iron would cause your bliss to suddenly drop to mild euphoria, you would withdraw your hand to gain the greater pleasure. Thus, your body is protected from injury by differential pleasure, thereby eliminating the need for pain. Many theists charge that Hume's thought experiment is nonsensical because 'pleasure' only has meaning by virtue of its contrast with pain.

The appeal to meaninglessness lives on; witness Putnam's treatment of the sceptical worry that I am a brain in a vat. Putnam's causal theory of reference requires the speaker to have appropriate causal connections with his referent. But if I were a brain in a vat, I would lack this relation and therefore my utterance would fail to express the worry. Hence, if I am a brain in a vat, I cannot even think that I might be a brain in a vat!

Those with no love for verificationism can sometimes achieve a similar dissolution with an appeal to unknowability. Many thought experiments have a hidden epistemic component; that is, in addition to assuming the possibility of the predicament, they assume the possibility of the character's knowing his predicament. This assumption is challenged by one analysis of the Death case. It concedes that the traveler's predicament is possible but denies that the traveler can realize that he is in the predicament.[19] The reasoning is that a chooser cannot pick an option that he knows to be inferior to the other alternative. So although outsiders can know that the traveler will meet Death either way, the traveler inevitably interprets the situation differently.

Since impossibility comes in different grades, there are varying grades of unknowability. This opens the possibility of mixed solutions to paradoxes. Consider a hypothetical that arose from the policy of mutual assured destruction ("If they nuke us, we'll nuke them"). Suppose that the only way to deter an evil is to threaten a fiendish retaliation. This threat of a counterstrike must be sincere because your adversary detects bluffing. Is it moral to form the conditional intention to do the immoral deed? The question arouses mixed feelings; for the intender seems virtuous in that his intention prevents evil yet vicious in that he is prepared to do evil. So which is it? David Lewis holds that there is no determinate answer to this speculative question.[20] Our moral code could be precisified to come down on either side of the issue. The shock caused by either answer could be explained away as a pseudoabsurdity generated by the peculiarity of the scenario. This may seem an evasion. Doesn't the existence of the current U.S. policy on nuclear warfare show that the scenario is realistic? Granted, our information is more limited in the real world. But can't these uncertainties be set aside with minor idealizations? Lewis answers that our ignorance is intractable. There are surprisingly deep reasons why the decision-makers could not know enough about how much damage had been sustained during a first strike or what the effect of a retaliatory strike would be. So Lewis adopts an impossibility solution for "realistic" variants of the hypothetical and a pseudoabsurdity solution for the more far out possibility where the decision-maker does have the knowledge.

C. Summary of Necessity Refuters

I close with a thought experiment that illustrates four of the five resolutions of a *necessity refuter* thought experiment: the St. Petersburg paradox. The mathematical theory of rational choice developed alongside advances in probability. At the time it seemed that if rational choice were amenable to mathematical treatment, rational agents must maximize their expected value. For instance, if you are offered a prize of $10 on the condition that a coin lands heads, then the offer is worth $5, because you have a probability of $1/2$ of gaining $10 and $1/2 \times \$10 = \$5$. An offer of a $24-dollar prize for a die yielding a six has a lower expected value because $1/6 \times \$24 = \$4$. Hence, the wise chooser takes the coin offer rather than the die offer because he should maximize expected value. But now suppose you receive a more complex offer. A fair coin will be tossed until one head results. You will then be paid $\$2^{n-1}$ where n equals the number of tosses. Hence, the expected return is $(1/2 \times \$1) + (1/4 \times \$2) + (1/8 \times \$4) + \ldots + (1/2^n \times \$2^{n-1}) + \ldots$. Since each addend equals 50¢ and there are infinitely many of them, the sum is infinite. Thus, someone who maximized expected money should be willing to pay any amount of money for this bet. Yet common sense tells us that the offer is not worth much at all. Few people would pay $1,000 for the deal. So the belief that rational choice is amenable to mathematical analysis seems to lead to an absurdity.

The St. Petersburg paradox led to a rich set of reactions. A few accepted it as a cogent refutation of the idea that free choice has a mathematical structure. Some mathematicians defiantly declared that a prudent man would risk everything on the bet. They claimed that the result was a pseudoabsurdity; it only looks absurd because we are only familiar with bets offering finite gains.

Condorcet and Poisson rejected the thought experiment on the grounds of impossibility. If heads does not come up by the hundredth toss, the gambler should receive a profit equivalent to a mass of gold bigger than the sun. So the deal cannot be kept. Others extended this criticism by insisting that money is essentially finite because money has exchange power only to the extent that it is scarce.

Daniel Bernoulli challenged the counterfactual 'If an expected utility maximizer were offered, then he would be willing to pay any finite price to participate'; that is, Bernoulli denied that the principle of maximizing expected utility implied that the deal was of infinite value (even if infinite money were possible). He pointed out that doubling one's cash holdings from a million to two million does not really double the value to you. Each new dollar tends to have less influence on your welfare than the preceding dollar. Bernoulli's insight is enshrined in contemporary economics as the law of the diminishing marginal utility of money. The rate of diminution resists precise calculation, but Bernoulli persuasively argued that it was a logarithmic function which prevents the sum from being infinite.

D. Possibility Refuters

A minority of thought experiments vary from the format introduced at the beginning of the chapter. Instead of connecting the modal source statement to a necessity, they connect it to a possibility statement. And instead of affirming the possibility of the thought experiment's content, they say the possibility statement implies its copossibility with the content. The two revisions yields a new quintet of jointly inconsistent propositions:

i $\ $ S

ii' $\ $ S $\supset \Diamond$I $\ $ *Possibility extractor*. This proposition draws a possibility consequence from the source statement.

iii $\ $ (I & C) $\Box \rightarrow$ W

iv $\ $ $\sim\Diamond$W

v' $\ $ $\Diamond$I $\supset \Diamond$(I & C) $\ $ *Content copossibility*. This asserts that the statement extracted at ii' is true only if it is compatible with the content of the thought experiment.

Three of the five members are unchanged from the necessity refuter schema. They do not need further explanation. But the change at steps ii' and v' betokens an important shift in tone. Necessity refuters say that the source is too closed minded, that it rules out genuine possibilities. *Possibility refuters* say that the source is too open-minded, that it saddles us with spurious possibilities.

1. *Bad source statement.* We can show how possibility refuters are supposed to work with the help of a standard objection to polytheism.

i $\ $ Polytheism: more than one god exists.

ii' $\ $ If polytheism is correct, then two omnipotent beings can coexist (because gods are perfect and perfection implies unlimited power).

iii $\ $ However, if there were two omnipotent beings and they were to have a shoving match, an irresistible force would meet an immovable object. (For one would have the power to move anything and the other would have the power to resist any movement.)

iv   But it is impossible for an irresistable force to meet an immovable object: either can exist, but it is contradictory to say they coexist.

v′   If it is possible for two omnipotent beings to exist, then it is possible for them to have a shoving match.

Something has to go! Most theologians dump polytheism, thereby agreeing that the number of gods is either 1 or 0. Thus, the major battle is between monotheism and atheism.

Purveyors of possibility refuters often engage in overkill; they make an already absurd consequence more absurd. Mach is responsible for a charming example of this magnification. His thought experiment demonstrates the physical transitivity of the equal-mass relationship, that is, that two masses that are equal to a third mass are equal to each other. Suppose that A, B, and C are elastic bodies that freely move along a smooth, rigid ring (Figure 6.3). Further suppose that the principle fails to hold for these three objects. In particular, assume that A has a mass value equal to B upon collision and that B has a mass value equal to C but that C has a greater mass value than A. If we then impart a velocity to A, A will pass the velocity to B, and then B to C. But since C has a greater mass value than A, its collision will give A more velocity than A initially had. This surplus will then be carried to B, then to C, whereupon a further boost will be given to A. So on each cycle, the bodies go faster and faster without limit. This endless amplification violates the first law of thermodynamics, which says that energy is neither created nor destroyed. So Mach concludes that 'equal mass' must be transitive. But notice that even one cycle around the ring is precluded by the conservation of energy.

Economists use the money pump argument with the same zeal to prove the transitivity of indifference: if a rational agent is indifferent between A and B and indifferent between B and C, then he is also indifferent between A and C. Suppose indifference were not transitive. It would then be possible for a rational agent to be indifferent between A and B and indifferent between B and C yet prefer A to C. But if this were so and if the agent, Sap, were offered the following trades, he would be turned into a money pump.

	Sap	Trader
0. Initial Holdings:	B, C, $100	A
1. Sap trades C and $3 for A	A, B, $97	C, $3
2. Sap trades B for C and $1	A, C, $98	B, $2
3. Sap trades A for B and $1	B, C, $99	A, $1

Notice that Sap has traded for a net loss. Continued trading in this pattern eventually results in all of Sap's money's being lost to the merchant. The agent

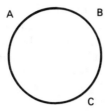

Figure 6.3

will have been turned into a "money pump." But no rational agent can be turned into a money pump. Yet if it is possible for a rational agent to have intransitive indifference, it is possible for this preference structure to occur in the trading context. Hence, rational indifference is transitive. Ducky, but once again notice that a single cycle suffices to demonstrate Sap's irrationality.

Since 'redundant' is relative to belief systems, there are thought experiments that legitimately amplify absurdity. One targets microphysicists who maintain that at the quantum level, things are in an indeterminate state until the observation is made. Erwin Schrodinger ridiculed this idealist stance by imagining a cat that is sealed in a box containing a capsule of poisonous gas. The gas will be released just if an emission of a radioactive source triggers it within a certain interval. After the period elapses, we wonder whether the cat survived, so we believe 'Either the cat is alive or dead'. But given that the particle from the radioactive source only acquires definite properties upon being observed, it follows that the cat itself is in an indeterminate state, neither dead nor alive! Looking inside the box settles its fate, but it has no fate until we peek. But obviously, the only indeterminacy here is epistemological. We do not know whether the cat is dead or alive, but it is surely one or the other. Many microphysicists eat this absurdity. So Eugene Wigner hardens the bullet by substituting a conscious human observer for the cat. "Wigner's friend" may be unneeded for the persuasion of some physicists but necessary for the persuasion of those with a different background or temperament. Although our sense of absurdity has the biological uniformity suggested by Mach, there are also systematic variations.[21] Thus, effective thought experimenters throw a wide net.

2. *Misconnection.* Archytas of Tarentum attacked Aristotle's doctrine of a finite universe with a thought experiment that influenced cosmologists for two thousand years. If the universe is finite, then it is possible for a man to travel to its edge. If he were then to throw a spear, it would rebound or vanish. But that's crazy. So the universe *must* be infinite. Einstein attacked the second step of the thought experiment: finitude does not imply a boundary, because space could be curved.

For a contemporary illustration, consider the reasoning behind those who say the existence of a universal speed limit should not shock us because the opposite possibility is at least as absurd.[22] If there were no maximum speed, an object could travel at infinite speed. If this infinitely fast object made a grand tour of every spot in the universe, it would be everywhere simultaneously. But an omnipresent object is absurd. The trouble with this thought experiment is that the absence of a maximum does not imply an infinite. (For example, there is no largest fraction of the form $n/(n + 1)$ even though none of them is infinitely large.)

3. *Erroneous counterfactual.*   One common objection to time travel is that time travelers could change the past; for suppose that Tina the Time Traveler goes back to the days when her father was a toddler. Now it seems that Tina could grab his throat and strangle him. But if he dies, Tina would not get born and so could not be around to do the dirty deed. Thus, the objection is that if time travel were possible, the time traveler could depart for the past in 1992 and then alter what happened in 1942. But since what has already happened cannot be changed, time travel is impossible.

Time travel buffs reject the counterfactual. They agree that the past cannot be changed but deny that this would make the time traveler *miraculously* impotent; for if time travel is possible, then Tina the Time Traveler is already part of the past; her contribution is preembedded in the course of history. Hence, when Tina's fingers curl about the throat of her toddler father, something will prevent the patricide: he runs away, Tina's grandfather enters, she has a change of heart, or some such. We do not know exactly what prevents it. But we know that something must prevent it, because we know the toddler lives to sire Tina the Time Traveler.

4. *Pseudo-absurdity.*   The fourth move is to deny that the apparently absurd consequence is really absurd. As with the necessity refuter schema, there are four submoves corresponding to the four ways of defending yourself against a charge of wrong doing.

First one can deny the appearance of absurdity. Consider Mark Johnston's objection to Robert Nozick's closest-continuer theory of personal identity.[23] According to this theory, the closest among a field of sufficiently close continuers of an individual is that individual. Johnston objects that this theory would make one's identity dependent on future events. Suppose that a machine has transferred the psychology of a person who began with body A into body B. The person in the B-body gets up, walks about, and says to himself "I am A. I did not just come into existence." He then notes that the machine is reading the A-psychology into body C and will soon make the person with the C-body a better continuer than B. So he turns the machine off, thereby making himself the best continuer. According to the closest-continuer theory, he has thereby made his past thought "I am A" true. But Johnston stresses that the relevant facts were settled *before* the B-body person could decide whether to turn off the machine. Some defenders of the closest-continuer theory respond that this consequence only appears absurd as long as we overlook the fact that the future determines the past just as much as the past determines the future.

Another possibility is to divide and conquer: agree that the consequence is strange under a seductive misinterpretation, and then direct attention to the correct reading. Sheldon Krimsky applies this strategy to Stevin's chain of balls (discussed in chapter 3).[24] Krimsky's contention is that Stevin confuses permissible and impermissible sorts of perpetual motion. Contemporary physicists distinguish between three kinds of perpetual motion. Perpetual motion of the first kind, such as Mach's ring, violates the first law of thermodynamics, which says that energy is neither created nor destroyed. The second law of thermodynamics says that no device can transfer heat from a cooler body to a warmer body as its sole effect. A steamship that operated by extracting heat from the ocean would defy the second law. The century of analysis devoted to exorcising Maxwell's demon is testimony to the commitment of physics to the principle. However, there is a third kind of perpetual motion that only violates the principle that dissipative effects are omnipresent. For instance, a tuning fork cannot endlessly produce sound because the air resistance eventually consumes its energy. A hockey puck eventually slows to a halt because of friction with the ice. Krimsky grants that Stevin's endless chain does produce perpetual motion of this third kind but stresses that violations of this third principle are legitimate in thought experiments that stipulate away friction. Stevin cannot selectively apply the ban against frictionless environments within the same thought experiment!

Interestingly, Mach himself is largely responsible for the distinctions between kinds of perpetual motion. In *History and Root of the Principles of the Conservation of Energy*, Mach points out that the tuning fork can vibrate infinitely in a vacuum. Another one of his nice examples is a spinning top in a vacuum. Although its angle of rotation varies continuously, it produces no work and so does not violate the first law. Mach even applies the point to Stevin's chain of balls.[25] Yet in his later work, *The Science of Mechanics*, Mach accepts Stevin's thought experiment. Krimsky speculates that Mach might have come to believe that the first law was being violated because the balls are continuously overcoming gravity and are therefore doing work. However, endless work need not violate the law. A pendulum in a vacuum raises the bob over distance and so performs work, but no new energy is created. The pendulum just endlessly converts potential to kinetic energy and back again. Doesn't Stevin's chain merely engage in the same endless energy volley?

My guess is that Mach first interpreted Stevin's absurdity as merely that of perpetual movement but later interpreted the absurdity as perpetual *acceleration*. A bias in favor of motion in either direction would ratchet up the speed with each cycle. This accumulation of kinetic energy would be a genuine violation of the first law. Unlike Mach's ring, this magnification of the absurdity would be no rhetorical flourish!

Second, one can respond to the charge of absurdity by "biting the bullet," the most popular response to possibility refuters. For example, the twin paradox was one of the startling consequences of Einstein's theory of relativity. If time is relativized to a frame of reference, then things accelerated relative to that frame change more slowly than unaccelerated things. Consequently, a clock sent off in a speeding spaceship lags behind an earthbound clock. Many people found this incredible. The absurdity was given a human face by substituting twins for the clocks. If acceleration was sustained long enough, the speedy twin could return to find his brother an old man. Relativity theorists treated the absurdity as a veridical paradox. They have since gained empirical support from the fact that atomic clocks sent on speeding planes lag slightly behind ones left on ground, from the increased lifetimes of highly accelerated mesons, and from slight inaccuracies in Doppler's original formula for the wavelengths of light emited by rapidly moving atoms.

Richard Dedekind's definition of 'infinite set' embodies a pseudoabsurdity response to a paradox discussed by Galileo. When we say that the number of As equals the number of Bs, we mean that each A can be uniquely paired with each B—that there is a one-to-one correspondence. However, cases involving infinite sets suggest that this account of equality is too broad; for if it were true, then an infinite subset of a set would be as big as the set itself. As Galileo noted, the natural numbers can be set off in a one-to-one correspondence with those that are squares:

Natural numbers 0, 1, 2, 3 . . . n . . .

Squares   0, 1, 4, 9 . . . n^2 . . .

However, the squares are included in the natural numbers. Moreover, the squares get rarer and rarer as one progresses down the number line. In any case, it's absurd for a set to have a proper subset whose members can be put in a one-to-one correspondence with the set itself. However, Dedekind accepted the "absurdity" and even enshrined this paradox as the *definition* of 'infinite set'!

Third, one can reply to a charge of absurdity by saying that the absurdity is inherited from the thought experiment's weird scenario. This is how hedonistic utilitarians tame utility monsters. Suppose that an individual was able to convert resources into pleasure much more efficiently than anyone else, at a nondiminishing rate, and without limit. Would we then be obliged to turn over all our assets to this lean, mean pleasure machine? Yes, says the hedonistic utilitarian, but the absurdity can be completely traced to the weirdness of the utility monster. If you ask a funny question, expect a funny answer.

5. *"Compossibility" denial.* One class of thought experiments misinfers a possibility statement by means of the following any/all fallacy:

1. <u>Any F is possibly a G.</u> $(x)\Diamond Fx$

2. It is possible for all Fs to be Gs. $\Diamond(x)Fx$

The invalidity of this argument form is demonstrated by its invalid instances. Any New Jersey lottery ticket is possibly a winner, but it is impossible for all of the tickets to be winners. Now consider a thought experiment designed to show that personal identity does not require consciousness. We first assert that no thought is essentially a conscious thought. Then we reason that since any thought could be unconscious, all of one's thoughts could be unconscious. But yet you would still exist. The suspicious step here is the *any*-to-*all* inference. Perhaps an unconscious thought is just one that has a low degree of accessibility as compared to the average accessibility of one's thoughts. Then although each thought could be unconscious, it would be impossible for all of them to be unconscious—they cannot all be below average!

A second example concerns a thought experiment designed by Michael Scriven to refute predictive determinism. Predictive determinism says that every event can be predicted beforehand. Scriven has us suppose that there is an agent whose dominant motivation is to avoid prediction (Avoider). Should this countersuggestible fellow learn that Predictor says he will drive to town, he will stay home. Therefore, to predict what Avoider will do, Predictor must conceal his prediction. Unfortunately, Avoider has enough data, laws, and calculating capacity to duplicate Predictor's reasoning and thus his prediction. Hence, Avoider is unpredictable and therefore a counterexample to predictive determinism.

David Lewis and Jane Richardson object that Scriven has falsely assumed that Predictor and Avoider can simultaneously have all the needed data, laws, and calculating capacity:

> The amount of calculation required to let the predictor finish his prediction depends on the amount of calculation done by the avoider, and the amount required to let the avoider finish duplicating the predictor's calculation depends on the amount done by the predictor. Scriven takes for granted that the requirement-functions are *compatible*: i.e., that there is some pair of amounts of calculation available to the predictor and the avoider such that each has enough to finish, given the amount the other has.[26]

Lewis and Richardson maintain that this compatibility assumption is only tempting as long as we are under the spell of the ambiguity cf 'Both Predictor and Avoider have enough time to finish their calculations'. Parsing the sentence one way yields a truth—that against any *given* avoider, the predictor can finish and that against any *given* predictor, the avoider can finish. But the compatibility premise claims that they can finish against *each other*.

E. Summary of Possibility Refuters

The paradox of the stone illustrates four resolutions of the possibility refuter thought experiment. If God exists, then it is possible for an omnipotent being to exist. If there were an omnipotent being and he tried to make a stone so big that he himself could not lift it, then he would succeed (because he can do anything) and not succeed (because success implies a contradiction). But no one can both succeed and not succeed at a task. Yet if an omnipotent being is possible, he can try to make unliftable stones (because omnipotence ensures he *can* try to do anything).

Atheists are free to reject the source statement and conclude that God does not exist. A few theists (such as the Boston personalists) challenge the modal extractor by denying that God is omnipotent. Descartes is the most famous of a small band of thinkers who take omnipotence to mean that God's power is not even limited by the laws of logic. Members of this group might therefore deny that the contradictory feat is absurd.

However, the most popular theological response is to deny the counterfactual. In particular, theologians reject the following element of the larger counterfactual: 'If an omnipotent being were to try to make a stone too large for him to lift, then he would succeed'. Their case proceeds by first getting the promoter of the paradox to agree that contrary to Descartes, omnipotence does not imply the ability to bring about contradictory states of affairs. Otherwise, there is a much simpler objection to omnipotence: no one can make a round square. Omnipotence has to be understood as the ability to do whatever is *logically* possible. Hence, God is omnipotent if He can bring about any consistent state of affairs. We can make unliftable stones because the stone's weight only has to exceed our finite lifting ability. So 'unliftable stone' is a consistent term relative to *us*. However, it is not consistent when relativized to an omnipotent being. Our anthrocentricism misleads us into thinking we have specified a consistent task for God. Hence, the answer to "Can God make a stone so big that He Himself cannot lift it?" is "No, but so what?" *Omnipotence* does not imply that He could lift it.

III. The Identity Conditions for Thought Experiments

Objection: My analysis underestimates the flexibility of thought experiments. For many thought experiments are *modified* to make new points and to meet objections. These changes can only be formally reflected by replacing a proposition in the set constituting the thought experiment. But since a set is defined in terms of its members, you no longer have the same set. Therefore, we are forced to say that any alteration of a thought experiment yields a new thought experiment. However, we can easily demonstrate that looser standards prevail.

For instance, Bertrand Russell's five-minute hypothesis has been mounted by different thinkers to make different points. Russell introduced it in 1921 to refute 'If there are memory-beliefs, then there were past events'.[27] By supposing that the universe popped into existence five minutes ago complete with fossils, "memories," and so forth, Russell showed how it is possible for memory-beliefs to exist without a past. This boosted scepticism about our knowledge of the past. One might think the universe could be shown to be older than five minutes by applying standard dating techniques. Cut down a tree. Count ten rings. Conclude that the universe is at least ten-years-old because a new ring forms once a year. But this begs the question. The five-minute hypothesis grants that things *look* like they have been around for many years. Dating methods *presuppose*, rather than prove, a past. Hence, Russell's thought experiment propels an interesting sceptical issue. But others retread the five-minute hypothesis for purposes not envisioned by its inventor. Peter Unger has used Russell's brainchild to attack the causal theory of reference.[28] The causal theory says that names refer to their bearers by virtue of appropriate causal chains leading from the event of naming the bearer to the speaker using that name. Hence, if the universe popped into existence five minutes ago, names would only appear to refer to their bearers. Since Unger finds this consequence absurd, he rejects the causal theory of reference. Causal theories of knowledge, meaning, and intention have attracted parallel criticisms. The five-minute hypothesis has also been deployed against etiological accounts of biological function. These theories say that an organ's function is the effect for which it was selected. For instance, the function of a human heart is to pump blood because human beings survive because of that effect. But if the universe popped into existence five minutes ago, organs lack histories of selection. Etiological accounts therefore imply that our organs would then lack functions! Russell's five-minute hypothesis is also used in ethics to contrast "forward-looking" utilitarian theories of justice with "backward-looking" retributivist theories. The utilitarians say that an act is right exactly when it maximizes goodness. Since we can only affect the future, this amounts to saying that right actions maximize good consequences: the past is irrelevant. Therefore, the truth of the five-minute hypothesis would be morally irrelevant. Deontological ethicists hold retributivist theories of morality saying that an injustice is done whenever someone is punished for a crime they did not commit. If the universe popped into existence five minutes ago, then so did the inmates of Attica. The prisoners would be the (albeit unwitting) victims of injustice because their punishment is undeserved.

The only proposition that remains constant through this variety of applications is the C part (the content saying that the world just popped into existence). Yet we are not misspeaking when we say that Russell's thought experiment has been reused for different purposes. In ordinary usage, content continuity appears to be a sufficient condition for identity: same scenario, same thought experiment. However, content continuity is not a necessary condition. In the sixth century, Simplicius quotes Archytas' spear-throwing objection to finite space and asks what would happen if the man at the edge of the universe stuck out his hand. Since there would be no resistance, the man could take another step and stick his hand out again. Since the peripatetic frontiersman could continue this hand extending process without limit, the universe is infinite. Simplicius' scenario is a bit different from Archytas', but many commentators insist that it is the same old thought experiment.

We freely revise the content of thought experiments to overcome sticking points and count ourselves as amending rather than replacing the original thought experiment. Witness how news that a scenario is logically impossible is rarely treated as lethal. The normal inference is that the thought experiment has developed a glitch, so that we need to tinker to bring it back on-line. Thought experiments that run afoul of recent science readily adapt to the new climate of opinion. So instead of being delicate, set-theoretic butterflies, thought experiments are as tough as subway rats.

Often, the original thought experimenter is not in the best position to do the repair work. An outsider may take over because he has the training, insight, or time to make modifications. As other thinkers join in, the thought experiment becomes a collective enterprise. Peter Galison has documented how the high price of apparatus drives high-energy physicists toward collective experiments.[29] Apparatus-free thought experiments show how collectivization can arise just through the need for intellectual division of labor.

Frequently, the thought experimenter will even revise the content to meet objections he foresaw. This occurs when the thinker uses the dialectical process as a vehicle for exposition. Indeed, the typical thought experiment undergoes a combination of "canned" revisions and external modifications. Consider Anthony Quinton's Two-Space myth. Its target is the metaphysical principle that there can only be one space. If this principle is true, then it is necessarily the case that every place can be reached from every other place by traveling through intermediate places. Our confidence in this necessity underlies the joke about the confused guide who, after several false starts, concludes, "You can't get there from here!" Quinton challenges this strand of common sense with the following scenario:

> Suppose that on going to bed at home and falling asleep you found yourself to all appearances waking up in a hut raised on poles at the edge of a lake. A dusky woman, whom you realise to be your wife, tells you to go out and catch some fish. The dream continues with the apparent length of an ordinary human day, replete with an appropriate and causally coherent variety of tropical incident. At last you find yourself awaking at home to the world of normal responsibilities and expectations. The next night life by the side of the tropical lake continues in a coherent and natural way from the point at which it left off. And so it goes on. Injuries given in England leave scars in England, insults given at the lakeside complicate lakeside personal relations. . . . Now if this whole state of affairs came about it would not be very unreasonable to say that we lived in two worlds.[30]

The key counterfactual in our regimented version of Quinton's case is that if all places are mutually accessible and you were to have these experiences, then you would not be in a position to describe yourself as living in spatially isolated worlds. You would be instead obliged to discount at least one set of your experiences as delusory (but which?) or would have to persist in believing that there is some hidden route from England to the lakeside (despite your most exhaustive geographical investigations). Since Quinton thinks this epistemic fate absurd but the sequence of experiences possible, he concludes that space is not essentially singular.

So far, so good. Quinton's thought experiment snugly fits the necessity refuter schema. The trouble is to accommodate its dialectical development. An early objection to the Two-Space myth was that no one would believe your report of a double life. Your audience's incredulity would give you ample reason to discount the veridicality of at least one line of experience. Quinton therefore amended the scenario so that most of the members of your community report similar experiences. Indeed, their experiences correlate: each person in England has exactly one counterpart in the lake district. (Perhaps they identify each other with self-portraits or by making secret appointments to meet in the other world.) A second objection was that the two-space explanation of your experiences requires locating your body in two spots at a single time, namely, a British bed and tropical boat. To avoid saddling your explanation with the burden of this heresy, Richard Swinburne adds the friendly amendment that folks who report double lives vanish from their beds soon after falling asleep and reappear just before waking.[31] He also helps by weakening the rival explanation of there being a hidden route between the worlds by further stipulating scientific differences between the world. For example, in England gravity follows an inverse square law while in the lake district it follows an inverse cube law. Swinburne's kibbitzing continues for a full chapter.

Evidently, the thought experiment can survive the revisions even if every component and subcomponent of the original set of propositions is replaced. It owes its survival to the resemblance between the new components and the old as well as to its historical connection with the old thought experiment. Learning that William Newcomb invented Newcomb's problem after hearing the prisoner's dilemma buttresses the case for their identity. The identity thesis is further supported by David Lewis's logical demonstration that the prisoner's dilemma is two Newcomb's problems "put side to side."[32]

The question 'When is a thought experiment the same thought experiment?' has many of the complexities of other issues of identity: When is a ship the same ship? When is a person the same person? Sometimes, as in patent law, the question has practical significance; the significance in our case is theoretical. Regardless of the niceties involved in specifying content continuity, propositional resemblance, and historical connection, we can see that the customary standards of thought experiment identity are much more lenient than the set-theoretic conception I have proposed.

The force of this objection is dissipated by pointing out that the same difficulties arise for the logician's definition of 'argument': an argument is a set of propositions of which one, the conclusion, is supported by the rest. Commentators on the design argument, the private-language argument, and the paradigm case argument do not treat any component as essential to the identity of the argument. History and resemblance are as relevant here as they are to thought experiment. Indeed, virtually all set-theoretic definitions are embarassed by the same flexibility.

The definitions are saved by the relativity of individuation: what counts as the same F varies with one's purpose. The "conflict" between our ordinary usage of 'same thought experiment' and our regimented usage is only an illusion induced by equivocation. Relative to the purpose of establishing authorship and priority, historical criteria dominate. Relative to the purposes of theoretical unification and heuristics, propositional resemblance is more significant. Since the purpose at hand is theoretical classification, maximally fine-grained criteria are appropriate. Hence, instead of allowing a thought experiment to survive a slight amendment, we say that there are two thought experiments, one being a slight variation of the other. Two factors reinforce this strictness. First, unlike concrete objects, the abstract original can exist simultaneously with its altered version. Although a car owner cannot park his prepainted and postpainted automobile side by side for comparison, a poet can compare his original sonnet with its modified descendent. This difference makes us more apt to treat the modified abstract object as nonidentical to the original. Another factor favoring the set-theoretic definition of thought experiment is that ordinary speakers gravitate toward this fine-grained standard when pressed by the challenge 'Is this thought experiment really *identical* to that one?' This tendency toward pickier standards holds for all absolute terms (*flat, free, certain*), as noted in chapter 4. Since speakers usually overlook their slide to different standards, they often have the impression that the finest-grained standard is the uniquely correct one. This privileging illusion accounts for some of the persuasiveness of scepticism and hard determinism. My classification scheme will enjoy the same illicit boost! But honesty is the best policy. Therefore, I abjure this shadowy force and reiterate my claim to have found only a scheme suited to theoretical purposes, not the all-purpose truth of the matter.

IV. An Extension to Ordinary Experiments

Since most thought experiments could be turned into regular experiments by actually bringing about the envisioned scenario (the C part), we should expect a subset of normal experiments to be analyzable as paradoxes. Let us illustrate with Blaise Pascal's demonstration that air is something, rather than nothing. He weighed an empty balloon, inflated it, measured a small weight gain, and concluded that air was a substance. We regiment it thus:

1. Source Statement: Air is nothing.
2. Modal Extractor: If air is nothing, then it must be weightless.
3. Counterfactual: If air is weightless and only air were added to a container, then the container would not grow heavier.
4. Absurdity: It is impossible for the inflated container to have not gained weight.
5. Content possibility: It is possible that only air was added to the container.

As before, the necessity refuter scheme lines up the possible challenges and indirect usages of the experiment. For example, one might reject (5) and take the experiment as evidence that an invisible impurity rushed in with the air.

Some critics might complain that the necessity refuter formulation *dilutes* the force of the experiment. Instead of (5), we should insert the stronger statement from which it was derived: only air was added to the container. However, the experimentalist should use the least controversial assumption necessary for the result. The *actual* insertion of pure air should be cited only as a lemma.

Another critic might object that proposition 4 *overburdens* the experiment: Pascal only tried to show that a particular balloon did in fact weigh more after inflation, not that it *had to* weigh more. My reply is that this critic is also confusing salient intermediates with the key players. Scientists have little professional interest in facts that are only accidentally true. Even their simplest enumerative inductions are intended to establish the *necessity* of the concluding generalization:

1. Animal 1 is a platypus and has a bill.
2. Animal 2 is a platypus and has a bill.

. . . .

100. Animal 100 is a platypus and has a bill.
101. All platypuses have bills.

If the conclusion is read merely as a universal generalization (not as a low-level law), then it will not support the counterfactual 'If animal 101 were a platypus, it would have a bill'. Without such counterfactuals, scientists have no hope of explaining why things are the way they are, rather than some other way. Pascal's experiment fails to refute the source statement if the balloon merely happened to weigh more after inflation. It is only effective if understood as showing that the inflation *made* the balloon heavier—that the balloon *had to* get heavier. As Leonardo remarked, "Necessity is the guardian of nature."

This logical similarity between a thought experiment and its executed counterpart is compatible with the existence of epistemological differences. Executing an experiment gives you better evidence for the fourth and fifth components of the necessity refuter scheme. As stressed in the chapter on armchair inquiry, both experiments and thought experiments are designed to make you an authority, but they try to do it in different ways. An experiment makes you an authority by enhancing your perceptual opportunities. An effective thought experiment amplifies your nonperceptual resources—such as recall; the transformation of one kind of knowledge into another; internal redelegation of cognitive tasks; and the elimination of cognitive obstacles, hindrances, and hang-ups. Since a regular experiment can also employ these nonperceptual strategies along with its perceptual strategies, one might expect it to dominate thought experiment. But this overlooks the advantage of specialization. Powered planes and gliders can both glide, but the different design emphasis leads us to expect gliders to glide better.

V. The Big Picture

My goal in this chapter was to present a taxonomy of thought experiment. By treating a thought experiment as a stylized paradox, we mature the idea that it reveals inconsistencies. We also expose the structure of our ensuing ambivalence, as well as the structure of resolutions. These benefits are achieved at the price of artificiality. Often, we must reshape thought experiments to make them fit the mold. Since this regimentation perturbs the customary exposition of many thought experiments, the result will strike us as unnatural and even perverse. However, the logical merit will be preserved even if there is some aesthetic and rhetorical disfigurement.

My emphasis in this chapter has been on "global" aspects of thought experiments. I craved a general framework that could subsume the particulars in a systematic fashion. But I am also eager to detail the classification scheme so that it can guide a fine-grained analysis of special kinds of thought experiments. In particular, I wish to apply the framework in a way that will do full justice to the sort of cases that intrigued Kuhn. Hence, chapter 7 opens up the mechanism by which Kuhnian cases materialize on the intellectual landscape.

7

Conflict Vagueness and Precisification

Creative workers need drink at night, 'Roses and dung'. (Or: mathematicians read 'rubbish'.) An experimentalist, having spent the day looking for the leak, has had a perfect mental rest by dinner time, and overflows with minor mental activity

J. E. Littlewood

Now I want to pinpoint the property that excited Kuhn's interest in thought experiments. This property, conflict vagueness, often generates inconsistent beliefs but is not itself inconsistency. Although it is absent from most thought experiments, a substantial portion of the most provocative thought experiments do spring upon this species of vagueness; for they motivate conceptual reform by touching a nerve of indeterminacy. Hence, study of conflict vagueness reveals the ways thought experiments restructure our conceptual scheme.

I. General Features of Vagueness

A word is vague if and only if it has at least one borderline case. For example, 'bald' is vague because there are men who are neither clearly bald nor clearly nonbald. The uncertainty created by borderline cases is irremediable. No amount of observation, experiment, or conceptual analysis can answer the question. Borderline cases are inquiry-resistant.

If a term has an *actual* borderline case, it is *extensionally* vague. 'Food' is extensionally vague because coffee is an actual borderline case of food. If a term has a *possible* borderline case, then it is *intensionally* vague. 'Mermaid' is intensionally vague because it is possible for a part woman–part fish to be too fishy to qualify clearly as a mermaid yet too womanish to qualify clearly as a nonmermaid. Since whatever is actual is possible, any extensionally vague term is also intensionally vague—but not vice versa: that is the message behind 'mermaid'. The extension of a term is the set of things to which it applies. The extension of 'vowel' is {a, e, i, o, u} and that of 'human' contains you, me, Manuel Noriega, and the rest of us.

167

The intension of a term is the rule governing the application of the term. The rule for 'mother' is to apply the term to exactly those individuals who are female parents. The intension of a term matches our intuitive idea of what a definition should be. Once we know the intension of a term, we understand its meaning. Does a person ignorant of the term's definition know its meaning? Don't answer! First distinguish knowing *how* from knowing *that*. In one sense, a competent speaker of English understands *cousin* even if he cannot define it. To understand in this sense is simply a matter of knowing *how* to use the term, of being able to sort standard from deviant usages. But in another, more peripheral sense, the speaker does not understand *cousin* because he does know *that* it means an equally distant but indirectly related codescendent.

The definer may be stumped in two ways. First, he may lack candidate definitions. If you ask the average person to define 'capon', he will be without a clue as to how to proceed. The opposite difficulty occurs when the definer has too many candidate definitions. Sometimes he escapes this embarrassment of riches by scaring up subtle objections that disqualify candidates. In other cases, the definer shows that the definitions are compatible. For instance, there is no rivalry between 'A brother is a male sibling' and 'A brother is a son of the same parents'. They give the same results. But even if the definer reconciles some definitions and eliminate other contenders, he must often be left with a "tie." If the tie must be broken, it must be broken arbitrarily.

II. Dueling Definitions

This semantic excess suggests an alternative definition of *vagueness*. A term is vague if and only if there is an unbreakable tie between rival definitions of it. This gives the same result as saying that vagueness is a matter of having borderline cases.

A. The Psychology of Conflict Vagueness

The advantage of the rivalry definition is that it spotlights *conflict vagueness*. Normally, the convergence between the definitions is extensive enough to make the area of divergence a surprising discovery. Indeed, the speaker normally conceives of himself as operating with only *one* definition. Like the woman in the movie *Dead Ringers* who unwittingly dates twins, the victim of conflict vagueness has a false sense of unity. For example, a private may be unaware that there are two readings of 'clean' until his sergeant orders him to clean a rifle that the private has already cleaned. On the effect reading, cleaning is matter of eliminating dirt and disorder. This reading makes the sergeant's command impossible to obey. However, 'clean' also has an action reading that merely refers to the movements that tend to eliminate dirt and disorder. Under this action reading, it is possible to clean an already clean rifle because one can perform the characteristic movements. But since the private is not sure which reading is correct, he will be uncertain whether an already-clean rifle can be cleaned. The private's uncertainty is the result of a semantic insight. He knows more about cleaning because he now realizes that there are two rival definitions of 'clean'. This double-trouble illustrates how borderline cases can be conceptually edifying.

When thinkers confront these hidden borderline cases, they say the concept has "come apart." The expression "conflict vagueness" conveys this tension. Under my analysis, all vagueness involves rivalry between candidate rules. Hence, to avoid trivialization, I confine *conflict vagueness* to cases of *unsuspected* rivalry. The mark of conflict vagueness is surprise and ambivalence.

Indeed, there is a rich psychology of vagueness that is partly anticipated by ethnographic research into the bizarre "conflict behavior" of animals.[1] For example, herring gulls have a drive to remove all red objects from their nests. They also have a drive to retrieve any egg that rolls away from the nest. Ornithologists stimulate conflict behavior by placing a red egg in a vacant nest. When the herring gull returns, she pushes the red object out, then rolls the egg back in, then pushes it out again, only to retrieve it once more. Similarly, borderline cases make speakers vacillate between incompatible descriptions.

Alternation is only one form of conflict behavior. A cat *averages* the conflicting drives to fight and flee by retreating with its front paws while advancing with its back feet (giving rise to the "Halloween" arched back). Speakers compromise between clashing linguistic tendencies when they describe paradoxical phenomena with oxymorons such as 'blind sight' and when they invent blend words like 'brunch'. Conflicted conversationalists delight proponents of many-valued logic by saying that 'A hankerchief is an article of clothing' is partly true and partly false. Redirection is a third form of conflict behavior. An angry robin aborts its attack on an intimidating foe and completes it on some hapless bystander or even an inanimate object. Now recall my hostility toward the conceptually inconvenient midget in chapter 5. Perhaps this tendency to project vying semantic tendencies onto the stimuli that activate them is one source of belief in vague objects.

Unlike animals, humans volunteer introspective reports of the conflict. The gestalt switch induced by the Necker cube is just alternation experienced from the inside (Figure 7.1). Your visual system automatically switches back and forth between rival hypotheses as to which edge is outermost. Thus, the ethnographic lessons learned from the conflict behavior of animals hold for to expressionless thought experimenters.

However, intellectual conflict behavior is not a sufficient condition for conflict vagueness. Imagine heating a ring. What happens to the size of the hole? Although most people dither, 'The hole expands' is not borderline. It's definitely true. To see why, mentally uncurve the ring to form a bar. As you apply heat, the bar becomes thicker, but it also becomes longer. Hence, when you recurve the bar back into a ring, both the inner circumference and the thickness are greater.

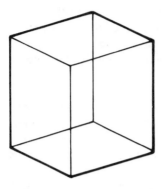

Figure 7.1

B. How Conflict Vagueness Creates the Mirage of Local Incoherence

Conflict vagueness is easily confused with inconsistency. A speaker who has not distinguished the two rules acts as if the term means the *conjunction* of the rival definitions. Although the rivals are consistent relative to each other, their conjunction is not consistent with the proposition that one of their conflict cases is possible. For example, one broad conception of 'naked' defines the condition as an absence of clothing, while a narrower one defines it as an absence of covering. Although no contradiction can be derived from the conjunction of these rival definitions, a contradiction can be derived once we add the assumption that a person can be covered with nonclothing. Thus, the conjunction of the conceptions implies an impossibility theorem: no woman can cover her body with only a net. Since this theorem is absurd, we must reject the conjunction of the rival rules. But the innocence of the excluded situation leads us to overlook its participation in the paradox. Hence, we erroneously infer that the rules are in *direct* conflict. Continued conviction that the conjunction of the rules is the correct definition of the term leads some people to conclude that the concept is contradictory. Like Kuhn, they marvel at the semantic epilepsy: the concept works for the most part but occasionally lapses into incoherency.

Pockets of support for the resulting notion of local incoherence can be detected in the "solution" to certain problems and riddles. The case discussed in the previous paragraph is inspired by a story featuring a daughter pleading for the life of her father. The king says that he will spare her father if she returns the next day both naked and not naked. When she enters wrapped in a fish net, the king concludes that he must release her father. A second example is a Renaissance riddle: "When is a man both rightside up and upside down?" The official answer is "When he is at the center of the earth." The answer exploits the fact that we have no more reason to relativize 'up' to the direction his head points than to the direction his feet point. Under one relativization he is rightside up, and under the other he is upside down. Nevertheless, the answer is fallacious because under no relativization is he simultaneously rightside up and upside down.

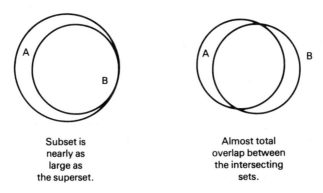

Subset is
nearly as
large as
the superset.

Almost total
overlap between
the intersecting
sets.

Figure 7.2

We can readily see how the rival meanings can pass undetected if there is only a *slight* difference in the intensions of the two terms, for then the applications of the rules strongly converge (Figure 7.2). However, if the two rules never agree on a positive case, it might seem that the rivalry between the rules would always be manifest. However, there is another way for distinct rules to blur together. Instead of having the rules agree on most cases, have them only slightly and inconspicuously disagree on their positive cases. Consider the difference between the opening and filler readings of 'window'. Since the readings place the window in almost the same spot, the divergence in meaning makes little behavioral difference when we are asked to point to the window. Sometimes the readings pick out contiguous regions or even divide between a visible cause and an invisible effect (as with 'knock'). In yet other cases, the two readings do not disagree on spatial locations, because one reading is concrete and the other abstract. 'President', 'pawn', and 'dollar' each has one reading in which it means a role and another designating the concrete thing filling the role. Since roles do not have locations, there is no divergence in what you point to. Another example is 'surface' in which one reading selects the outermost layer of the object, while another designates an abstract boundary between the object and its environment.

C. Extensional Conflict Vagueness

When empirical discoveries pose a conceptual challenge, unanticipated extensional vagueness is usually at play. The platypus was puzzling because it is a hairy, warm-blooded egg layer and hence was a borderline case of 'reptile' and 'mammal' (as used in eighteenth-century English; 'Mammal' was subsequently "precisified" by zoologists to include the platypus and other monotremes).

Experimentalists love to *create* the conflict cases. In 1732, Hermann Boerhaave distinguished plants from animals by the location of their nutritional organs: plants have exterior roots and animals interior roots. This distinction was well received by eighteenth-century naturalists in Geneva. So Abraham

Trembley hoped it would illuminate the mystery posed by the freshwater polyp.[2] This tiny "being" had some marks of an animal and some marks of a plant, for the polyp moved like an animal yet reproduced by budding and regenerated into distinct polyps after being cut into several pieces. Eventually, Trembley observed a polyp ensnare a tiny eel and stuff it into its central cavity. Hence, on Boerhaave's criterion, the polyp counts as an animal. Yet the simple, sacklike shape of the polyp led Trembley to suspect arbitrariness. What if the polyp was turned inside-out? With preternatural patience, Trembley inverted the polyp and found that it continued to feed and reproduce as before!

D. Intensional Conflict Vagueness

Actual cases are sometimes near misses of interesting borderline cases and so inspire the invention of hypothetical cases that "correct" for the missing part. Thus, news that some zygotes fuse after dividing into twins leads a metaphysician to imagine them fusing later, in midlife. Would the fused man be identical to both twins, either twin, or neither?

J. L. Gorman explains a voting puzzle as a case of conflicting criteria. Suppose three issues are voted on as follows (1 means *yes* and −1 is *no*):

Voter	Questions		
	A	B	C
1	1	1	1
2	1	1	1
3	1	−1	−1
4	−1	1	−1
5	−1	−1	1

The election is democratic in the sense that the questions are resolved by a *yes-no* majority vote. However, a majority of the voters vote in the minority on a majority of the issues. All the questions pass by a three to two majority but a three to two majority of voters opposed a majority of the questions. This majority is frustrated in the sense that it loses over half of the issues. To put the point more paradoxically, we have democratic frustration of the majority. The theoretical interest of this voting pattern lies in the fact that "we have two possible and conflicting criteria for saying 'the majority has its will fulfilled'. . . . Should there be more successful votes than unsuccessful votes for each question? Or should there be more successful votes than unsuccessful votes for the majority of voters?"[3] Since the two criteria yield the same verdict in almost all situations, we are not apt to notice that there are two rules of decision.

In 1835 Gaspard Gustave de Coriolis published a paper on how the Earth's rotation distorts the motion of fluids. According to Newton's first law, objects continue to move in the same direction unless acted upon by a force. Coriolis

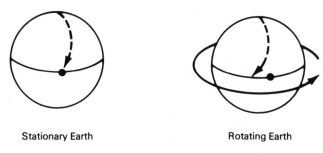

Stationary Earth   Rotating Earth

Figure 7.3 Stationary Earth and Rotating Earth

said that objects on earth, however, will follow curved paths. Suppose that a missile on the North Pole is launched against Kisangani on the equator (Figure 7.3). If the earth were not rotating, the missile would go directly to target. But since the earth is rotating, Kisangani will rotate with the earth toward the right. Thus, the "straight" missile curves to the left of Kisangani. The Kisanganians are good Newtonians and so joyfully infer that some force acted on the missile to account for the deviance from a straight flight. But an astronaut far away from the earth detects no deviation. Who is right?

The paradox reveals the conflict vagueness of 'straight line'. The *dynamic* reading defines the concept in terms of Newton's first law: a straight line is the path followed by an undisturbed projectile. The *geometrical* reading takes it to be the shortest distance along the surface in question. The missile went straight according to the dynamical definition but curved according to the geometrical. The Kisanganians and other earthdwellers find that the geometrical conception suits their terrestrial purposes and so prefer the geometrical definition. That's okay. We can accommodate their preference by postulating a force that curved the flight of the missile. However, this "Coriolis force" has to be understood as a fictitious force, that is, one that depends on a frame of reference. The dynamic alternative is more austere because it does not require the postulation of a force.

Is there any reason to prefer one conception of a straight line over the other? There is no *linguistic* reason. To decide *whether* we should choose (instead of just leaving the term vague) or *which* reading to choose demands evidence of expedience. The best supplier of this evidence is normally someone with expertise in the pursuit of the goals in question. Hence, there is a division of labor in cases involving conflict vagueness. The role of the thought experiment is merely to show that we have a choice. Whether we choose and how we choose is settled pragmatically, by the aims and ingenuity of speakers.

Kuhn's cases illustrate extensional and intensional conflict vagueness. Piaget's car races plop children into a quandary over 'faster'. The borderline case activates the two sufficient conditions, the goal-reaching criterion and the perceptual blurriness criterion, for two competing objects. Likewise, the participants in Galileo's dialogue have their criteria clash with the medieval necessary condition that faster objects have higher mean speeds. Once we adopt the view that conflict vagueness is a matter of superimposed candidate definitions, we can explain how thought experiments generate internal conflict. Since the two definitions almost always coincide, they speak with one voice most of the time. Where they diverge, we begin to hear two voices, hence our ambivalence.

III. Application of the Quintet Schema

Thought experiments that reveal conflict vagueness conform to the schema introduced in chapter 6. The simplest kind of case exploits the transitivity of synonymy:

1. 'Decapitate' is synonymous with 'make headless' and synonymous with 'cut a head off'.
2. If so, then 'make headless' is synonymous with 'cut a head off'.
3. If they really are synonymous and someone were to cut a head off of a two-headed man, he would be decapitated and not decapitated.
4. It is impossible to be both decapitated and not decapitated.
5. It is possible for someone to cut a head off a two headed man.

The source statement, (1), assumes that there is no friction between the two conceptions of decapitation. The second statement notes that this assumption implies a semantic equivalence, which is then refuted by the hypothetical scenario described in the counterfactual (3). The same pattern would emerge if we regimented the thought experiment for 'naked' and 'fulfillment of the majority will'.

Another pattern emerges when the thought experiment reveals vagueness about what to relativize to. 'Tributary' seems tame until we consider the possibility of a thick river branch, B_1, joining a long thin one, B_2. Now we wonder which is the tributary because B_1 is smaller in volume and B_2 is smaller in length. We can formalize our puzzlement as follows:

1. 'Tributary' means the smaller of two river branches.
2. If so, then all river branches that are smaller in length are tributaries and all river branches that are smaller in volume are tributaries.
3. If smaller length and smaller volume each suffice for being a tributary and B_1 was a longer but less voluminous river branch than B_2, then B_1 and B_2 would be tributaries of each other.
4. It is impossible for two river branches to be tributaries of each other.
5. It is possible for a river branch to be longer but less voluminous.

The action takes place at the second step. In the absence of any specification of what to relativize to, (2) counts both kinds of *relata* as sufficient for the application of the term. The counterfactual, (3), then serves up a semantic surfeit because the relation is asymmetric (as noted at (4)). Yet, as (5) assures us, the situation triggering the sufficient conditions is certainly possible.

For the purposes of illustration, I favor examples of conflict vagueness that we have little stake in resolving. Our intuitions about high-stakes conflict vagueness are less reliable because our commitments prompt subterrean precisifications. Nevertheless, the analysis is just as applicable to substantial cases. Recall how the flagship example of the identity issues, the Ship of Theseus, pits the spatiotemporal contiguity condition against the reassembly condition.

IV. Conceptual Reform

Thought experiments turning on conflict vagueness can be resolved in the five ways discussed in chapter 6. Sometimes recognition of the vagueness is enough to tell us which member of the inconsistent quintet is false. On other occasions, the recognition narrows down the alternatives to two. The ensuing decrease in credence in the pair may then diminish our dissonance to a tolerable level. But often, recognition of the vagueness does not settle the matter. Sometimes the problem is that too many members of the quintet remain compelling, so that we are still haunted by a "subparadox."

Since vagueness brings inquiry resistance, the residual mystery is irresolvable. Unlike the thought experiments in chapter 6, we may not be able to *discover* a false member. But we do have an ersatz response. Instead of searching for the false member, we could eliminate the vagueness by refiguring the key concept. Once this precisification is completed, we have a verbally similar quintet that can be solved by discovery; that is, we can *invent* our way out of inconsistency by abandoning the original concept in favor of its more accessible replacement. Old inconsistent beliefs will just fade away because our credence can now be realigned around the rectified terminology.

The most straightforward way of breaking a tie is to arbitrarily stipulate one contender as the winner. When read under this precisification, the source sentence comes out false. Precisification resolves the paradox by redefining the term. But brute precisification is rare. Typically, we search for a reason to stipulate a winner. This must be distinguished from searching for the real winner. When we look more closely at the contenders with an eye to detecting the winner, we do not regard the contest as a tie. When we look at the contenders with an eye to an optimal tie-breaking scheme, we admit that there is a tie and so are free to invoke a new set of criteria for settling the matter. These new considerations tend to be utilitarian and opportunistic, rather than ontological and principled. We change the issue from 'Which alternative is correct?' to 'Which alternative is most profitably counted as if it were correct?'. Since stipulations have no truth value, our final verdict has no truth value. What applies generally to tie breaking, applies to the particular case of resolving conflict vagueness.

A couple of factors lead us to overlook the distinction between reasons for 'Let's stipulate that x is F' and reasons for 'x is F'. The first cause of confusion is just the close resemblance between the two. The second is that stipulation (like flattery) is most successful when not recognized as such. People are more apt to acquiese in your usage of a term if they believe that it is already common practice. They appreciate the foolishness of bucking established usage; but if the usage is presented as a stipulation, then the audience feels free to consider the pros and cons of adopting the convention. They will not believe that they are committing an error by diverging. Hence, whether sincerely or insincerely, stipulators tend to present their innovations as old news. Just as politicians legitimate their authority by acting as if their authority is already legitimate, speakers legitimate their innovations by treating them as if they are already accepted. Thus, much stipulation is passed off as description. The bluster behind this subterranean stipulation is reminiscent of the posturing that occurs in setting grading standards. As J. O. Urmson notes, persuasive speakers blur the distinction between *proposing* criteria and *applying* already accepted criteria.[4]

Once we decide to stipulate, we need not stick with the old candidates. We can fashion a new candidate. The newcomer can be cannibalized from the originals or be made from scratch.

One way to keep both candidate rules is to stipulate two meanings for the word, one corresponding to each of the rivals. This invention of an ambiguity is different from the discovery of one. However, we should expect that this distinction will often pass unrecognized because of our general problems with the distinction between reasons for something being the case and reasons for stipulating it to be the case.

Once the ambiguity is declared, we can reject the modal extractor; for the conditional expressing the extractor will now have an ambiguous antecedent. It will be analogous to 'If you stand beside a bank, then you are near a financial institution and near the side of a river'.

There are two main kinds of ambiguity. The more familiar one is *polysemy*: multiplicity of *senses*. When we look up a polysemous word in a dictionary, we expect a number of entries. For example, 'rake' has an entry for a gardening implement and another for a lewd man. Since dictionaries include only the more prevalent senses of a word, we might say that our expectation is that a polysemous word would have many entries in an ideally complete dictionary. When we stipulate that a word is polysemous, we are encouraging others to talk in a way that would lead the word to gain a new entry in a future, complete dictionary.

William James's famous example of a verbal dispute illustrates how conflict vagueness is handled by stipulating polysemy. The borderline case features a squirrel on a tree trunk. She hides from a circling man by always keeping the tree between them. So although the man always faces the squirrel's belly, he is successively before, right of, behind, left of, and then again before the squirrel. Does the man go around the squirrel? James's solution is to bifurcate 'go around' into two senses. In one sense, it means moving so that each side is successively faced. In another, it means moving in a path that encloses the object. However, James presents his solution as the *discovery* of an ambiguity rather than the proposal of one. The same goes for the famous riddle "If a tree falls in an uninhabited forest, does it make sound?" Philosophers and scientists assure us that the issue is verbal because 'sound' means an auditory sensation in one sense and a physical wave in the other. But 'sound' is vague between these alternatives, not ambiguous. The scholars are right to condemn debate over the question, but they are not quite right about the nature of the defect.

Interesting verbal disputes rarely involve the detection of a preexisting ambiguity. The ambiguity is constructed. This creative aspect of dispute diagnosis is self-masking; for the arbitration is more likely to be accepted if presented as a *discovery*, and once the stipulation is commonly accepted, the ambiguity comes to really exist. This self-masking feature of ambiguity invention explains why our intellectual ancestors seem so prone to equivocate and to fall into simple verbal disputes such as the *vis viva* controversy between Leibniz and Cartesian physicists. Relative to our vocabulary (which incorporates the multiple senses prompted by these past perplexities), there is a straightforward ambiguity. But relative to the old vocabulary, there was only vagueness and conditions that made one stipulation subtly more propitious than the others.

A word is relative when it has a multiplicity of *relata*. For instance, 'north' is relative because New York is north of Florida but not north of Maine. Consider the ambiguity of 'Fred and Karl are brothers'; are they brothers of each other or just brothers of some people or other? This subtler ambiguity is at play in the sophism in Figure 7.4. There is a strained reading of the conclusion that makes the argument sound. But the natural, specific reading invalidates the argument.

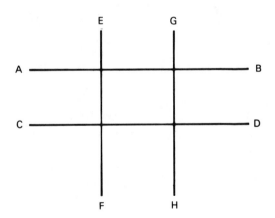

AB and CD are parallel lines.
EF and GH are parallel lines.
AB and EF meet.
___
Some parallel lines meets.

Figure 7.4

Usually a term is described as relative in order to alert the audience to *hidden* variables. When we overlook those variables, we "drop a variable" and become susceptible to equivocation or meaninglessness. For example, a physics riddle asks "Does a conductor in a train move if he walks 1 mph east on a train going 1 mph west?" The physicist diagnoses our uncertainty as a failure to relativize 'move' to a frame of reference. We are ambivalent because there is no frame of reference obvious enough to be presupposed. So we vacillate between relativizing 'move' to the inside of the train and to the surrounding countryside. Once we recognize the relativity of 'move', we can explicitly ask 'Which do you mean?'. Unless we relativize to something, the question is meaningless.

This suggests a strategy for handling borderline cases: assimilate them to the relativity puzzles. The idea is to invent another variable for the puzzling relation and outlaw utterances that do not plug something into the new slot. Einstein applies this strategy to a problem that arises from the constancy of the speed of light. Suppose a train travels six-tenths the speed of light. One observer is stationed on the middle car while a second observer stands on an embankment. Lightning bolts now strike each end of the train, leaving burn marks on the train and the ground. The light from the bolt striking the locomotive and the light from the bolt hitting the caboose reach the ground observer at the same time. Hence, the events look simultaneous to him. To be sure, the ground observer measures the distance to each of the burn marks on the ground and verifies that he was standing exactly midway between the two events. Now consider the observer on the train. Since he is traveling toward the light emanating from the locomotive and away from the light of the caboose, he sees the locomotive bolt before the caboose bolt. He, too, is situated exactly midway between the burn marks left on the train. So he infers that the bolts were not simultaneous. Who is right? Einstein amplifies the query by supposing that another bolt hits the locomotive so that it is measured as simultaneous with the caboose bolt by the train observer but as later by the ground observer. We have a number of options here. We might draw the moral that light should not be assigned a central role in our tacit definition of simultaneity. We have assumed that two distant events are simultaneous if the light emanating from them reach their midpoint at the same time. However, Einstein urges us to stick with this aspect of the definition and instead relativize 'simultaneous' to reference frames. Thus Einstein's answer to 'Who is right about whether the bolts struck simultaneously?' is 'Bad question!'.

Lawyers share the scientist's affection for relativization but are prone to relativize in a more higgledy-piggledy fashion following history's palseyed finger of precedent. Witness how they handle the problem of determining 'next of kin' for the purpose of intestate inheritance.[5] Suppose that in the family depicted by Figure 7.5, O dies and A, C, D, and E are already dead. Is B or F the next of kin of O? The source of ambivalence has been traced by legal scholars to two ways of determining next of kin. To be next of kin is to be the fewest degrees removed. But what counts as a degree? Under the "gradual scheme" a step of ascent or descent count as a degree, yielding five degrees

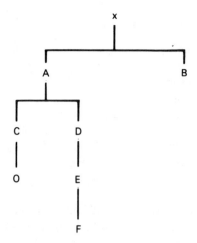

Figure 7.5

from O to F and four from O to B. According to the "parentelic scheme," you start from the person and exhaust his descendents, then exhaust descendents of his ascendent, and so on. So since O has no descendents, we try C. Since C has no descendents, we ascend further to A and find that he has a descendent D. But since D is dead, we look to D's descendent, E who is also dead, leading us finally to F. Although the schemes are equally reasonable and agree most of the time, they yield conflicting results in this case. In law, the conflict is resolved by relativizing 'next of kin' to types of inheritance. The gradual scheme is given dominion over "personality" inheritance and the parentelic over realty.

The mathematical drive increases the precision of scientific vocabulary at one level but introduces new indeterminacies at another; for if we define the term as a quantity, we raise the question 'Quantity of what?'. Usually, there are many potential units that give converging results for the most part but diverge in a few cases. Evolutionary theory predicts that the behavioral and structural parts of an organism are designed to maximize survival value—but the survival of that individual, its offspring, or its group or species? Utilitarians agree that right actions maximize goodness—but average or absolute goodness?

Here the problem is vying definitions of the *definiens* of the key term. This deeper-level conflict vagueness can be qualitative. 'Weight' refers to a tendency of things to fall to the center of the earth. But 'center' might mean a material portion of the earth, or it might mean a mathematical point in space that just happens to be occupied by the earth's core. In the fourth book of *De caelo*, Aristotle brings out the difference by supposing that the earth is moved from its position at the center of universe. If 'weight' is understood as an attraction to the mathematical point, then rocks will fly from the earth toward the point. But if 'weight' means an attraction to the earth's core, then the rocks remain. (Aristotle opted for the abstract version.)

V. Eliminative Reactions to Conflict Vagueness

Thus far, we have concentrated on replacement resolutions of conflict vagueness. A replacement response *modifies* the vague word so that the paradox can be resolved by rejecting a member of the (reformed) quintet. Although the replacement method changes the original meaning of the word or even gives it a new meaning, the method is still conservative. The idea is to reform the concept by explication, to change the letter but not the spirit. Eliminative responses do not aim at preserving the basic meaning of the old word. They deal with the conflict vagueness by avoiding the word or even banning it. The word is treated like a defective tool that is not worth repairing. By discarding the word, you relieve yourself of its aggravations.

Most scientists are happy to use the problematic term in its unproblematic contexts. Physicists believe that plenty of things are hot, but they don't say so in their official language.[6] This can be seen by the way they respond to the riddle "Which of is hottest—a 200-degree dime, a 300-degree turkey, or a 400-degree cake?" Physicists reject the question on the grounds that 'hot' is too vague. They try to satisfy you by saying that the cake has the highest *temperature*, the turkey has the greatest quantity of *thermal energy*, and the dime has the greatest rate of *heat exchange*. But physicists will not give a direct answer. The coy physicist does not *deny* the existence of hot and cold, but neither does he affirm it. He just refuses to talk in that terminology. Scientific vocabulary resembles the staff of a baseball team. Managers, coaches, and batboys meander through the warm-ups and time-outs, but only the pros stay on the field when the game is officially in play.

VI. Tolerating the Vagueness

Suppose I can't answer 'What is the capital of Delaware?' even though I can identify Dover as the capital when presented as an alternative on a multiple-choice test. Do I remember? Alternatively, suppose I cannot pass the test on capitals but I am able to relearn them easily because I memorized the capitals as a boy. These conflict cases reveal three readings of 'remember': recall, recognition, and enhanced relearning. Psychologists resist attempts to break the tie but wish to retain 'remember' as part of the official vocabulary. The same pattern holds for technical terms such as 'gene'. In the 1950s Seymour Benzer demonstrated that the smallest unit of mutation (having a distinct functional affect) need not be the smallest unit of recombination.[7] So he coined the terms 'cistron', 'muton', and 'recon' to distinguish between the units of function, mutation, and recombination. But the scientific drive for precision has not retired 'gene' from active duty.

So we should distinguish desertion of a concept from resignation to the concept's vagueness. Few biologists bother to precisify 'organism' to eliminate the borderline case of viruses. They have decided to live with the vagueness

because none of the "cures" are better than the disease. Scientists in general are wary of premature precisification. A good stipulation is like a good piece of legislation: unless you have both the factual background needed to predict the consequences and the support of those affected, your declaration is apt to do more harm than good.

VII. Extending the Analysis

There are two ways of extending this analysis to cases that do not strictly involve vagueness. The first is to "play along" with an erroneous theory that implies conflict vagueness. Consider the trouble dead babies cause for 'innocence' within Christianity. If innocence is merely a negative matter of being sinless, babies are innocent and go to heaven (except for Christians who are gung-ho on original sin). But if innocence is a substantive property (like submission to the will of God), then babies are no more innocent than daisies. Since Christians believe that everyone goes to either heaven or hell, the latter construal heartlessly condemns babies to hell. But airlifting them to heaven also seems objectionable because they get in too easily. Hence, some Christians back up and say that babies go to a third place: limbo. An atheist can analyze the dead baby dilemma as if it involved conflict vagueness by assuming Christian premises. However, the atheist does not really think that babies are borderline cases of 'innocents'.

The second way to extend the conflict vagueness analysis is to restrict one's domain of discourse. Calling x a borderline F amounts to saying that there is no possibility of discovering x to be an F or discovering x to be a non-F. By limiting the set of possibilities considered, one increases the appearance of indeterminacy. We can illustrate this type of thought experiment with Henri Poincaré's disk world.[8] The thought experiment has two phases: one intended to demonstrate the conceivability of non-Euclidean geometry and one to establish the conventionality of geometry. Poincaré's world is a two-dimensional disk that is hot at the center and gradually cools to absolute zero at the circumference. (More exactly, the temperature is $R^2 - r^2$ where R is the radius of the disk and r is the distance of a point from the center.) Two-dimensional physicists live in the disk and try to ascertain the intrinsic geometry of their world with rigid rods. Unbeknownst to them, their rods contract uniformly with the decreasing temperature so that at the edge of the disk, the rods shrink to zero. Thus, the surveyors conclude that their world is a Lobachevskian plane of infinite extent. This refutes Kant's contention that only Euclidean geometry is conceivable.

The second phase of Poincaré's thought experiment begins with a conjecture from a two-dimensional heretic. He points out that all the experimental data also fits the hypothesis that the plane is a finite Euclidean disk that shrinks measuring rods. But how do the two-dimensional inhabitants discover the truth? Poincaré maintains that they must treat the matter as indeterminate and then decide in favor of one geometry on the basis of convenience. The point of the parable is that we should not ask whether Euclidean geometry

Figure 7.6

corresponds to the actual world. We should restrict ourselves to the question of whether it is a better convention than the other geometries.

The analogy between us and the disk inhabitants breaks down when we press the question of indeterminacy. Since from our three-dimensional perspective we can see that the heretic is right, there is a fact of the matter. However, for the purposes of the thought experiment, we are to restrict our attention to what is determinate relative to the capacities of the disk inhabitants.

In some thought experiments, disanalogies are initially kept in the shadows and then later spotlighted. Early sceptics about witchcraft told stories in which various diseases, eccentricities, and coincidences make women look just like witches. The ability of these factors jointly to mimic witchcraft makes it impossible to know that a woman is a witch, rather than a pseudowitch. However, the point of the analogy is not that the witches and pseudowitches are equivalent; for the thought experimenter recommends tolerance over perse-

cution. This is the point at which the disanalogies are stressed. The pattern is also evident in the utilitarian's critique of fairness. He has us imagine a quiet utilitarian community in which goods are distributed in accordance with the principle of diminishing marginal utility. Since this utilitarian policy tends to give people equal shares, outsiders naturally interpret the community as fair-minded. However, the convergence between utilitarianism and more complicated ethical systems does not hold universally. And the utilitarian does not permanently suppress divergences; he eventually parades them as surprising moral truths. The point of his thought experiment is that much of the admirable behavior that seems to issue from fairness could just as well have its source in the principle of diminishing marginal utility. Therefore we have no reason to postulate fairness as a fundamental moral force.

These cases illuminate Einstein's elevator thought experiment (Figure 7.6). The history behind this hypothetical is powered by unhappiness with gravity.

Even Newton was reluctant to admit that this kind of action at a distance is a real force that pervades the universe. Yet unlike occult, animistic theories saying that objects are psychologically attracted to each other, gravitational theory is precise and empirically fruitful. Hence, a physicist cannot flatly deny gravity. Einstein's solution to the dilemma was to assimilate gravity to fictitious forces such as the Coriolis force. He noted that gravity is mimicked by acceleration in an elevator; you feel heavier going up, lighter going down. Standing on a scale during an elevator ride substantiates this impression. However, people on terra firma outside the elevator do not feel the forces you feel against your feet as the elevator goes up. Hence, our geocentricism leads us to dismiss the felt forces as fictitious like the centrifugal force we feel when speeding round a curve. Now suppose that the elevator ride occurs in empty space, so that we are not tempted to relativize to earth. A demon who pulled the elevator just right could convincingly fake the force of gravity. From within the elevator, you would not be able to perform an experiment that would distinguish between the hypothesis that you are accelerating and the hypothesis that you are being pulled down by gravity. For example, if you release a ball, it will drop the same way under both hypotheses.

The elevator scenario does not demonstrate a genuine indeterminacy. The demon can know that acceleration, rather than gravity, is at work. There are even some subtle tests that could be performed from within the elevator. In one, the rider releases *two* balls from the ceiling. Under the acceleration hypothesis, they follow parallel paths. If the force is gravitational, the paths of the balls will converge toward the object's center of mass. Nevertheless, we can appreciate the resemblance to indeterminacy by temporarily ignoring these subtle tests and the outside world. Once we see how well we can along without the classic conception of gravity, we have inductive grounds for taking its rival seriously. Maybe the classic conception will be the one refuted at the points of divergence. And indeed, Einstein's account of gravity prevailed over the classical physics by correctly predicting that light from a star could be observed bending around the sun during an eclipse in 1919.

VIII. Sunder, Enlighten!

Scientists often discover interesting properties but then misdescribe them. Kuhn's discovery shows that what goes for the scientist goes for the historian of science. Kuhn characterized the interesting property as a peculiar type of inconsistency. I have argued that the property is conflict vagueness. Thought experiments powered by this vagueness are disquieting but edifying because they shatter our false sense of unity. They are also provocative, because our drive to reduce inconsistency joins with our craving for unity to yield proposals aimed at smoothing out the rift. These precisifications or eliminations are conceptual reforms that are sometimes performed openly but more often under the guise of semantic discovery.

I think that conflict vagueness does full justice to the thought experiments that concerned Kuhn. In any case, I have nothing further to say on this development of the inconsistency strand of the cleansing model. It has given us a good idea of what some of the more interesting thought experiments are and how they work. Curiosity now turns to other questions: What are thought experiments, in general? How did they arise? What is the relationship between thought experiment and physical experiment? Happily, these ontological, historical, and comparative questions have a unified answer.

8

The Evolution of Thought Experiment

Whether it is a parable or a fable depends on oneself.

Henry S. Haskings

Thesis: thought experiments evolved from ordinary experiments by a process of attenuation. This evolutionary story paves the way for the conclusion that thought experiments are limiting cases of experiment just as circles are limiting cases of ellipses.

I. 'Experiment' Defined

An *experiment* is a procedure for answering or raising a question about the relationship between variables by varying one (or more) of them and tracking any response by the other or others. For the sake of simplicity, most experiments are designed around two variables. The one you directly manipulate is called the independent variable and the one you try to affect indirectly through these manipulations is the dependent variable. For example, Benjamin Franklin answered "Which colors absorb the most heat?" by laying out squares of cloth on the snow. After a few hours in the sunshine the squares had sunk into the snow at depths that increased with the darkness of each square.

A. Stereotypical Features of Experiment

When one thinks of apples, one tends to think of *red* apples. Even though green apples are plentiful, there is some temptation to treat redness as a defining condition of 'apple'. Likewise, experiments seem more and more like experiments to the extent that they involve manipulation, instruments, hypothesis testing, scientific content, and public access. Although these features affect the degree to which a procedure is a representative, typical, or a paradigmatic of *experiment*, they are not essential features of experiment or (for that matter) *good* experiment.

186

1. *Testing hypotheses about nature.* We usually think of experiments as confirming or disconfirming controversial theories. But many merely establish constants such as the freezing point of oxygen. Other experiments just display puzzling effects. Luigi Galvani showed others how electricity causes a dead frog's leg to twitch even though there was no antecedent question 'Will a dead frog's leg twitch if electrically stimulated?'. The experiments were intended to *raise* the question of why they twitched. Likewise, early demonstrations of magnetism, Brownian motion, and the photoelectric effect gave us answers to questions that we never had reason to ask. Still other experiments only aim at confirming a preexisting consensus. A social psychologist may grant that it is commonly and correctly believed that one's rate of drinking is influenced by the rate of drinking of one's coimbibers but still buttress this bar lore with a controlled study.

Galileo said that experiments put a question to nature.[1] But even experiments that do involve hypothesis testing need not be testing a hypothesis about *nature*. Sometimes the scientist puts the question to those who put the question to nature. These metaexperiments test the adequacy of experimental methods. Robert Rosenthal's experiments on experimental bias are well-known contemporary examples. Antoine Lavoisier's refutation of Johann Eller provides a historical illustration. In 1766 Eller presented experiments designed to prove that water could be transmuted into both earth and air by fire, or phlogiston. Two years later, Lavoisier responded with a series of metaexperiments on water intended to vindicate his suspicion that Eller's experiments really derived glass from the vessels holding the water through a leeching effect. The procedure was to weigh both the water and the apparatus before and after a three-month heating. The precipitate was found to weigh about as much as the weight loss of the apparatus.

Indeed, behind every published experiment lies a harem of metaexperiments. Some of them help the scientist adjust his apparatus. Others were needed to enlarge the effect or to simplify the procedure.

2. *Instruments.* Experiments are associated with test tubes, bunsen burners, and caged rats. However some experiments are performed without any equipment. Phoneticists prove that the *p* in *pat* is pronounced differently than the *p* of *spat* by having you say each word into your moistened hand. A puff of air is felt only with *pat* because the *p* is aspirated. Physiologists demonstrate how behavior can arise from the simple fact of circulation by having you cross your legs. After one minute, the hanging foot bobs slightly in syncopation with the heart.

3. *Scientific content.* Experiments are so closely associated with science that it is tempting to view every experiment as an act of science. This generalization is prey to three classes of counterexamples.

First, there are experiments that predate science. Although the ancient Greeks had social and metaphysical views that made them neglectful of experiment, their reputation as purely a priori thinkers is exaggerated. The Pythagoreans determined the lengths of strings emitting harmonic notes by varying the position of a bridge on a monochord. Ptolemy initiated a whole set of experiments to answer questions about the refraction of light. Some ancient experiments took place outside of Greece. The efficacy of prayer is tested experimentally in the Old Testament story of the contest between Elijah and the priests of Baal. Herodotus reports that Pharaoh Psammetichus (664–610 B.C.) conducted an experiment to determine whether the Egyptians or the Phrygians were the most ancient race.[2] The method was to see which language would be spoken by two children deprived of all linguistic contact. Eventually, one child uttered a sound that was taken to be the Phrygian word for bread. So Psammetichus admitted that the Phrygians had greater antiquity.

Second, experiments occur in fields outside of science such as art, history, and music. Even philosophy has a few. Some are possibility proofs. Compare Socrates' demonstration that Meno's untaught slave boy can recognize geometrical theorems (see p. 88) with S. L. Miller's demonstration that amino acids can be produced from an electrified mixture of water vapor, methane, and hydrogen (all gases believed to have existed in the early atmosphere). Witness how Frank Jackson argues for the heavy contribution to understanding of conventional implicature. "If you are unconvinced of the importance of words like 'but', 'yet', 'nevertheless', 'anyhow', 'even', 'however', and so on in aiding the assimilation of information and furthering understanding, I suggest the following experiment. Take a philosophical paper with which you are unfamiliar; get someone to blank out all these words and their kin; then try to follow it. It is not easy."[3] Another group of experiments alien to science are the ones conducted by pseudoscientists. For example, one group of East German researchers concluded that the soul weighs 1/3,000th of an ounce after weighing 200 terminally ill patients before and after death.

Lastly and most importantly, many everyday acts count as experiments. Changing the amount of sugar in lemonade to discover the tastiest proportion is experimentation. Ditto for the fisherman varying his bait and the merchant's varing prices in search of the best overall return. Experiment is topic-neutral.

4. *Causal manipulation.*   The definition also permits experiments in fields associated with a priori reasoning. Getting to know a program by feeding it a wide variety of input values counts as experimentation. The same goes for those who get a feel for a formula by working it through with various values and expert puzzle busters who manipulate Rubik's latest toy in order to obtain a general solution. If varying properties required *causally* changing the things having those properties, then such a priori experimentation would be impossible. But our ordinary usage of 'vary' shows that we are willing to count relational changes as variations.

'Vary' has a second subtlety; it does not require the agent actively to change the values of the variable.[4] Aristotle's classic embryology experiment (reported in *Historia animalium* 6. 561a3–562a20) illustrates one sort of passive variation. He got the experimental design from the Hippocratic work *On the Nature of the Infant*: "Take twenty eggs or more, and set them for brooding under two or more hens. Then on each day of incubation from the second to the last, that of hatching, remove one egg and open it for examination. You will find that everything agrees with what I have said, to the extent that the nature of a bird ought to be compared with that of a man." Although there is no suggestion that the author of this passage executed the experiment, Aristotle checked it out. At the end of twenty days, Aristotle had an excellent record of the embryonic stages of the chick's development. A scandalous example of passive experimentation is the Tuskegee syphilis study in which treatment was withheld in order to plot the natural course of the disease. Any use of a control group involves variation by omission.

The distinction between experimentation and observation is best drawn within action theory. The experimenter *allows* the independent variable to acquire certain values or *makes* them acquire those values. For observers, the question of acting or refraining fails to arise. Astronomers neither make eclipses nor allow them to occur. One can only allow what one can control.

5. *Publicity.*   Experiments can be private affairs. Children make silent wishes to see whether wishing so can make it so. Francis Galton discreetly conducted the first experiments on ultrasonic sound while touring zoos and city streets. He had made a brass whistle that emitted sound above the maximal range for humans. The whistle was concealed in a walking stick and was powered by a bulb in the handle. If the animal in question pricked its ears after Galton squeezed the bulb, Galton inferred the animal could detect the sound. Another example is phosphene experimentation.[5] Phosphenes are hallucinated light displays that occur when there is a lack of external stimulation. They can also be induced by pressing your fingertips against closed eyelids, by certain drugs, and by electric shocks. In 1819 Johannes Purkinje published the results of some self-experimentation. He was able to stabilize his phosphene images by putting one electrode to his forehead, another in his mouth, and then rapidly making and breaking the current with a moving string of metal beads. Since the phosphenes are mental events, they are only observable to the hallucinator. Other people can have the same *kind* of hallucination, but the particular events are private.

B. A Cognitive Aim Is Essential to Experiment

Early speculation about the origin of fossils included the hypothesis that they are the remains of God's experiments. That is, the Creator had to run through a number of prototypes before perfecting the blueprints for contemporary organisms. My definition of 'experiment' provides immediate grounds for rejecting this hypothesis. An omniscient being cannot intend to learn anything. Hence, God can only experiment to edify others. But according to the hypothesis, the experiments predate any audience.

Like God, we cannot experiment about what is obvious to us. As William Whewell noted, we cannot experiment concerning the hypothesis that $7 + 8 = 15$ because a favorable outcome would not strengthen our conviction and a negative outcome would merely lead us to blame the equipment, our perception, or memories. Whewell claimed that this untestability is also exhibited by fundamental principles such as "The pressure on the support is equal to the sum of the bodies supported." He inferred that the principles are therefore necessary truths. However, any closed-mindedness defeats experiment, not just the firmness that comes from knowledge. According to the sceptic David Hume, nature has made human beings absolutely dogmatic concerning the hypothesis "Bodies exist." So although Hume denies that we know the hypothesis, he puts the matter beyond experiment.

As Mach emphasized, there is nothing essentially human about experiments. Extraterrestrials may experiment as I write. The question whether machines can experiment is held hostage by the controversy over whether they can have intentions. The Viking landers were described as performing experiments to test for life on Mars. But when directly asked whether the *landers* performed the experiments, most people attribute the experiments to the designers and controllers of the craft. People are even more opposed to attributing *thought* experiments to a machine. There is a computer program called HYPO that generates and manipulates the sort of hypothetical cases that lawyers love.[6] But users are inclined to analyze HYPO's "thought experiments" as thin imitations that only give the illusion of being backed by real thought.

We attribute intentions to animals with much more confidence and so are friendlier to the possibility that they experiment. Horses tap ice to check whether it will support their weight, penguins push other penguins off ice floes to test for killer whales, chimps vary the length of twigs when fishing out termites from rotten logs. Still, none of this trial-and-error behavior is *obviously* experimentation because it is difficult to discern the *specific* content of the animal's intentions. This inaccessiblity is more severe in the case of thought experiment because there are fewer behavioral clues and because (as shall soon be shown) thought experimentation requires an extra layer of intention. Given the higher complexity, it is natural to ask how thought experiments could have evolved from ordinary experiments. The question might also be framed developmentally. Attributing ordinary experiments to children is easier than attributing thought experiments to them. How might a child graduate from one to the other?

II. Execution Is an Optional Part of Experiment

The instruction–execution distinction suggests an answer to the question of how thought experiments could arise from ordinary experiments; maybe thought experiments are slimmed-down experiments—ones that are all talk and no action. There may be a progression of simplifications that connects experiment to thought experiment in the way self-propelled aircraft are linked to gliders by a process of attrition and compensating elaboration of the surviving parts.

Most experiments are planned *actions*. However, the execution component is inessential, just as physical realization is inessential to diets and drills. Each can exist merely as a set of instructions. Of course, the instructions are usually designed to be followed. But sometimes a procedure fulfills functions independent of its execution. Once these by-product services are appreciated, procedures will be designed without the intention of being followed. If this general point applies to experiments, we are on our way to showing how experiments can shuck their execution element and become thought experiments.

To show that the general point applies, we need to uncover benefits that experimental plans produce autonomously. Then we must show how some of these advantages could have been noticed by astute individuals who were as yet innocent of thought experiments. Pacing the goodies forms a trail that could have led our forebears toward thought experiments, step by myopic step, without knowledge of their ultimate destination.

One benefit autonomously bestowed by experimental plans is advance warning. Although scientists are often surprised by the results of experiments, they usually have an informed opinion about what will happen. Probability of success is one of the criteria used by review committees that allocate expensive laboratory time or that regulate ethically questionable experiments. At the informal level, we find scientists betting on the outcomes of interesting experiments. Some rushed researchers save time by faking experiments that they think will work. Like traders in the stock market, scientists revise their opinions before the event takes place. For news of probable news is itself news. Hence, an experiment can affect the probability of a hypothesis prior to execution.

In 1818 the French Academy of Sciences offered a prize for the best treatment of diffraction. Augustin Fresnel's wave theory contribution was referred to a commission dominated by proponents of the particle theory of light: Laplace, Poisson, and Biot. Poisson deduced a peculiar consequence from Fresnel's theory. Place a circular obstacle in a beam of light diverging from a point source on the axis. Under the assumption that light has a wave motion, the waves should arrive in phase at the periphery and so recombine at the center of the shadow to form a bright spot. Poisson (and nearly everyone else who learned of the consequence) was confident that this strange effect would not really occur and urged that an experiment be done to seal the fate of the now-more-doubtful wave theory. François Arago (and probably Fresnel) did just that, and the news that the spot really does form became a colorful feather in the cap of the wave theorist.[7]

Execution of Poisson's experiment altered opinion by introducing fresh information. But probabilities were revised before this, as soon as the experimental plan became known. This first revision was not due to new evidence. It was, instead, due to new inferences from old evidence. The prospect of the experiment stimulates the sort of thinking described in the chapter on armchair inquiry. At a minimum, scientists are cued into analogous situations. Assembling these reminders puts them in a better position to generalize.

This early evidence thesis explains how experimental results conforming to a theory's predictions sometimes *weakens* support for the theory. Even satisfactory results may be less satisfactory than was expected. Strangers to the stock market are puzzled when news of high profits fails to increase the value of the stock and sometimes even decreases it. The explanation is that the news of profits was "discounted." Those studying the company anticipated the earnings and bid up the price of stock. Thus, the news of profit was reflected in the price of the stock before the profits were officially announced. If profits are just as predicted, the expected return on the stock is unaffected by the official reports of high profits. If the profits are high but less than predicted, then the official reports will lower the expected return, causing a sell-off and lower prices. The phenomenon is also familiar to political scientists. Politicians who are predicted to win primaries by large margins are counted as losers if they win by only small margins.

Further evidence of the evidentiary value of experimental plans comes in the form of secret experiments. The insiders try to keep both the experimental procedure and the results of following that procedure secret. If the experimental plan was not scientifically informative, then we lose one of the standard reasons for keeping the plan secret.

III. The Progression from Experiment
to Thought Experiment

The notion of discounted performance suggests a progression of informative experimental plans. A plan may have no discount value because it is immediately implemented. Many casual experiments fall into this zero-lag category. But since scientific experiments tend to be complicated schemes, grandiose experiments have a lag time that allows the experimental plan to be disseminated and digested. We are still waiting for some experiments to begin. In the early 1960s Leonard Schiff extracted a remarkable prediction from the general theory of relativity: there will be a slight torque on rotating balls in orbit around the earth. Schiff has since died; but in the late 1960s another Stanford University physicist, Francis Everitt, began designing an experiment to be conducted by the National Aeronautics and Space Administration in the 1990s. Other experiments have begun but have yet to be completed. Over a hundred years ago, Lord Kelvin designed an experiment to demonstrate the very high viscosity of some substances. He built a twenty-inch-high staircase and put a block of pitch at the top. Over the years the pitch has flowed over the treads like very slow lava. The pitch still flows at Glascow University. In the case of finite lag, the plan has evidential value because it is an indicator of a future event. As in all *futures* trading, the indicator value of the experimental plan is exhausted by the occurrence of the event.

Some experiments are canceled. These events are, contrary to intention, never completely performed: infinite lag. Many experiments are never executed because they lack promise. Mach was once asked to perform an experiment to show that vivid visual ideas caused an image to form on the retina.[8] The reasoning was that since sensations are formed by images on the retina, the converse may also be the case. Mach was unimpressed and dismissed the experiment as hopeless. Other procedures may be only partially executed because unanticipated obstacles force truncation or approximation. For example, experiments on bird migration are enfeebled by the casualties inflicted by the trauma of transportation to the release site.

Some promising experiments are not even partially executed. In 1914 the Freundlich expedition to the Crimea was canceled by World War I. Members of the team were arrested and their equipment impounded in southern Russia. This delayed confirmation of Einstein's general theory relativity for five years. The cause of cancellation can be a dramatic event such as the 1986 Space Shuttle disaster or something inconspicuous like procrastination. Scientists who learned of the plan will have revised their opinions in anticipation of the performance. When they learn that the experiment was canceled, they do not revert to their old views. Scientists retain the revised opinion—a legacy of the experimental plan. When asked to defend their opinions, they may cite the experiment even though it is common knowledge that the experiment neither was, nor will be, performed. Hence, the evidential value of infinite lag experiments is not parasitic on the information value of some future event. It draws its news value from a merely possible event.

Molyneux's question illustrates how infinite lag can be caused by unforeseen complications. He stated his question in a letter to John Locke in the seventeenth century:

> Suppose a man born blind, and now adult, and taught by his touch to distinguish between a cube and a sphere of the same metal, and nighly of the same bigness, so as to tell, when he felt one and the other, which is the cube, which is the sphere. Suppose then the cube and sphere placed on a table, and the blind man made to see; query, Whether by his sight, before he touched them, he could not distinguish and tell which is the globe, which the cube? To which the acute and judicious proposer answers: 'Not. For though he has obtained the experience of how a globe, how a cube, affects his touch; yet he has not yet attained the experience, that what affects his touch so or so, must affect his sight so or so'[9]

Although Molyneux expected that his answer would eventually be supported by execution of the experiment, no clear execution has taken place. Sight has been restored to hundreds of congenitally blind people by cataract removals. However, these operations disturb the optics of the eye and so require a waiting period to give the eye a chance to settle down. This waiting period makes the restoration of sight gradual. But Molyneux's experiment can only be clearly executed with a subject who suddenly regains sight. Hence, psychologists have had to settle for subjects who only approximate Molyneux's requirements. But even if Molyneux's experiment is never properly executed, its substantial influence on theories of vision will survive.

Since they were all intended to be performed, none of the lag cases is a thought experiment. But they do form a progression that invites further steps toward thought experiment. Many innovations mimic and extend good effects of accidents. Early fishermen may have discovered that a hook inadvertently barbed worked better and so used it as a model for making future hooks. Even without knowledge of why the barb helps, fishermen would develop better and better hooks by choosing the most successful hooks as their new models. Representative art may have begun similarly. Leon Battista Alberti in his treatise on sculpture, *De statua*, suggests that art begins with touch-ups. Our protoartist stumbles upon a tree trunk or lump of clay that happens to look like some natural object. This fortuitous resemblance is accentuated by a trivial addition or subtraction. The adjustments gradually become more ambitious. As time goes on, the artist needs less and less of a running start from nature and can bring the final product into closer and closer conformity to the model. At the bottom of this shapely, slippery slope sits a professional sculptor carving a precise likeness out of a block of marble.

Experimental designs also improve by accident. Developmental psychologists have long explored the role of social learning in morality by designing experiments in which subjects are induced to perform good deeds. In one experiment, ten- and eleven-year-olds who won at a miniature bowling game were encouraged to donate part of their winnings (5¢ gift certificates) to orphans.[10] Children assigned to the "guided rehearsal condition" practiced putting gift certificates into the open charity box. The second group played with an adult model who made lots of donations. The third merely watched an adult donating, while the fourth group were merely told about the charity box. During the first session, the greatest altruism was exhibited by the rehearsers. But in the second session, several days later, the children donated less and even began stealing from the charity box (especially the rehearsers). The surprised experimenter had strengthened the experimental design by inadvertently including immoral options.

Robert Millikan's ultimate design for his oil drop experiments exploited a whole sequence of serendipitous improvements.[11] Much of the everyday "fine-tuning" of an experiment resembles the fishhook story. In many cases, the experimentalist does not really know why an adjustment to the procedure makes the experiment "work."

The same happy-go-lucky ignorance holds at the level of general experimental design. Galileo engineered his methodology with the pragmatism of a physician who changes his treatment of a patient.[12] Galileo found that nuisance variables could be eliminated by creating a highly artificial setup. Some biases could be canceled by others. Small disturbances could be ignored. Hard-to-reach values could be extrapolated by a trend from a series of easier-to-reach values. All of these are empirical lessons learned about experiment, not a priori methodological principles. When things get working, there is often a residue of uncertainty as to how the changes succeeded, what the *specific* problem was, or whether a particular adjustment even had any real impact at all. Hence, experimental design evolves in the uneven, half-blind manner of medical lore. Theorists can come in later and rationally reconstruct many aspects of laboratory craft. They will dismiss other features as vestigial or as superstitious pieties. Thought experiments are equally dependent on their track records. After all, an idealization is supposed to be an *effective* simplification—a shortcut that gets you most of the effect by concentrating on just a few variables.

The transition from experiment to thought experiment turns on a change in how the design is used rather than on any immediate change in the design itself. When an experimenter notices experiments yielding an epistemic return without the price of executing the procedure, he will seek another free ride. He will design executionless experiments in the same spirit that others design cordless telephones. More specifically, our innovator enters the realm of thought experiment when he presents experiments that purport to work by supposition rather than performance. He might make the transition with one revolutionary leap. Maybe straw votes were invented this way. Someone noticed that a vote invalidated by a technicality nevertheless helpfully exhibited the preferences of the group. When the need for the information arose again, he may have suggested that the group deliberately vote without official effect. Thus, the procedure is remodeled into a preference-revealing device.

Then again, pioneers may have inched over the threshold to thought experiment by many small, unreflective steps. Experiments composed of individually useful trials invite such gradualism. In these cases, failure to perform a particular trial is not fatal. Most are even designed to tolerate missing parts; that is, some trials are discretionary. Now consider someone who uses a standardized experimental design that makes some patently infeasible trials mandatory. The experimenter might nevertheless proceed, inviting us to guess what the results of the missing trials would have been. These missing mandatory trials are miniature thought experiments. As we increase the proportion of thought trials to executed trials, the experiment looks more like a thought experiment. Upon reaching the extreme, where none of the trials are intended to be performed, we have a clear case of thought experiment.

Missing elements are rarely random omissions. Normally, there is a systematic limitation, such as our inability to reach extreme values. In Galileo's experiment with the law of equal heights (discussed in chapter 1), he made smooth, double inclined planes and allowed a ball to roll down one side and up the other. By varying the angles of the plane, Galileo demonstrated that the ball rolls (almost) up to the same height whence it rolled down. He then asked what would happen if the right-hand plane were made perfectly smooth and horizontal. Galileo's answer is that it would continue move horizontally at a constant speed indefinitely because the ball could never regain its original height. If we divide the experiment into subexperiments, this extreme case is a thought experiment. But we can also view the enterprise as one large hybrid experiment—part ordinary experiment, part thought experiment.

Galileo's continuation illustrates both the power and limits of thought experiment. On the plus side, the thought experiment directs us toward a fundamental law of nature, the law of inertia: unless disturbed, bodies at rest stay at rest and bodies in motion continue in motion with the same speed in a straight line. On the downside, we should note that Galileo did not get this law quite right. Lingering geocentrism led him to think that 'horizontal surface' meant a surface that is everywhere perpendicular to the radius of the earth. This is apparent from his discussion of another thought experiment intended to make the same point. Picture a ship on a calm sea that is progressively losing resistance. When the ship loses all resistance, it will continue around the earth indefinitely. Unlike Newton, Galileo does not think that the circular motion is in need of explanation. Since the law of inertia demands movement in straight lines, the Newtonian says that the ship's own weight keeps it from flying off the earth on a tangent. Thus, Galileo imagined the ball in the inclined plane experiment moving in a giant circle around the earth—not straight, as dictated by the law of inertia. Luckily, Galileo was willing to use the law of inertia as a simplification. Since he realized that a small segment of a huge circumference very nearly equals a straight line, Galileo accepted the straight-line movement as a close approximation for terrestrial purposes.

Thought experiments are not the only experiments that are designed without the intention of being performed. Some experiments are designed by students as part of their laboratory instruction. Prototype experiments serve as stepping-stones to an improved experimental design. Some experiments are designed with the intention to fake performance. None of these unexecuted experiments is a thought experiment because none of them is presented as an abstract instrument of rational persuasion. Thought experiments must be presented with the public aim of answering or raising the experimental question without the benefit of execution.

Bureaucratic requirements compel the design of experiments that will never be executed. For example, a scientist whose job requires him to design safety experiments that could be performed by survivors of an all-out nuclear war may actually believe that all-out nuclear war is not survivable. The experiments he designs are not thought experiments even if he appends the note, "These experiments will never be executed."

The requirements generating this sort of example need not be bureaucratic or alien. Thinkers impose requirements on themselves as part of their methodology, pedagogy, or style. For instance, ancient Greeks favor "Begin at the beginning," that is, demonstrations that march from the initial state to the goal state, rather than solutions that proceed indirectly or by reasoning backwards from the goal state. They used this psychologically insensitive style of reasoning even when other paths were the ones actually and more easily taken. Methodologies that stress the need for experimental validation (such as Newton's) will naturally pressure adherents into filling in the required experiments. An instance is the French physicist André-Marie Ampère, who was one of the founders of electrodynamics and a stout Newtonian. In his main work, Ampère presents his research as if it had been conducted in strict conformity to the inductivist rules of natural philosophy established by Newton:

> I have consulted only experience in order to establish the laws of these phenomena, and I have deduced from them the formula which can only represent the forces to which they are due; I have made no investigation about the cause itself

assignable to these forces, well convinced that any investigation of this kind should be preceded simply by experimental knowledge of the laws and of the determination, deduced solely from these laws, of the value of the elementary force.[13]

But (as Duhem points out) Ampère did not practice what he preached. Many of Ampère's experiments are vague, and others were afterthoughts arranged to make his inquiry look as if it squared with Newtonian scruples. Ampère himself admits that some of the experiments he describes were never carried out. He did not have time to construct the necessary instruments. Nonetheless, Ampère was sure about how these experiments would turn out because of his other results. He was not intending to persuade the reader by mere contemplation of the experimental design. The experiments were instead presented in the hope of showing how a good scientist would methodically arrive at his results.

IV. Classifying Thought Experiments by Grounds for Inaction

The picture of thought experiments as attenuated experiments invites us to classify them in accordance with three reasons for inaction.

A. Unimprovables

The first reason for doing nothing is the absence of evidential gain. The usual point of the experimental procedure is to answer a question. Yet sometimes merely thinking about the procedure answers the question; the thought sometimes renders the action superfluous. Aristotle's principle that velocity equals force divided by resistance makes projectile motion puzzling because nothing is pushing the moving object. One popular solution was that the air displaced by the arrowhead rushed to the tail to prevent the formation of a vacuum. Jean Buridan objected that if the air was pushing from behind, sharpening the back end of an arrow should greatly shorten its flight. Buridan thought the consequent too absurd to merit actual trials but did not think the shortened flights to be logically impossible.

There are notorious cases in which the thought experiment is executed and the "absurd" is proven true. One of these ill-fated thought experiments was part of the battery of arguments Aristotelians used to show that the earth was at rest. If, as Galileo maintained, the earth rotated from west to east, unattached objects such as clouds and birds should move east to west like the sun—at about a thousand miles per hour. Second, the rotation hypothesis conflicts with the fact that rocks dropped from towers fall straight down. If the earth were rotating, the rock would land far west of the tower. The Aristotelians explained this consequence by drawing an analogy between a moving earth and a moving ship. If a rock were dropped from the mast of the ship, it would land at the rear because the ship continues to move after the rock is released. In

1632 Galileo challenged the counterfactual in his *Dialogue Concerning the Two Chief World Systems*. He maintained that the tower argument and the thought experiment featuring the moving ship, overlooked the law of inertia. When the stone is released from the mast of a moving ship, it has the same motion as the ship. This motion will be conserved in its descent, so it will keep pace with the ship and land at the foot of the mast, not the rear. Likewise, a stone dropped from a tower on a moving earth will have the same motion as the tower and so fall to the foot of the tower. The law of inertia also explains why the clouds and birds can move freely. Galileo was just assuming that gravitation did not interfere with the horizontal velocity. He was vindicated in 1640 when Pierre Gassendi carried out the experiment.

The execution of an experiment can also be rendered pointless by conceptual barriers. Actually "sending" a person through a teletransporter would not resolve the questions asked by personal identity theorists. They are not in doubt about the empirical facts. Their doubt is about whether to describe the event as one of personal survival or one of death and replacement by a replica. These questions of description are raised equally well by actual and hypothetical events. Hence, carrying out the procedure would be redundant, an uninformative duplication of data.

The irrelevance of physically instantiating these experiments leads some people to deny that the instantiation constitutes execution of the thought experiment. They say that these thought experiments are executed by reflecting on the design—by figuring out how you would describe the situation. Since many philosophical thought experiments concern questions of description (Quinton's Two Space Myth, the Ship of Theseus, Searle's Chinese Room) one might even conjecture that the two types of execution distinguish scientific thought experiments from philosophical ones.

One of the themes of this book is *gradualism*: there are no qualitative differences between philosophical thought experiments and scientific thought experiments, only differences of degree. Hence, the conjectured demarcation on the basis of execution styles provokes a search for counterexamples. The first counterexamples are philosophical thought experiments that are more than indirect reports of linguistic conventions. Several of these made an appearance in my negative answer to the question of whether all thought experiments are appeals to ordinary language (chapter 4). But let us consider another counterexample that challenges the principle that distinct pains must have distinct locations. The procedure is to manuever two victims of phantom limb pain so that their phantom feet overlap. Since it is obvious that their pain would continue, we conclude that two pains can exist at the same time and place. But suppose a bored (but philosophical) nurse executes the procedure. She discovers that the two victims no longer feel pain! When the phantom limbs are repositioned, the pain returns. Further study by physiologists substantiates the "interference effect," and they hail a new painkiller. This scenario shows that the philosophical thought experiment could be empirically disconfirmed.

The alleged asymmetry is also refuted by scientific thought experiments that only concern the description of events. Consider the thought experiments cosmologists use to demonstrate that universal isotropy implies universal homogeneity, but not vice versa. An isotropic system is a setting that looks the same in all directions. For instance, an astronaut who splashes down into the ocean can rotate around without seeing anything new. Thus, at that point the ocean is isotropic. A homogeneous system is a setting that looks the same at any point. Think of a man walking along railroad tracks in the desert. The view does not change as he travels forward. Yet the scenery does change if he turns: he sees sand instead of tracks. Hence, the system is homogeneous but not isotropic. It can also be shown that a universe that is homogeneous at every point need not be isotropic; for suppose the man is now in an infinite field of arrows that all point north. However far he treks, the scenery remains the same. But if he turns south, he sees tails, instead of heads, of arrow. Now suppose that the man is in a totally isotropic universe. From where he stands, each point in his circle of vision is indistinguishable from the rest. So if he moves out to one of these equivalent points, he will once again find himself surrounded by indistinguishable "reference points." Since this process can be extended indefinitely, the universe must be homogeneous by the transitivity of 'indistinguishable'. Notice that the lesson taught by this last pair of thought experiments is entirely conceptual.

My next complaint against 'Execution is irrelevant in all, and only philosophical thought experiments' is that it equivocates between different relativizations of *execute*. The preferred reading of 'The thought experiment was executed' is a report of physical instantiation. There is a secondary reading that only reports competent reflection on the experimental plan. When a thought experiment "goes in one ear and the other," one has not executed it even under this lenient reading. Hence, a teacher could complain that his lazy student did not execute the assigned thought experiments or note that his toddler is too young to execute thought experiments. But under the primary reading, none of the students executed the thought experiment. Describing thought experimenters as "executing" their experiments is misleading because the reading under which it is true is also trivial. People will construe the reading under the impression that you are abiding by the maxim "Be informative." Although the reading is not analytic, it has little content, because people almost always think about what they are asked to think about. Occasionally, the audience cannot or will not reflect on a thought experiment because it is arcane, unpromising, or disturbing. But usually, the audience's compliance is an involuntary mental reflex. They have difficulty *refraining* from execution in this sense. It is like trying to entertain but not calculate $(5 \times 6) + 3 = x$. Since virtually all experiments are executed under the weak reading, there is no way of executing an experiment that is peculiar to philosophy.

B. Unaffordables

The second reason for inaction is that the gains are outweighed by the losses. *Loss* should be interpreted broadly. Opportunity costs count. A procedure might be a thought experiment by virtue of a more efficient alternative to execution. Many of the thought experiments appearing as exercises in text-books are in this category: What would happen if a fly in a bottle resting on a pan balance were to start to hover within the bottle? Would the registered weight increase, decrease, or stay the same? The ignorant are not intended to capture a fly and carry out the instructions. They are to quiet their curiosity by reviewing the relevant part of the text. *Loss* is also intended to include legal, prudential, ethical, and aesthetic considerations. Would pushing a suitless astronaut into space cause him to explode? If a wound and an unwound watch were dissolved in acid, what would become of the wound watch's potential energy? Here the point is not learning by doing.

Let *unaffordable* include cases in which mild inconvenience prevents the execution of the experiment. For example, Galileo mentions an experiment designed to show that projectiles are slowed even by thin media.[14] Fire a bullet into the stone pavement from a gun poised two hundred feet up and another bullet from the same gun from only two feet above. The bullet shot from the lower distance will be more badly smashed if thin media (such as air) impede projectiles. Galileo (in the guise of Salviati) admits that he has never carried out this experiment but still regards it as added assurance for his theory. Since this is an easy experiment, his reason for inaction is only the mild inconvenience.

C. Impossibles

The third reason for inaction is impossibility. To say that something is impossible is to say that it violates a constraint. *Impossible* is relative to one's means. Since one's means vary over time, what was once impossible can become possible. Low-gravity experiments began as impossible experiments because we did not have the means to reduce the influence of gravity. Once we determine that the deed cannot be done, we ask, "How impossible?" We do this to determine the relevance of the conditional. The relevance of the conditional is determined by its resemblance to our stock of accepted conditionals. If the infeasibility is due to a minor obstacle, then the corresponding counterfactual is similar to other conditionals in our stock of beliefs. If the infeasibility is due to a major divergence from actual conditions, there is less resemblance between the counterfactual and the conditionals about the actual world. Textbooks have plenty of thought experiments whose execution is theoretically possible but impossible in practice: A hole bored through the center of the earth. What happens to the weight of a stone that is tossed in? What happens to its mass? Suppose that an apple hovers a hundred kilometers over the earth and that the earth's volume suddenly expands so that its radius is extended by a hundred kilometers. Would the gravitational attraction between the apple and the earth change? Many casual definitions of 'thought experiment' imply that all thought experiments are like these technologically impossible scenarios. This overgeneralization tempts one to infer that some kind of impossibility is essential to the very concept of a thought experiment. For example, Kathleen Wilkes emphasizes that "Such forays of the imagination are called *thought* experiments precisely because they are imaginary, they cannot be realized in the real world."[15] Since her scepticism about philosophical thought experiments is generated by making heavy weather about the difficulty of judging counterfactuals with impossible antecedents, the error cascades through her reasoning. The mistake is just a cousin of the one most people initially make about counterfactuals: inferring that the antecedent of counterfactual must be false because it is *counter*factual. A thought experimenter is free to be neutral on the question of whether the procedure has ever been executed.

The depth of the impossibility is proportional to how different the world would have to be for the procedure to be executed. Difference in detail is not much difference. So a procedure rendered infeasible by our position in space or time is still a "realistic" case. The transition from these shallow impossibilities to deep ones is gradual; for once we see that an experiment can be evidence even if its execution is precluded by minor technological limitations, we naturally ponder major technological barriers. Thus, puzzlement about Saturn's rings may prompt an astronomer to suppose that a bridge is built around the earth. He then supposes that the bridge's pillars are simultaneously destroyed. Would the bridge fall down? The global bridge is infeasible, but supposing its construction summons relevant counterfactuals.

As we pass to procedures calling for altered planetary orbits, acceleration to near light speed, and the performance of infinitely many operations, we ease ourselves into scenarios that violate natural laws. There are fundamental reasons why we cannot "turn off" forces such as gravity, natural selection, and subscription to moral codes. Does this spoil the evidential value of experiments that violate these laws? Having countenanced thought experiments whose execution is precluded by basic technological limitations, we are committed to the further extension of law contravening thought experiments. Interestingly, there is a whole set of "nonperformable" operations in crystallography that can only be studied by thought experiment. Certain operations, such as those under the category of rotational symmetry, can actually be performed on objects. A cube, for instance, can be rotated so that its lattice is the same as when the operation began. Reflection symmetry is a nonperformable operation on solid objects because it cannot be performed in three-dimensional space even though it could be done in a space of a higher dimension.

Some experimental procedures are even logically impossible to execute because they eliminate preconditions of physical experimentation. For instance, Max Black attacks the principle of the identity of indiscernibles (distinct things must have different properties) by means of a thought experiment that is incompatible with the existence of observers: "Isn't it logically possible that the universe should have contained nothing but two exactly similar spheres? We might suppose that each was made of chemically pure iron, had a diameter of one mile, that they had the same temperature, color, and so on, and that nothing else existed. Then every quality and relational characteristic of the one would also be a property of the other.[16] Neuman defended absolute space by eliminating everything except Jupiter. Russell's five-minute hypothesis cuts out the past. These procedures make one wonder; if it is common knowledge that the procedure is logically impossible to execute, what can be gained by having people suppose the procedure is executed?

V. The Immigration of the Supposition Operator

"Nothing!" is the short answer of those who suspect we have attained the acme of academic inanity. They condemn experimenter-precluders as inconsistent and dismiss them as epistemically impotent.

Here we connect with a long tradition that draws the boundary of imagination along the horizon of perceptibility. Goethe, for example, denied that anybody believes in his own mortality on the grounds that no one can imagine his own death. He granted that you might *seem* able to imagine your own death. For example, you might visualize your head being lopped off by a samurai. But Goethe would then point out that you visualized the spectacle in color and from a particular height and angle, usually that of a well-positioned spectator. According to Goethe, this shows that you really imagined yourself *viewing* the death of a lookalike.

Some Newtonians championed absolute space with a thought experiment featuring a single ball accelerating through empty space. Since the ball is not moving relative to any other object, its movement must be in relation to space itself. Berkeley objected that the thought experimenter smuggles himself into the scenario:

> We tend to be deceived because when in imagination we have abolished all other bodies, each of us nevertheless assumes that his own body remains . . . no motion can be understood without some determination or direction, which cannot itself be understood unless, besides the body moved, we suppose our own body, or some other, to exist at the same time. . . . Since, then, absolute space never appears in any guise to the senses, it follows that it is utterly useless for distinguishing motions.[17]

Gesticulative conductors of the Newtonian thought experiment "point" to the moving ball while the audience "follows" its path with eyes (like baseball fans viewing a home run). Berkeley would find this behavior particularly damning evidence of "peeking."

Berkeley is better known for his use of this point against an obvious objection to his principle "To be is to be perceived." In his dialogue between Hylas and Philonous, Hylas tries to refute the principle by imagining a solitary tree in an uninhabited wilderness. Philonous' reply is that such an exercise in imagination is self-defeating because one can only picture the tree if one imagines observing it.

The best rejoinder to Berkeley is to distinguish between external and internal supposition. An external supposition has the form 'SBp' for 'Suppose you bring about p'. An internal supposition has the form 'BSp' for 'Bring it about that you suppose p'. This is the difference between a supposition that you are following a procedure and following a procedure of supposition. If I externally suppose that I am following an experimental procedure, then I must suppose that I am doing various things. Part of the supposition will be that I exist, in good working order and in the sort of circumstances conducive to investigative activities. These assumptions rarely interfere with down-to-earth thought experiments; but they mix poorly with suppositions about me never being born, everyone going mad, or the earth increasing in temperature a thousandfold. However, there is no problem if we interpret the experimental procedure internally, as a sequence of directed suppositions; for there is a difference between supposing you are perceiving and just supposing. When asked to suppose that I am dead, I visualize my inert body in a coffin from a particular angle, in color, with stereoscopic vision. But this does not imply that I am supposing that I am *perceiving* my dead body. It is one thing to imagine a tree in an uninhabited forest and another to imagine myself discreetly viewing a tree in an otherwise uninhabited forest. (And so we must also distinguish between BSp and SBSp.) Compare visualization with paintings. The painter's position, aesthetic talent, and access to technology can be inferred from his paintings, but the paintings do not *depict* those facts. Just as the painter's intentions influence what his painting represents, so the imaginer's intentions influence what his visual images represent.

We are especially tempted to abrogate the distinction between SBp and BSp if we believe that all supposing is pretended perception. For if I am pretending to see things, I am one of the characters in the story; hence, detached imagination becomes as impossible as pretence without a pretender. Raw empiricism inspires one line of reasoning for the premise that all supposing is pretended perception: if all of what I know arises from *my* experience, then my imagination is limited to this personal experience. When I imagine, I only cut and splice stored reels of perception. Regardless of how the film is reworked, I remain in the story as the perceiver just as I always remain in the story when the film is run unedited, that is, when I recall the past. Ryle endorses the analogy between imagining and recalling: "I recall only what I myself have seen, heard, done and felt, just as what I imagine is myself seeing, hearing, doing and noticing things; and I recall as I imagine, relatively vividly, relatively easily and relatively connectedly. Moreover, much as I imagine things sometimes deliberately and sometimes involuntarily, so I recall things sometimes deliberately and sometimes involuntarily."[18] But notice, contrary to Ryle, that even recollection goes beyond personal experience. I recall but never witnessed any of the following: Kepler stole Tycho Brahe's data, there is no largest prime, birds sometimes get sucked into jet engines. Even events that I remember because they were objects of my perceptions can be recalled without any ability to recall features of the perception that sired them.

One might challenge the notion of detached supposition with a weaker premise than brute empiricism. Assume, instead, that supposition is pretended assertion. As Ryle notes, people find pretended adoption of positions useful in the way boxers find sparring useful. Usually, the speaker will signal that "his intellectual tongue is in his cheek" with words such as 'if', 'suppose', 'granting', and 'say'. If the audience takes him to be asserting rather than supposing, he will explain that he is not committing himself, that he was trying out the thought, "practicing." Ryle goes on to make a point that buttresses my claim that experiment is logically prior to thought experiment:

> Supposing is a more sophisticated operation than ingenuous thinking. We have to learn to give verdicts before we can learn to operate with suspended judgments.
>
> This point is worth making, partly for its intimate connection with the concept of imagining and partly because logicians and epistemologists sometimes assume, what I for a long time assumed, that entertaining a proposition is a more elementary or naive performance than affirming that something is the case, and, what follows, that learning, for example, how to use 'therefore' requires first having learned to use 'if'. This is a mistake. The concept of make-believe is of a higher order than that of belief.[19]

I agree so far. Indeed, Ryle's insight about the sophistication of make-believe helps the evolutionary theory of thought experiment by knocking out an attractive rival: the view that experiment is thought experiment plus execution. This reversal of my position falsely implies that one can suppose that x is done prior to understanding what *doing* x is. Abnormalities may prevent some thought experimenters from performing an ordinary experiment. The cosmologist Stephen Hawking, for example, has been nearly totally paralyzed by a progressive motoneuronal disease. But the handicap does not alter the fact that anyone who grasps the concept of a thought experiment also understands what it is to conduct an ordinary experiment.

Despite this point of agreement, there is an influential application of Ryle's idea that would doom detached supposition. David Lewis has argued that storytelling is pretend reporting: "The author purports to be telling the truth about matters he has somehow come to know about, though how he has found out about them is left unsaid. That is why there is a pragmatic paradox akin to contradiction in a third-person narrative that ends '. . . and so none were left to tell the tale'."[20] Kendall Walton treats the pretence of knowledge as just the tip of the iceberg.[21] He analyzes fiction as a cooperative game of make-believe between speaker and audience. By studying the roles of both parties, we gain an understanding of our puzzling emotional reactions to fiction such as fearing that which we know cannot really hurt us. Despite these successes, the Lewis–Walton account of fiction is incomplete. Some stories depict the author as a well-informed character, but other stories depict no such epistemological relationship. Indeed, a work of fiction can preclude the existence of knowers. If none yet exist, the literary lacuna can be filled forthwith:

Granite World

Once upon a time there was a universe that contained three hunks of granite and nothing else. Sometimes they came close to hitting each other. But they always missed. So just as nothing ever preceded their existence, nothing new came after them. The end.

To handle my story, a theory of fiction must allow for detached supposition. We are free to say that the Lewis–Walton type of story temporally and even logically precedes detached stories. Indeed, the transition from SBp to BSp for thought experiments may be intimately linked with the development of this derived sort of fiction.

The transition from supposing that one follows a procedure to following a procedure of supposing is implicit in idealization. Exclusion of extraneous variables ordinarily requires intricate precautions. Thought experiments offer control by *stipulation* once we incorporate suppositions directly into the procedure. One can simply declare the surfaces frictionless, the bales of hay equally appetizing, the agents perfectly rational. Thus, demonstrations by means of mere experimental plans tend to be simpler. The setups are Spartan with just a few things to track. In contrast, running an ordinary experiment is like running a small business. Hundreds of details must be monitored and the demands of principle and practicality must be balanced. Construing the procedure as directing us to *suppose* the conditions are so, rather than as a procedure that makes them so, spares us the dreary details needed for control.

Since the preconditions of ordinary experimentation can themselves be extraneous variables, the quest for control coupled with desire for completeness will inspire procedures free of those preconditions, as well. At this stage internal supposition is more than a laborsaving device. It is mandatory, because the external construal leads to the inconsistency Berkeley fastened upon. Thus, the pain of contradiction forces the supposition operator to migrate from SBp to BSp. In its new home, it supports the most powerful form of thought experiment.

Since any sequence of suppositions is a story, the internal interpretation makes the experimental procedure a story. The freedom to eliminate extraneous variables through stipulation, rather than detailed control, strains the resemblance between this sort of thought experiment and ordinary experiments. The thought experiments are almost always far briefer, sketchy, and strange. Their resemblance to minimalist science fiction should lead us to expect enlightenment from theories of fiction.

VI. A Definition of 'Thought Experiment'

A *thought experiment* is an experiment (see p. 186) that purports to achieve its aim without the benefit of execution. The *aim* of any experiment is to answer or raise its question rationally. As stressed in chapter 6, the *motives* of an experiment are multifarious. One can experiment in order to teach a new technique, to test new laboratory equipment, or to work out a grudge against white rats. You could be doing it for the sake of fame, fortune, or fun. (The principal architect of modern quantum electrodynamics, Richard Feynman, once demonstrated that the bladder does not require gravity by standing on his head and urinating.) The distinction between aim and motive applies to thought experiment as well. When I say that an experiment "purports" to achieve its aim without execution, I mean that the experimental design is presented in a certain way to the audience. The audience is being invited to believe that contemplation of the design justifies an answer to the question or (more rarely) justifiably raises its question. The thought experimenter need not actually believe what he is inviting others to believe. He could be insincere. Perhaps the thought experimenter is a devious lawyer who thinks he can trick the jury into acquitting his client by walking them through a fallacious hypothetical. Or perhaps the bad faith thought experimenter is a popularizing scientist who privately regards his imaginary scenario as hopelessly oversimplified but wants to give his readers the illusion of understanding frontier research.

A. The Selectivity of the Definition

My definition of 'thought experiment' resembles George Dickie's institutional analysis of art: "A work of art in the classificatory sense is (1) an artifact (2) a set of the aspects of which has had conferred upon it the status of candidate for appreciation by some person or persons acting on behalf of a certain social institution (the artworld)."[22] Likewise, what makes an experimental design a thought experiment is the way it is presented to the audience—as a design that aims to convince or puzzle in its own right. Of course, not any mode of presentation makes the design a thought experiment. If the curator of the Guggenheim presents it as an object of appreciation, then the experimental design would qualify as a work of art (not necessarily as a *good* one). Experimentalists sometimes present especially elegant designs in this aesthetic fashion. Just as Duchamp's urinal is art relative to one presentation but not relative to another, an experimental design can be a thought experiment in one setting and an ordinary experiment in another. For example, one textbook thought experiment asks what will happen if a gun is pointed at a squirrel that drops from its tree just at the time of firing. The scenario is also a popular classroom demonstration with toys playing the role of gun and squirrel. Since the substitution of toys for a real gun and squirrel is physically irrelevant, the demonstration is a physical instantiation of the design used in the thought experiment. Indeed, the experimental design could be a thought experiment on Monday, an executed experiment on Tuesday, and a macabre conceptual artwork on Wednesday. Compare the design with a slab of wood that is used as a door, then a table, and finally a lid. My theory is adverbial in the sense that I deny that thought experiments are new kinds of entities—they are old kinds of entities used in new ways. It's not in the meat, it's in the motion.

My definition of 'thought experiment' is intended to rule out many cases that casual definitions overlook. Fraudulent experiments are examples of unexecuted experiments that are rationally persuasive (to the victims) but are not thought experiments. The reason is that the faker presents the (putative) execution rather than the abstract element as the evidence. Plainly, Cyril Burtt's fraudulent twin studies are not thought experiments.

Nevertheless, one eminent historian of science fails to draw the distinction between thought experiment and fraudulent experiment. Alexandre Koyre is well known for his thesis that the scientific literature of the seventeenth century "is full of these fictitious experiments."[23] In particular, he argues (with as much diplomacy as he can muster) that many of Galileo's experiments are only "imaginary" because he did not really execute them. Galileo only worked them out in thought and gave readers the impression that they were reports of real experiments. Koyre uses this thesis of secret nonperformance as a premise for his conclusion that Galileo, rather than being an archempiricist, was actually a Platonist who triumphed by underground reliance on thought experiment. My complaint against Koyre is that he characterizes Galileo's *published* experiments as thought experiments. Koyre has been deluded by his own politeness. He is so worried about slandering Galileo that he uses the euphemism "imaginary experiments" and then assimilates them to thought experiments. Koyre should call a spade a spade.

A close relative of the fraudulent experiment is the experiment that is *secretly* intended to do the persuasive work without execution. This type of experiment is rationally persuasive and may even produce knowledge but is not a thought experiment because its intended manner of persuasion is not common knowledge between the speaker and his audience. Closely related are those experiments that are only half-heartedly intended for performance. In the course of explaining sailboat design, James Trefil warns "If we want to increase the height of the mast or column, we have to worry about the fact that the column may buckle when its height reaches a certain value, even if no new weight is added to it. You can convince yourself of this fact by doing a simple experiment. Take a few drinking straws. A single straw will easily support the weight of something like an orange, but a column made by joining the straws together will quickly collapse."[24] Although Trefil tells us to learn by carrying out the experiment, he must realize that nearly all readers will be too convinced to bother. Another close case is created by half-hearted imaginative scenarios that are not seriously presented as evidence. Einstein once responded to a reporter's plea to summarize his theory by saying: "If you will not take the answer too seriously, and consider it only as a kind of joke, then I can explain it as follows. It was formerly believed that if all material things disappeared out of the universe, time and space would be left. According to the relativity theory, however, time and space disappear together with the things."[25] If Einstein had presented the scenario seriously, he would have been conducting a thought experiment. But given his disclaimers, we know that he is only trying to convey a vague, alluring idea of what his theory is about.

Daydreams and fantasies are excluded by the definition because they are not presented as raising or answering questions. We can also distinguish recreational speculation from thought experiment. Many people take intrinsic pleasure in working out the consequences of hypothetical events that would dramatically impact their lives: What would I do if I won the lottery? How would I survive if I were the last man on earth? These questions are not asked to prove a point or raise an issue. They are asked because people just enjoy entertaining the propositions. Supposition arouses feelings and emotions in the way that fiction does. So a person may suppose certain propositions in order to have those experiences. This drives curiosity about recherché aspects of established literature: Who was the greater detective, Sherlock Holmes or Hercule Poirot? Why didn't Superman defeat the Nazis? Of course, thought experiment offers many of the same delights as idle speculation. The difference is that the *aim* of a thought experiment is enlightenment, rather than fun. Moreover, the enlightenment has to proceed in the right way. "Granite World" is a concise piece of fiction invented for the purpose of proving a philosophical thesis. But it is not a thought experiment. "Granite World" persuades by *exemplifying* the predicate that I claimed to be consistent, namely, 'story that precludes the existence of knowledgeable storytellers'. There is no manipulation of the question's variables.

Irrationally persuasive imagery also fails to count as thought experiment. Having a cat lover imagine Dan Quayle biting off the heads of live kittens may convince her that Quayle is vicious. But since this exercise in imagery is not presented as a device of *rational* persuasion, it is not a thought experiment.

B. Stereotypical Features of Thought Experiments

Many of the heuristics used to identify procedures as experiments are also used to identify thought experiments. Thus, the typical or representative thought experiment scores high on scientific content, hypothesis testing, and manipulation. But other stereotypes are inverted. A procedure seems more like a thought experiment to the extent that it is private and isolated from the external world. Thus our tendency to call a procedure a thought experiment waxes with the richness of its mental imagery, its depth of introspection, and its distance from actuality.

1. *Autonomy*. Whereas the stereotypical experiment teems with apparatuses that bridge abstraction to particulars, the stereotypical thought experiment is free of tangible intermediaries. When we think of the thought experimenter, we picture an intellectual in a pensive pose, lost in abstraction. To the extent that we clutter the setting with laboratory paraphernalia, we lose the Olympian quality of aetherial detachment. This makes us prone to misclassify prop-laden thought experiments as executed experiments.

Conversely, experiments that are restricted to the self-manipulation of mental states are apt to be mislabeled as thought experiments. For example, voluntaristic theories of belief (ones implying that we literally *decide* what to believe) are met with a challenge: try to believe there is a jellyfish on your head. Your failure to form the belief is evidence that belief is involuntary. But the evidence is obtained by executing a psychological experiment, not a thought experiment. The same applies to an experiment that addresses the mystery of why we have emotional reactions to events that we know to be fictional (such as the death of Anna Karenina). Just study your reaction to the supposition that your child's head is transformed into a rat's head.[26] Mere contemplation of this possibility is distressing. The point of the experiment is that the puzzling phenomenon is not peculiar to fiction. The classic empiricists specialized in mental experiment. Berkeley criticized Locke's recipe for abstract ideas by challenging readers to follow it. Hume casts doubt on the concept of self by reporting his inability to summon up a corresponding idea and by asking his audience whether they had any better luck. These experiments are philosophical, but they are not thought experiments.

2. *Mental imagery.*   The grandmaster of *Gedankenexperiment*, Albert Einstein, was deeply impressed by the role of visual images in thinking. His autobiographical remarks and extensive interviews with the gestalt psychologist Max Wertheimer reinforce the natural association between thought experiments and mental imagery.[27] Einstein's Twin Paradox, Shrodinger's Cat, and G. E. Moore's Two Worlds strike us as especially clear instances of thought experiment because they arouse rich mental imagery. We seem to peer into another realm. Hence, they manage to strongly resemble executed experiments while remaining aloof from actual phenomena. In contrast, the St. Petersburg paradox arouses little or no imagery and so tends to be misclassified as "merely a problem."

Since the amount of imagery can vary from person to person, some disputes over whether a procedure is a thought experiment can be traced to mental differences between the parties. S. M. Kossyln persuasively argues that people can solve a problem either through imaginal or propositional processes.[28] Often, the two types of operations proceed in parallel. Since we use whichever answer pops out first, the two processes race against each other. The propositional process is faster when the question does not require much retrieval or inference. Imaginal processes prevail when questions become more complex. For example, when subjects take a true/false test made up of "Lions have fur," "Trucks have big wheels," and "Donkeys have long, furry ears," little imagery is reported for the simple sentences; but progressively more imagery is reported as the adjectives increase. As we become more familiar and practiced, complex tasks become simple because more is accomplished in automatic chunks. Hence, imagery prevails when a question is novel and unprecedented but fades away as we become more skilled at handling the task. Thus, pupils hearing Einstein's thought experiments have rich mental imagery. The teacher, in contrast, has given the relativity lecture umpteen times and so has "nothing in his head." The beginning philosophy student describes his desert island scenario as a thought experiment because he has a vivid mental picture of the situation. His professor describes it as "just a hypothetical question" because familiarity with social isolation scenarios has dampened his mental imagery of them.

As you answer the following questions, consider whether you are performing a thought experiment. How many windows does your house have? Identify the breeds of dog whose ears stick up above the head: German shepherd, beagle, cocker spaniel, fox terrier. Is the square ■ the same shape as the diamond ◆? Most people use a mental tour for the first question, inspection of mental snapshots for the second, and mental rotation for the third. They have a strong impression about how the mental imagery helps them answer: a mental model is formed and manipulated, and the results are noted. The analogy between the mental model and the manipulations then gives them the evidence needed to answer the question.

Given the accuracy of this introspective impression, there has been no thought experiment. Instead, a psychological experiment was *executed*, and its results used to complete a simulation. The 'thought' in 'thought experiment' is an excluder term.[29] It informs us that the experimenter is not using data generated by performance of the procedure; he is relying solely on the experimental design to make his point. Thus, the three cases named, although rich in mental imagery, run against the fundamental contrast between execution and reflection.

3. *Bizarreness.*   A fanciful setting makes a thought experiment easy to recognize as such; for in addition to conveying heady independence from the actual world, bizarreness signals that execution is not intended.

Judith Jarvis Thomson assembles a wide array of hypotheticals in "A Defense of Abortion."[30] But her reputation as a thought experimenter was made on her three most surreal cases. The first is intended to show that the right to control your body sometimes prevails over another's right to life. Imagine that the Society for Music Lovers kidnapped you and hooked you up to a comatose violinist. Their plan is to have you lie in a hospital bed for nine months while your kidneys clean the poisons from his circulatory system. Since only you have the correct blood type for this treatment, unplugging yourself will kill the musician. Are you obligated to let him use your kidneys? Thomson's second strange case cuts against the assumption that a person is entitled to do only what a third party may do. Imagine that you are trapped in a tiny house with a growing baby. The baby is growing real fast—so fast that he will soon burst through the house and walk away a free man. A bystander might say that he cannot intervene and kill the child before it crushes you. But does it follow that you are prohibited from killing the child in self-defense? Lastly, consider Thomson's *people seeds*. If you are not careful, they may waft in through your window and take root. You could prevent this by always keeping your windows shut. But since you like fresh air, you install screens that filter out the human pollen. Unfortunately, your screen has a rare defect and a people seed has taken root in your carpet. Has it acquired a right to stay by the fact that you are partially responsible for its rooting? No, says Thomson, you are free to Hoover it away.

VII. Verbal Disputes over 'Thought Experiment'

I have already trumpeted the importance of relativizing thought experiments to the intentions of their presenters. It is also important to relativize them to a standard of complexity.

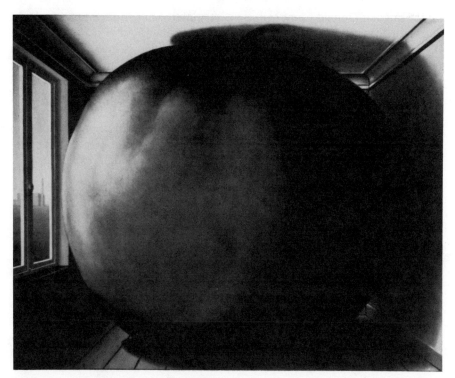

Figure 8.1 The paintings of René Magritte exemplify the same bizarreness of stereotypical thought experiment. *The Listening Room*, like Thomson's growing baby, elicits a feeling of claustrophia by simply exaggerating the size of a familiar object.

Experiments are procedures. Procedures must have a number of steps. Thus, sticking your big toe in a pool does not count as an experiment because this temperature check is too simple. Testing whether a suit fits by picturing yourself wearing it does not count as a thought experiment because there are not enough intermediate steps. Procedures are complex acts. Of course, 'complex' is a relative term; hence, one can make an act count as a procedure by adopting a standard that encourages fine-grained discriminations. For example, the efficiency expert scrutinizes workers for wasteful movements and so would analyze toe dipping into a long sequence of subactions. So relative to ergonomic standards, toe dipping is a procedure. Another case is the computer programmer who must flow-chart all sorts of simple deeds (alphabetizing, face recognition, ordering pizza) as richly structured projects. Therefore in addition to being relative to the speaker's aims, 'thought experiment' needs to be relativized to standards of complexity. This dual relativity makes 'Is x a thought experiment?' a potent source of verbal disputes. Debaters beware!

The foregoing suggests another evolutionary slippery slope. Thought experiment could have evolved *indirectly* from experiment through a kind of hypothetical act that is too simple to count as experiments. A "thought trial" (in the lingo of Donald Campbell) is an imaginative substitute for physical trial and error.[31] Instead of testing a hypothesis by acting it out in real life, you test it "in thought" by relying on your beliefs about what is possible. Here, you trade reliability for speed. However, any trial-and-error process—even one performed in thought—takes a long time when there are lots of alternatives. To cope, we complicate the hypothetical act so that it can rule out a whole class of possibilities. Complexity is also prompted by our frequent failure to have any direct intuition about whether a hypothesis fails; for we can respond to the gap by building a link to another, apparently irrelevant intuition. Once the supposition attains a degree of complexity, it qualifies as a thought experiment.

VIII. Five Theses Recapitulated

First is origin: thought experiments evolved from experiments by attenuation of the execution element and elaboration of the design element. Thus, thought experiments developed from experiments in the way a spider's spinners evolved from legs.

The evolutionary thesis may appear to conflict with chapter 4's emphasis on parallel evolution. I there maintained that the need to overcome similar obstacles, hindrances, and snags led experiment and thought experiment to develop similar precautions and antidotes. Since these coping mechanisms constitute many of the characteristic features of experiment and thought experiment, we should expect a strong resemblance between the two. But doesn't the claim that B evolved from A undercut the claim that A and B underwent *parallel* evolution? Not if we are talking about different stages. Many species began a process of coaccommodation with their parent species after separation.

Second is the objection that the evolutionary thesis is unverifiable. Darwin's theory of the evolution of organisms has been richly documented by fossil records; field studies of systematic variation between isolated groups of organisms; and corroborating geological, physical, and biochemical findings. Thought experiments, on the other hand, do not fossilize. My account places their origin in our preliterate past. A search through ancient texts will reveal early thought experiments, but the technique was matured by the time of the written word. The spread of civilization dashes hope that we can visit isolated groups of preliterates in the hope of pulling "living fossils" from the lips of neophyte thought experimenters.

I agree that the evolutionary thesis about thought experiments is much tougher to verify than ones about morphology. But this is just a special case of the more general problem of verifying evolutionary claims about behavior. Ethnographers grumble about their sparse evidence, but they have sleuthed up "historical pathways" of animal behavior.[32] For instance, paleontologists have used fossils and analogy with contemporary animals to infer much about dinosaur behavior, such as their gait, methods of attack and defence, and daily cycle of activity. Scientific ingenuity may surprise us with details about prehistoric thought experiments. For instance, paleontologists have been intrigued by Nicholos Humphrey's hypothesis that the development of our big brains was driven by spiraling social competition rather than environmental challenges.[33] Humphrey lays particular stress on the advantages that accrue of the ability to replicate another's thinking. So perhaps wily paleontologists will unearth evidence of prehistoric *verstahen*.

In any case, the empirical elusiveness is due to *contingencies*, not an accumulation of ad hoc qualifications that evasively shield the thesis from empirical refutation. Moreover, the problem is one of degree. There are faint empirical consequences because the evolutionary thesis about thought experiments is linked to my second, developmental thesis: children learn to thought-experiment after they learn how to experiment. By studying the order in which children learn the skills, we gain evidence about the order in which their ancestors acquired the skills leading up to thought experiment.

We should compare the evolutionary thesis to the remaining alternatives. The first is the reversal of my thesis. This says that experiment evolved from thought experiment. Ryle's point about the parasitic nature of supposition eliminates this possibility. The other option is to maintain that thought experiment and experiment have independent lineages. This is made unlikely by the striking resemblances between the two. Coincidence is a poor explanation of detailed convergence.

Third is a classification claim: thought experiments are (limiting cases of) experiments. This proposition does not deductively follow from the evolutionary thesis because Fs can evolve from Gs without being Gs. Indeed, there is some positive reason to think the evolutionary process eliminated an essential feature in the case of thought experiments. Recall that an experiment was defined as a procedure for answering or raising a question about the relationship between some variables by varying one of them and tracking any response by the other or others. If the procedure is to vary what one *supposes* to be the case, the experimenter is not really acting on anything. So perhaps 'thought experiment' is a misnomer like 'kangeroo rat' (which is neither a kangeroo nor a rat).

The worry is that we are being hornswoggled by the surface grammar of 'thought experiment'. On the surface, the expression is analogous to 'chemistry experiment' and 'costly experiment'. These expressions refer to subclasses of experiments. But it is naive to assume that all expressions of the form '_____ experiment' select a class of experiments. Behold counterexamples: *pseudo-experiment, dream experiment, fictional experiment, hallucinated experiment*. There is no short way to quell the suspicion that 'thought experiment' is a systematically misleading expression. However, there is a long way—a way best deferred to chapter 9.

Fourth was the justification thesis: thought experiments justify beliefs in the same fashion as unexecuted experiments. I am not saying that thought experiments justify beliefs in the same way as *any* experiment. Thought experiments are a subset of unexecuted experiments and only executed experiments justify beliefs by introducing *new* empirical information. Since, unexecuted

experiments can only work with old information, there is a significant difference in the way thought experiments justify beliefs and the way fully executed
experiments justify beliefs.

Even so, the comparison between thought experiments and executed experiments is fruitful. The fact that executed experiments bring fresh information
does constitute a striking and substantial difference between them and thought
experiments. Nevertheless, the cognitive value of executed experiments is not
exhausted by this new information. Like thought experiments, they have an
organizing effect. Comparing executed experiments with thought experiments
will uncover further functions of experiment that have been overshadowed by
attention to the empirical element. Thus, the comparison is *mutually* illuminating.

Thus, my final point is methodological: thought experiments should be
studied as if they were experiments. This maxim can be applied to the objection
that thought experiments are arguments, rather than experiments. Instead of
replying with a direct denial, the maxim advises a parity thesis: thought
experiments are arguments if and only if experiments are arguments. The
parity thesis switches the burden of proof. The objector's claim amounted to
the assertion that a paragraph expressing a thought experiment is actually an
enthymeme. Since an enthymeme is an abbreviated formulation of an argument (in which the conclusion or premises are left unexpressed), "the enthy-
memetic ploy" gives you plenty of room to maneuver. As Gerald Massey points
out, it has been the instrument of choice for defending antiquated logical
systems from counterexamples.[34] This is how defenders of syllogistic logic
"handled" relational arguments, how pre-Davidsonians dealt with adverbs,
and pre-Masseyians coped with inferences involving multigrade predicates. It
is difficult to prove that a stretch of justificatory discourse does not express an
enthymeme. The parity thesis turns this problem on its head by letting us take
equal interpretative license. With determination and ingenuity, you can make a
thought experiment look like an argument. But determination and ingenuity
will also put in you in a position to make executed experiments look like
arguments. Hence, I doubt that the objector can shoulder his burden of proof
by showing that thought experiments are enthymemes in a way that ordinary
experiments are not.

I am trying to achieve a braking effect with the parity thesis. But I grant the
wisdom of first canvassing reductive possibilities. The attempt to assimilate
thought experiments to arguments is especially popular among philosophers
because they are argument experts and the shortest path to understanding is to
explain what you do not know in terms of what you do know. Many philosophers of science who deny that explanations are arguments would grant that
Hempel was right to try the reduction of explanation to argument and would
also grant that the attempt netted some insights. These nonreductionists would
grant that explanations *suggest* arguments, can sometimes be profitably
pictured as arguments, and can be *supplemented* with arguments. But they
deny that explanations *are* arguments. A similar reception awaits attempts to
reduce to arguments such intellectual instruments as calculation, interview,

mnemonic aids, observation, paradigms, research programs, and similes. Even nonintellectual entities such as jokes and dares can be scientifically illuminating (e.g., Archimedes' boast that he could move the earth if he had a long enough lever and fulcrum). Are they to be reduced to arguments, as well?

The methodological thesis is deductively independent of the other four theses because an F need not be G in order to be most profitably studied as if it were a G. Air is not water, but Evangelista Torrecelli made great strides by supposing himself at the bottom of a sea of air. Likewise, it might be profitable to study thought experiments as if they were experiments even if they are not. But this would be surprising, given the falsehood of my other four claims. The successful theoretical fiction is rare. Usually, things are most successfully studied under true descriptions. No special explanation is needed to explain why people are most successfully studied as animals, the Earth as a planet, and rocks as collections of atoms. Hence, in chapter 9, I shall back the methodological maxim with the taxonomic thesis that thought experiments are experiments.

9

Are Thought Experiments Experiments?

> In a railway carriage: A: what is that basket in the rack for? B: A mongoose; you see, I have a friend who sees snakes. A: But they are imaginary snakes. B: Oh yes, but it is an imaginary mongoose.
>
> A joke, common in 1913, as reported by J. E. Littlewood

Linguistic philosophers will be tempted to "just say *no*." Their training puts them on the prowl for verbal illusions and 'thought experiment' resembles misnomers that have misled naive thinkers. So let's first exonerate 'thought experiment' from the charge of being a systematically misleading expression.

I. Systematically Misleading Expressions

Perhaps the 'thought' in 'thought experiment' signals make-believe status. If a scientist experiments in thought only in the way that he places an apple before his mind's eye, then thought experiments are not experiments. 'Thought experiment' would then be in the bad company of 'Jones is an alleged murderer', 'Smith is a possible or probable Lord Mayor', and 'Robinson is an ostensible, or seeming or mock or sham or bogus hero'. Gilbert Ryle observes that "These suggest what they do not mean, that the subjects named are of a special kind of murderer, or Lord Mayor, or hero, or Member of Parliament. But being an alleged murderer does not entail being a murderer, nor does being a likely Lord Mayor entail being a Lord Mayor."[1] Ryle does not list actual mistakes caused by these expressions, but candidates come to mind. Free speech advocates accuse those who would ban pornography of regarding imaginary sex as a kind of sex on par with marital sex and safe sex. But imaginary sex is no more sex than a birthday suit is a suit. Likewise, imaginary friends, memories, and illnesses are not friends, memories, and illnesses. Wittgenstein hints that a parallel confusion is at play in thought experiment: "Suppose I wanted to justify the choice of dimensions for a bridge which I imagine to be building, by making loading tests on the material of the bridge in my imagination. This

216

would, of course, be to imagine what is called justifying the choice of dimensions for a bridge. But should we also call it justifying an imagined choice of dimensions?"[2] Justifying a proposition in your imagination is not like justifying it in your essay or in court.

'Imaginary' is not a veridical adjective. If V is a veridical adjective of the noun N, then 'x is a V N' is true only if 'x is an N'. For example, 'red' is a veridical adjective of 'pen'. If F is a falsidical adjective of noun N, then 'x is F N' holds only if 'x is not N'. Thus, from 'x is counterfeit money' we may infer that 'x is not money. Adjectives that are neither veridical nor falsidical are neutral adjectives. 'Alleged' is neutral in 'alleged murderer' because some alleged murderers are murderers and some are not. Likewise, 'toy' is neutral in 'toy block', and 'apparent' is neutral in 'apparent cause'. However, 'toy' is falsidical when modifying 'gun', neutral with 'spoon', and veridical with 'maker'.

It is tempting to think that some adjectives are falsidical regardless of the noun modified. But there are many counterexamples to the universal falsidicality of even the most promising adjectives. Sometimes debunking words just intensify a debunking judgment: phony baloney is baloney and false pretenses are pretenses. If a restaurant disguises pigeon meat as pheasant meat, patrons will squawk about the fake meat even though they grant it is meat. 'Fake meat' here is elliptical for 'fake pheasant meat'.

Counterexamples to the claim that there are universally falsidical adjectives show that there is no adjective F such that 'x is an F N' entails 'x is not an N'. There may be *inductive* counterparts of this inference. Given the truth of a statement of the form 'x is an F N', it is probable that 'x is not an N'. But, as with all inductions, the strength of the inference is sensitive to further information. We need to look at the expressions that resemble the one in question.

So what about 'thought experiment'? If 'thought' is a veridical adjective of the noun 'experiment', then it behaves like the adjectives *chemical, animal, controlled, sequential*. If 'thought' is a falsidical adjective in 'thought experiment', then it runs with a ticklish class of adjectives: *pretend, bogus, fictitious, make-believe*. Then again, it may be a neutral adjective; perhaps some thought experiments are experiments and some thought experiments are not experiments.

How can we tell whether 'thought' is operating veridically, falsidically, or neutrally? We might float a statistical argument based on expressions of the form 'thought [Noun]'. But the only instances I have detected are veridical: thought control, thought crime, thought pattern, thought police, thought process, thought transferrence. The thought control of totalitarian countries is indeed the control of thought. And thought crime in Orwell's *1984* antiutopia is the crime of having politically illegitimate thoughts. These cases support the view that 'thought' also operates as a veridical adjective in 'thought experiment'. However, this inductive support is limp because the sample is small.

We can increase the quantity of cases by compromising on the degree of resemblance. 'Mental' is a good case because 'mental experiment' is a synonym of 'thought experiment'. Sometimes 'mental' behaves falsidically: *mental voice, mental rotation, mental masturbation*. Perhaps these cases lead some to infer that 'mental' is universally falsidical. However, the generalization is abetted by a tendency to associate 'mental' with 'subjective' and 'subjective' in turn with 'unreal'. This chain of association colors Thomas Szasz's *Myth of Mental Illness*. In any case, 'mental' does have many veridical constructions: mental addition, mental torture, mental philosophy, mental breakdown. Although 'mental' behaves veridically more often than falsidically, the pattern is too faint to sustain a substantial conclusion about 'mental experiment'.

Given the absence of specific statistical grounds for thinking 'thought experiment' to be nonveridical, one could appeal to broad generalizations. Although no particular adjective is universally falsidical, there could be uses of words that are always falsidical. Many people believe that metaphors are always false because they think that a metaphorical interpretation is only triggered by the blatant falsehood of what is said. If I tell you 'Pigeons are rats with wings', you will be reluctant to ascribe a biological blunder to me and so infer that I am merely drawing attention to the ratlike qualities of pigeons. Someone who believes that metaphorical usage is universally falsidical and who sides with C. Mason Myers's belief that 'thought experiment' is metaphorical, will embrace the following syllogism:

All metaphorical expressions are falsidical.

'Thought experiment' is a metaphorical expression.

'Thought experiment' is falsidical.

However, there are counterexamples to the first premise: 'No man is an island', 'Mort Downey has a big mouth', 'The barn has tons of hay'. Therefore, let's revise the premise to 'Most metaphorical expressions are falsidical'. We must further understand 'most' to govern only *currently* metaphorical expressions. After all, a substantial portion of our vocabulary began as metaphor but is now literal and veridical. But given these qualifications, the second premise is dubious. Perhaps 'thought experiment' *was* metaphorical, but we have no more reason to think that it is now metaphorical than we have for 'double-blind experiment' or 'pilot study'.

II. Comparisons with Lookalikes

Another way to argue that thought experiments are not experiments is to find a synonym that does not denote a real class of experiments.

A. Imaginary Experiments

An obvious candidate is 'imaginary experiment'. Besides its intuitive appeal as a synonym, we can appeal to the fact that many commentators on thought experiments use 'imaginary experiment' and 'thought experiment' interchangeably.[3] Since 'imaginary' in 'imaginary experiment' is a fairly clear case of a 'falsidical' adjective, we have support for the view that the 'thought' in 'thought experiment' is also falsidical. Indeed, the basic argument can be laid out as a syllogism:

A. All thought experiments are imaginary experiments.

No imaginary experiments are experiments.

No thought experiments are experiments.

Since the argument is valid, I can only avoid the conclusion by rejecting a premise. I believe the second premise. Therefore I shall concentrate my refutation on the first premise.

1. *Not all thought experiments are imaginary experiments.* Some thought experiments feature a universe devoid of experimenters. None of the events in such a universe are experiments, because every experiment implies the existence of an experimenter. For instance, G. E. Moore tried to refute hedonism with a comparison between a universe devoid of sentient beings and one containing a lone deluded sadist who obtains pleasure from hallucinated cruelties. Since most people prefer the world without the bastard, they incline toward Moore's conclusion that there is more to good and bad than pleasure and pain. For our purposes, it is important to note that Moore is indirectly asking us to suppose that no experiments have really occurred. Hence, we are not imagining an experiment. Nor does a brain chemist imagine an experiment when asked to suppose that the universe is devoid of phosphorus as part of a thought experiment to show that phosphorus is a precondition of thought.

Most thought experiments are closer to being imaginary observations than imaginary experiments; for they generally present scenarios devoid of manipulative experimenters. Nevertheless, the recalcitrant thought experiments are not all imaginary observations. The typical thought experiment features scenarios free of any commitment to observers. Contrary to Berkeley, we are not imagining that we are *observing* a tree when asked to imagine a tree in an uninhabited forest. We can imagine scenes devoid of observers just as we can paint pictures of scenes devoid of observers.

True, if we conduct a thought experiment by visualization, we necessarily imagine an observation. But an essential property of the visualization need not be an essential property of the thought experiment. When I fasten two boards by nailing them together, nails are essential to the nailing but not to the fastening. (I could have used glue.) Likewise, when a microphysicist conducts a thought experiment by visualizing the atoms, he must picture the atoms as colored. But the coloring is only essential to the visualization, not the thought experiment. A necessary condition for a sufficient condition for X need not be a necessary condition for X.

2. *Not all imaginary experiments are thought experiments.* Note that children play scientists the same way they play cowboys and Indians. They pour

colored liquids from bottle to bottle, seal and shake them, and pontificate about their results. But their imaginary experiment is not a thought experiment. Actors portraying experimenters conduct imaginary experiments but not thought experiments.

3. *Are some thought experiments imaginary experiments?*   No. My motive for pressing the attack is the threat posed by an argument leaner than syllogism A:

B. Some thought experiments are imaginary experiments.

   No imaginary experiments are experiments.

   Some thought experiments are not experiments.

Since I accept the second premise of this valid argument and reject the conclusion, I must argue that 'thought experiment' and 'imaginary experiment' are contrary expressions such as 'bachelor' and 'bigamist'.

Let me first underscore points of agreement. I concede that there are imaginary thought experiments. Just visualize Einstein illustrating a theoretical scenario on a blackboard. But this visualization is not a thought experiment even though it is *of* a thought experiment. Certainly, some thought experiments *feature* imaginary experiments. In some cases, an experiment overlooking an important fact is presented to test the student's knowledge of the fact: "An experiment is performed to determine the lowest temperature at which a certain magma can exist within the earth by melting a sample of rock that has hardened from this magma in a furnace. How meaningful are the results of this experiment?"[4] Methodological discussions often appeal to hypothetical experiments to illustrate design flaws of experiments and verification problems:

> Unfortunately there are logical difficulties in trying to demonstrate that forgetting has occurred due to storage failure. Imagine an experiment in which you tell someone your address, and he or she is able to repeat it back accurately moments later. An hour later you meet that person again, and he or she cannot remember anything about it. You try a few cues such as the first letter of your street name, but none of these is effective. At this point you are tempted to conclude that information about your address is no longer in that person's memory. However, it might be that you have failed to provide the right kinds of retrieval cues. To prove unequivocally that failure to remember is due to storage failure, you would have to examine independently the contents of that person's memory, but this is impossible.[5]

Notice that the imaginary experiment attempts to answer the question of whether the individual remembers while the thought experiment demonstrates that there is no decisive test for forgetting. Other thought experiments aim at showing that a proposition *is* testable. Consider the debate over whether mental images are discursive (like sentences) or pictorial (like pictures). Some critics of the debate complain that the issue is not open to empirical resolution. Jerry Fodor replies that if mental images are pictorial, subjects would have more freedom in accessing information and thus have faster reaction times for certain tasks.[6] Imagine an experiment in which the subject is shown a red triangle and then asked to report what he has seen. If his response is based on an image, he should answer the question 'Was it a triangle?' as fast as he answers the question 'Was it red?'. But if the subject's response is based on the description 'It was a red triangle', the first question should be answered faster. Notice that whereas the imagined experiment is intended to show that images are pictorial rather than discursive, Fodor's thought experiment is intended to demonstrate the metastatement "The hypothesis 'Images are pictorial rather than discursive' is experimentally testable." The thought experiment is conducted by depicting the reaction time experiment but is distinct from it.

The difference between an imaginary experiment and the encompassing thought experiment is especially striking when the subject matter of the two sharply differ. Discussions of ethical research practices contain striking instances:

> A hospital doctor is convinced, on the strength of his clinical experience, that a certain drug, commonly prescribed for asthma but no more effective than others on the market, is liable to induce fatal heart attacks in those already suffering from hypertension. He has voiced his belief to his colleagues, only to be met with near universal scepticism. He therefore runs a clinical trial. Alternate hypertensive asthma sufferers are prescribed the suspect drug over a period of five years. Out of 50 patients in the experimental group, 16 suffer fatal heart attacks within six months of receiving the drug, 25 within a year and 32 within three years. Of the controls, treated with an alternative drug, the corresponding figures are 1, 3 and 8. These results the doctor triumphantly publishes in *The Lancet*. Is the man to be regarded as a benefactor to mankind, or as a monster? Do we condemn him for the murder of literally dozens of patients, or do we congratulate him for having saved many others from being inadvertently killed by their doctors in the same fashion?[7]

The thought experiment is within the field of *ethics*, while the imaginary experiment is a hypothetical *medical* experiment. When we imagine the medical experiment, we are imagining an experiment designed to answer a question about whether a drug induces fatal heart attacks. The thought experiment, on the other hand, is intended to raise ethical questions by showing that some clinical trials resemble a murderous sacrifice of the few to save the many.

Another sign of the distinction between the thought experiment and the imaginary experiment it depicts is the contrast between the variables manipulated in each experiment. It is one thing to imagine physical things being manipulated and another to vary what one supposes. The varied suppositions are real, the varied physical conditions imaginary. Epistemologists suppose that a thirtieth-century delusionologist is experimenting on your brain. This imagined experiment is *part* of the thought experiment, not the thought experiment itself. In other thought experiments, many experiments take place as part of the scenario but none of them is identical to the thought experiment. Usually, the experimental format is a stylistic flourish. Compare 'What would happen to a man who tumbled into a hole that ran through the earth?' with

'What would happen to a man who was pushed (by a curious physicist) into a hole that ran through the earth?'. Both thought experiments succeed in raising the question of whether the man stops as soon as reaches the center, "falls" out the other side of the hole, or oscillates until he reaches equilibrium at the center. Granted, most thought experiments can be *recast* to involve imaginary experiments. Equally, most imaginary experiments can be recast to depict situations that are not experiments. Once we distinguish between the thought experiment and the experiments depicted by the thought experiment, we have good reason to treat 'imaginary experiment' and 'thought experiment' as mutually exclusive.

B. Fictional Experiments

Whereas thought experiments are always presented to make a point, fictional experiments are usually presented for entertainment. The experiments featured in the adventures of Sherlock Holmes, "Bedtime for Bonzo," and "The Harrod Experiment" are designed to intrigue, amuse, and titillate. Other fictional experiments are intended to make a serious point but do not make the point in the way thought experiments do. In 1889 Paul Bourget wrote a novel, *Le disciple*, about a man obsessed with the scientific method. As part of a psychological experiment, he seduces a woman. After she learns of her unwitting contribution to science, she commits suicide. The author intends the fictional experiment to exemplify the moral irresponsibility engendered by scientific naturalism. Nevertheless, it is not a thought experiment.

Real experiments do make an appearance in fiction. For example, seminal experiments on nuclear fission work their way into *The Jesus Factor* as "immigrant events." But these experiments are only fictional in the way the burning of Atlanta is fictional in *Gone with the Wind*. What we normally describe as fictional experiments are completely made-up events. Granted, the depicted experiment may be *prompted* by an historical experiment. (Mary Shelly's *Frankenstein* was inspired by reports about the reactions of corpses to electrical stimulation.) But this causal connection does not constitute reference to a real experiment.

Inspiration flows both ways. Many of the early rocket scientists were motivated by the science fiction of Jules Vernes. Fiction has also inspired thought experiments: "Some science fiction stories refer to a mysterious sister planet of the earth, which shares the same orbit as the earth but is always opposite the sun and hence remains unobserved. Approximating the earth's orbit as a circle, qualitatively explain how a spacecraft could be sent from the earth to the sister planet."[8] The converse influence is evident from the title of R. Wilson's novel *Shrodinger's Cat*.

Literary scenarios can be used to make theoretical points. D. H. Lawrence's "Rocking Horse Winner" has been harnessed by epistemologists as an objection to 'Knowledge implies justification'. The story features a boy who has a mysterious ability to predict the winners of horse races while riding on his rocking horse. The boy's predictions are so reliable that even before his track record is established, the justificationless boy seems to know which horses will win. Had Lawrence told this story with the dominant aim of refuting 'Knowledge implies justification', it would have been a thought experiment. Absent this intention, it is just a story that happens to do the same work as a thought experiment. The similarity of result is exploited by teachers who blend science fiction with straight philosophy to make a more palatable introduction to the subject.[9]

These inadvertently enlightening stories are "natural thought experiments." Natural experiments may be compared to natural dams and natural bomb shelters. These "ready-mades" resemble artifacts but have not been intentionally produced. Since intentional production is a necessary condition for experiment, no natural experiment is an experiment. (Hence, no natural thought experiment is a thought experiment.) Yet typically, natural experiments are nearly as good as genuine experiments. A natural experiment can be defined as a systematic and observable change of an independent variable that is not part of an attempt to raise or answer the observer's question. For instance, the Jewish practice of circumcision was described as a natural experiment refuting Lamarck's theory of acquired traits. Natural experiments play an especially large role in epidemiology and social science because it is immoral or infeasible to perform the scientifically necessary operations on large groups of people. For example, criminologists gain evidence about the deterrent effect of capital punishment by studies of bordering states: they compare, say, Illinois (which has the death penalty) and neighboring Michigan (which does not).

Stories that are told as a means of answering a question will seem like thought experiments to the extent that they appear dedicated to answering that question. But stories seem less like thought experiments to the extent that they also serve other purposes. Consider Bas van Fraassen's story "The Tower and the Shadow" appended to his article "The Pragmatics of Explanation."[10] The philosophical point of the story concerns a symmetry thesis: explanation is just reverse prediction. A classic objection is that we can sometimes predict (but can never explain) the height of a tower by the length of its shadow. Van Fraassen's story presents an ingenious scenario in which the height of a tower is appropriately explained by the length of its shadow. (A man erects a tower on the spot he killed his lover. The height is determined by his bitter wish that with every setting sun, the tower should cast a shadow over the terrace on which he proclaimed his love.) But because the story is told at a leisurely pace with many superfluous details that only enhance the example's literary merit, it is a couple of pages long, rather than a short paragraph. Its extraneous services make us reluctant to describe the story as a thought experiment. Our reluctance grows with the length of the stories because the proportion of theoretically irrelevant detail grows; it becomes less likely that the principal aim of the story was to answer the question. If we are in an expansive mood, we might exclaim that Dostoyevski's *Crime and Punishment* is an extended thought experiment that pits Christianity against utilitarianism. But that is hyperbole. In a cool hour, we grant that the novel can be "read as" a thought experiment but deny that it is literally a thought experiment. The reason is that novels aim at much more than simply making the theoretical point by manipulating the question's variables.

Parallel reservations explain why elaborate, recreational demonstrations do not seem like experiments. When we read how Benjamin Franklin introduced Parisian society to American science by giving an electric shock to a row of monks, he strikes us as an entertainer or promoter, not an experimentalist.

Thought experiments mustn't be too fat or too thin. Recall that thought trials fail to qualify as thought experiments because they have insufficient detail. The cases just presented have the opposite problem: the details should support, rather than engulf, the experimental intention.

C. Mythical Experiments

The most famous experiment in the popular mind is Galileo's dropping two balls from the top of the leaning tower of Pisa in order to check whether heavier objects fall faster than lighter ones. This experiment has been taken as emblematic of the new empiricism. The Aristotelians are cast as dogmatic, a priori bookworms who refused to put their ideas to the empirical test. Galileo is cast as the antiauthoritarian revolutionary. Oliver Lodge recounts the confrontation between the two world views:

> So one morning, before the assembled University, he ascended the famous leaning tower, taking with him a 100 lb. shot and a 1 lb. shot. He balanced them on the edge of the tower, and let them drop together. Together they fell, and together they struck the ground. The simultaneous clang of those two weights sounded the death-knell of the old system of philosophy, and heralded the birth of a new. But was the change sudden? Were his opponents convinced? Not a jot. Though they had seen with their eyes, and heard with their ears, the full light of heaven shining upon them, they went back muttering and discontented to their musty old volumes and their garrets, there to invent occult reasons for denying the validity of the observation.[11]

Lodge's account is typical of a tradition that uses the Pisa experiment to legitimate empiricist views of science. Lodge's choice of the Pisa experiment is unfortunate because it probably never occurred and certainly never took place as a big showdown between the empiricists and rationalists. Galileo himself is not a clear case of an empiricist. Some of his writings assign experimentation a central role in science and so conform to the popular view of Galileo. But Galileo also writes in ways disdainful of experiment. In his discussion of what would happen to a ball dropped from the mast of a moving ship, Galileo has the Aristotelian Simplicio ask Salviati if he ever performed the experiment. Salviati (the advocate of Galileo's view) admits that he has not but denies that this should weaken his confidence. Salviati avers that he must be right regardless of experience: "Without experiments, I am sure that the effect will happen as I tell you, because it must happen that way."[12]

Thought experiments and mythical experiments are unexecuted experiments. But the resemblance stops there. Unlike thought experiments, mythical experiments purport to have been actually executed, to provide fresh information. Thought experiments are intended to persuade by reflection on the experimental plan. Mythical experiments must be associated with a tempting error (or else it isn't *myth*). Thought experiments can be completely free of misconception.

D. Models, Simulations, Reenactments

Sometimes we answer questions about the relationship between Fs and Gs by performing an experiment on Xs and Ys. The cogency of the experiment is based on the analogy between the F–G relationship and the X–Y relationship. We settle for the substitute variables because of the higher cost of manipulating the originals.

Costs are usually cut by downscaling. For instance, aircraft designers put model planes in wind tunnels to answer questions about the aerodynamic properties of large planes. When the properties are of very small things such as molecules, we upscale. However, descaling isn't essential to simulation. War games are as big as real battles—just less destructive.

Simulations are experiments "once removed." Instead of directly experimenting with the properties mentioned in the experimental question, you experiment on ersatz variables and then bridge the gap by analogy. We can tell that analogy is at work because the inference can flow both ways. If an eccentric engineer wants to learn how his model plane handles turbulence in a wind tunnel, he can fly the full-sized plane in rough weather. The procedure would be expensive but the inductive argument would be cogent. The reason why the argument tends to run from model F to genuine F is economic, not logical.

We cut costs even further by only conducting the indirect experiment in thought. Consider how a physics instructor explains melting. He has his students imagine that they are holding hands and then begin jumping randomly. The more violent the jumping, the more difficult it would be to stay linked. The hand-holding students are then compared to molecules of a solid. As energy in the form of heat is applied, the forces binding the increasingly active molecules are overwhelmed. This indirect thought experiment is better than a direct thought experiment because it concerns a more familiar situation; it lets us understand microscopic phenomena in terms of middle-sized objects. Thought simulations also let us understand gigantic phenomena in terms of middle-sized objects:

> Imagine putting a bowling ball on one end of a yardstick and a billiard ball at the other. If you move your hand underneath the stick there will be one point where the two weights will balance. . . . This point is known as the center of mass of the system. If we support the stick as shown and push on the weights, then the entire arrangement will rotate about the center of mass.
> The earth-moon system behaves in a similar way.[13]

Other thought simulations reduce the number of entities to be tracked. Consider Ronald Dworkin's analogy between legal scholars and a group of Dickens scholars engaged in a literary game in which David Copperfield is described as though he were a real person.[14] The legal scholars must analyze much more literature, and the nature of that literature is not as clearly stipulative as the Dickens corpus.

A fourth example is Paul and Tatiana Ehrenfest's illustration of how gas disperses in accordance with the law of increasing entropy. The situation to be modeled features two vessels: A, which is filled with gas, and B, which is empty (Figure 9.1). The two vessels are connected by a pipe with a valve. When we open the valve, we know that the gas will flow from A to B. The problem is to explain how this happens in light of the fact that the motion of the gas particles is random. The probability of a given particle moving from A to B equals the probability of a given particle moving from B to A. How, then, does vessel B net half of the particles beginning in A? The Ehrenfests solve the puzzle by having us imagine two dogs A and B, and n fleas. Dog A begins with all n fleas, which are numbered 1 to n. We also have a pail containing n small balls. A flea will jump to the opposite dog only when its number is called. After the flea jumps, its ball is returned to the pail, which is then thoroughly mixed for the next selection. Now what happens if the selections continue indefinitely? Initially, all of the fleas are on dog A so the flow of fleas is mostly from A to B. The flow is quite lopsided at first, with only an occasional trip from B back to A. But as the number of fleas on B increases, the net flow from A decreases. After a while, there will be no net flow because the number of fleas will be the same on both dogs. This division will only rarely be exact but will be very close to half, with only minor fluctuations. The same behavior should be expected in the case of gas flow. The fleas correspond to the gas particles and the dogs to the vessels. Although the probability of any given particle moving from A to B is the same as a particle moving from B to A, the initially greater number of A particles ensures that the probability of some particle or other moving from A to B is higher than the probability some particle or other moving from B to A.

To answer the question of whether thought simulations are thought experiments, we should first ask whether simulations are experiments. The answer is *no*: a simulation is a two-part demonstration composed of an experiment and an analogical argument. The question addressed by the simulation concerns

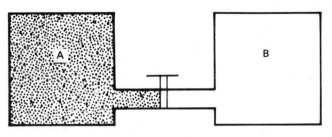

Figure 9.1

variables that are not manipulated in the simulation's experiment. Its component experiment manipulates variables that are analogous to the ones involved in the simulation's question. The job of the simulation's analogical argument is to support this connection. By parity of reasoning, a thought simulation contains a thought experiment but is not itself a thought experiment.

Thought simulations can be easily confused with simulations of mental processes. When I simulate a mental process, my simulation is actual, not hypothetical. The plan is to engage in a mental process that is similar to the mental process in question to give the basis for an analogy between two actual (mental) episodes. Simulations of mental processes are sometimes used by historians of science to convey psychological difficulties in theory acceptance. Rather than present the theory itself, they will ask the reader to imagine an analogous situation that produces similar problems of comprehension. Hanson gets his reader to appreciate the strain Copernicus put on the imaginations of his contemporaries with the following simulation of their mental process:

> Imagine yourself on the outer edge of a merry-go-round, sitting in a swivel chair. The constant rotation of the chair, when compounded with the revolution of the chair around the center of the merry-go-round, would generate—to say the least—complex visual impressions. Those impressions are compatible either with the motion as just described or with the supposition that it is the chair which is absolutely fixed and that all of the visual impressions stem from the motion of the merry-go-round about the chair-as-center and of a like motion of the walls of the building in which it is housed. The actual observations could be accounted for by either hypothesis. But what is easy to visualize in this example was extraordinarily difficult to comprehend in astronomical terms. That it was the earth that rotated and twisted, and revolved around the sun, seemed contrary to experience, common sense, and Scripture. Yet it was this simple alternative hypothesis that, for reasons demanded by astronomy, Copernicus espoused.[15]

Hanson's analogy is between two actual events: the seventeenth-century imaginings of the astronomers evaluating Copernicus' theory and the twentieth-century imaginings of his readers performing the mental exercise concerning the merry-go-round.

Thought simulations are closely related to ersatz visualization. This process presses the imagination into production of substitutes for real examples. Ersatz visualization is sometimes used to convey properties of an actual observation. Consider H. M. Smith's account of the synchronous flashing of Asiatic fireflies:

> Imagine a tree thirty-five to forty feet high thickly covered with small ovate leaves, apparently with a firefly on every leaf and all the fireflies flashing in perfect unison at the rate of about three times in two seconds, the tree being in complete darkness between the flashes. . . . Imagine a tenth of a mile of river front with an unbroken line of . . . trees with fireflies on every leaf flashing in unison, the insects on the trees at the ends of the line acting in perfect unison with those between. Then, if one's imagination is sufficiently vivid, he may form some conception of this amazing spectacle.[16]

Smith's strategy is to communicate his observation by having the reader engage in an exercise of visual imagery. In other cases, the visualization is only intended to provide the analogue of an illustration:

> The principle of counter-shading is simple and may be visualized by picturing a tennis ball resting on a table. The characteristic roundness of the ball will be easily identified by the fact that top of the ball will be lighter because of its greater illumination; and the bottom, partly in shade, will appear considerably darker. Therefore, by counter-shading the top of the ball with a dark tone, which gradates to lighter tones on the bottom, the three-dimensional aspect tends to disappear and the result has to some extent the semblance of a circular plane of a sphere.[17]

Unlike thought simulations, analogical visualizations are not used to answer or raise a question. They are instead devoted to producing substitutes for a phenomenon.

Another relative of the simulation is the reenactment. Thomas Settle dampened doubt that Galileo actually performed the experiments he says he did by reenacting one of them.[18] Settle recreated one of Galileo's inclined plane experiments using a primitive water clock. Since Settle already knew the answers to the questions Galileo's experiment was intended to answer, he was not experimenting in the way Galileo was. Settle was trying to answer a higher-order question about the feasibility of Galileo's experiment. Compare Settle's experiment with expeditions undertaken by Thor Heyerdahl to show the feasibility of long ocean voyages in ancient ships.

Higher-order questions do not preclude lower-order ones. Galileo describes another experiment involving wine and water. The procedure is to first take a glass globe with a small hole at the bottom and fill it with water. Then place the globe (hole downwards) in a bowl of red wine. The result, says Galileo, is that the wine flows into the globe while the water flows into the bowl. Alexandre Koyre took this amazing claim as proof that Galileo never performed the experiment:

> It is, indeed, difficult to put forward an explanation of the astonishing experiment he has just reported; particularly, because, if we repeated it *exactly as described*, we should see the wine rise in the glass globe (filled with water) and water fall into the vessel (full of wine); but we should not see the water and the wine simply replacing each other; we should see the formation of a mixture. . . . [Galileo] had never made the experiment; but, having heard of it, reconstructed it in his imagination, accepting the complete and essential incompatibility of water with wine as an indubitable fact.[19]

However, J. MacLachlan later defended the authenticity of Galileo's report by actually performing the procedure and obtaining the surprising result.[20]

III. The Analogy with Ordinary Experiments

Having contrasted thought experiments with lookalikes, I now nurture the analogy between thought experiments and experiments in general.

A. The Taxonomic Point of the Analogy

The purpose of the analogy is to show that thought experiments *are* experiments. We explain the analogy between whales and mammals by the hypothesis that whales *are* mammals. Ditto for the analogy between tomatos and fruits, alcoholic beverages and drugs, skin and organs. Analogy is the fundamental ground for classification. My comparison between regular experiments and thought experiments should be understood as evidence for a taxonomic thesis.

Actually, I can afford to be less demanding. A reader sympathetic to the themes of chapter 7 may wish to say that the analogy only provides good reasons for *stipulating* that 'experiment' encompasses thought experiments. This reader regards thought experiment as a borderline case of 'experiment' and so does not see any hope of *discovering* an answer to 'Are thought experiments experiments?'. He takes my claim that thought experiment is a *limiting case* of experiment as a sign that I am engaged in semantic legislation. He is willing to listen to my case but urges me to be more forthright and not to engage in the albeit (but illicitly) more persuasive practice of subterranean precisification. The objection that the analogical considerations are only reasons for *stipulation* might be right. Perhaps thought experiment is a borderline case of 'experiment'. Happily, there is no need to resolve the question whether I should describe or stipulate. If the reader believes that I have only given good reasons for stipulating that thought experiments are experiments, then that is good enough. Since analogical considerations are legitimate reasons for both stipulation and describing, the appropriateness of my arguments is not jeopardized by the distinction.

Our judgment as to whether an expression is systematically misleading should be based on the existence of *both* helpful and hurtful associations. Hence, I can grant that 'thought experiment' misleads some particularly vulnerable individuals. For example, the more reverential infer that our heads are private laboratories or that thought experiments are events accessible only to privileged scientific minds. I also agree that the expression 'thought experiment' leads people to believe that thought experiments are experiments. But I take this to be positive transfer, an inference to the truth.

Place 'thought experiment' with 'mind chess'. Mind chess is chess played without the benefit of chess equipment: each side merely states his move. Part of the awe inspired by these matches is deserved; only a skilled chess player with exceptional memory can play well without looking at the board. But part of the awe is due to the misleading associations conjured up by the expression 'mind chess'. People tend to believe that the players have private chess sets in their heads or that there is a Platonic common chess set to which the players have privileged access. But mind chess is just two-sided blindfold chess. In blindfold chess, one player has the handicap of not seeing the board. Mind chess is just chess where both sides have this disadvantage. So although 'mind chess' is a misleading expression, it is not misleading with respect to it being chess.

Some may suspect that my analysis makes 'Are thought experiments experiments?' as moot as 'Is meatless hamburger hamburger?'. Why not just note the relevant similarities and differences between ordinary experiments and thought experiments and leave it at that? This advice is as useless as 'Don't worry about what species of mushroom you pick; just eat the nutritious ones and discard the poisonous ones'. How am I to know what the relevant similarities and difference are? If I bracket out the question of what thought experiments are, I forego standards of relevance. Of course, I will never really live out this neutrality; I will only do implicitly what I disavow explicitly, namely, assume it be this kind of thing rather than that. Critical, frank classification will be displaced by bad faith informality. Little wonder that *lists* of features fail to satisfy our craving for generality! We are not looking for a *summary* of antecedently discovered resemblances and differences. We want the means to organize and *expand* the list, to see how thought experiments systematically relate to neighboring phenomena. The hypothesis that all thought experiments are experiments gives us what we hunger for. It suggests that thought experiments have a place in any complete classification of experiments and hence must be accommodated by theories generating the categories. Finding that place promises to illuminate both thought experiments and experimentation in general.

B. Points of Resemblance

The fruitfulness of the comparison between thought experiments and ordinary experiments is best demonstrated by commencing the harvest. Let us start with points of resemblance. Since ordinary experiments are the better-known phenomenon, insights about their properties should transfer to thought experiments. Some of the transfer is already embedded in the past chapters. Therefore, let us pursue novel points of resemblance.

1. *Parallel failures of naive inductivism.* Observation is popularly pictured as providing neutral data from which partisan theories develop. This data suggest theories whose predictions are then *tested* against further data. If the theory fits, we continue to infer observational and experimental consequences (perhaps with the benefit our new facts) and then see whether these predictions also conform to the real world. If the theory fails to fit, it must be rejected. Thus, observation is commonly thought to provide a perfectly impartial test of theory adequacy.

Historians and philosophers of science have demonstrated that this picture exaggerates the degree to which observation and experiment exert direct and impartial control over theory. Several of the standard themes in courses on philosophy of science target this "naive inductivism"—a doctrine associated with Francis Bacon but only safely ascribable to his brother, Eggsand.

a. *The failure of naive inductivism as applied to ordinary experiment.* The first element of this picture to go is the clarity of confirmation. If theory T has predictive consequence P, then the falsity of P gives us a deductive refutation of T by *modus tollens*. But if P is true, we cannot *deductively* infer the truth of T because a false statement can have true consequences. We can only say that the truth of P suggests an *inductive* argument for the truth of P. Hence, there is an asymmetry between confirmation and disconfirmation: confirmation is inductive and a matter of degree, whereas disconfirmation is deductive and absolute. One might console oneself with the thought that experiment still provides direct *negative* control over theory. We can still have crucial experiments which test the conflicting predictions of rival theories.

But the decisiveness of disconfirmation is challenged by Duhem's thesis. Pierre Duhem pointed out that theories only imply predictions when conjoined with background assumptions. But from (T & B) ⊃ P and the falsity of P, we can only deduce ~(T & B). Logic does not tell us to reject the theory, it only tells us that *either* the theory or the background assumptions is false. Therefore any theory can be consistently defended against an experiment if you are willing to ruthlessly edit background beliefs. The richness of the Duhemian maneuver tends to be masked by the retiring nature of this presupposed realm.

We tend to overlook the reasoning needed to connect the theory with its prediction. Otto von Guericke's seventeenth-century experiment is still used to "demonstrate" that sound is made up of waves. The procedure is to insert a noisemaker, such as a ringing bell, into a bell jar, then pump out air. As the air diminishes, so does the sound, until the bell is finally inaudible. One is supposed to infer that the silence is due to the absence of a medium for the sound waves. But the inaudibility is actually due to the difference in air density.

In addition to overestimating the degree to which experiment controls theory, we overestimate the impartiality of observation: perception is theory-laden because what you see is influenced by what you expect to see. Psychologists have an assortment of standard illustrations. In one study, subjects were instructed to identify flashed playing cards.[21] They did well with normal cards but poorly with doctored ones. These anomalous cards were initially misidentified. For example, a red ace of spades would be identified as a normal card such as the red ace of diamonds or the black ace of spades. Once the subjects realized that some of the cards were abnormal, their performance improved dramatically.

The lesson is that your ability to recognize a viewed F as an F depends on your beliefs about Fs, which are in turn influenced by your theory of Fs. Ptolemaic astronomy made it difficult for ancient astronomers to see meteors as burning stones from outer space; for they believed that the heavens were eternal and made up of etherial material. This view precludes extraterrestrial stones. Hence, although the ancients witnessed meteor showers and found some meteors in the ground, they did not recognize them as meteors. The astronomers saw meteors but did not see *that* they were meteors.

Following on the heels of theory-ladenness is a challenge to the universal priority of observation over theory. The objection is that the influence between observation and theory flows both ways. Sometimes we revise a theory to accommodate an observation, but we will also revise an observational statement in order to accommodate theory. Consider the judgment of the audience at a magic show. Rather than believe their eyes, they continue to believe generalizations such as "Sawing a woman in half kills her." The sort of visual experiences that lead gullible people to say they witnessed a miracle, Martians, or psychokinesis, lead most people to hedge their reports with *seems as if*, *looked like*, and *could pass for*. These hedges illustrate how theory restrains observation. The history of science contains many examples of theorists coaching experimentalists on how to improve the equipment or technique to get more accurate answers. Newton made an enemy of the astronomer John Flamsteed this way.

Theory-ladenness leads to further complexities. If the evidential situation were as simple as naive inductivism depicts, then failures of fit between fact and theory require urgent and drastic remedy. There should be zero tolerance of empirical or conceptual problems. But since the situation is much more complicated, the objections are not nearly as alarming. When things are complicated, malfunctions are to be expected; it would be hasty to blame the theory rather than our equipment, skill, or reasoning. Hence, the wise man waits for the dust to clear before addressing the difficulty. Indeed, he has a reasonable hope of the problem's disappearing without any direct intervention. Chapter 6 of *The Origin of Species* bears the frank title "Difficulties on Theory."[22] In addition to this anomaly toleration, we get a case for intellectual division of labor. Naive inductivism suggests that anyone can check out scientific claims for himself. But science is too complicated for such individualism. When Newton was deriving his laws from the data, he did not turn to Tycho Brahe's mountain of observations; he turned to Kepler's distillation in the form of Kepler's three laws of planetary motion. Likewise, modern scientists have come to rely on each other in the intricate way of citydwellers.

Naive inductivism has also been criticized for its overly restrictive view of the way in which experiment affects theory. Experiment is not limited to hypothesis testing. For instance, John Worrall has shown that some experiments merely prompt theories to greater precision. He notes that ever since Newton proposed the corpuscular theory of light, physicists have tried to detect pressure exerted by light;[23] for the finding would appear to constitute a crucial experiment. If light is made up of particles, one could get a very mobile object to move by merely training a beam of light on it. But if light is made up of waves, then it should remain unperturbed. However, the unfolding dialectic of experiment–analysis–counterexperiment revealed that this reasoning rested on an oversimplified conception of the contending theories. Early failures to detect pressure were eventually followed by apparent successes—and then debunkings of those positive results. At the same time, the empirical consequences of the theories became complicated by the development of different versions of the contending theories. Some corpuscular theories made the particles so small that their pressure would be indetectible. Some wave theories made the ether a solid and so predicted that light pressure would arise as stresses in this medium. Thus, the "crucial experiment" was an illusion generated by a monolithic conception of the two theories of light. Instead of *testing* the two theories, the experiments wound up as invitations to be more specific. Wave theorists accepted the invitations, and their accounts became more elaborate and detailed. Corpuscular theorists declined and their view stagnated.

These refutations of naive inductivism fuel scepticism about experiment. Thomas Hobbes rejected the elaborate experiments of the newly formed Royal Society on the grounds that contrary to advertisement, they were not open to the public or free from interpretation or political influence.[24] Hobbes compared experimentalists with the clergy (because they sought independent authority), lawyers (because they relied on artificial cases), and boys with popguns (because of their fascination with superficially impressive devices such as air pumps). For more recent scepticism, turn to Andrew Pickering's history of physics. Pickering argues that the theory-ladenness of observation prevents experiment from adjudicating between any theories. Experiments are epiphenomenal because "scientific communities tend to reject data that conflict with group commitments, and, obversely, to adjust their experimental techniques and methods to 'tune in' phenomena consistent with those commitments."[25] The global incommensurability that results from this freedom to select and reinterpret is embraced by many sociologists of science. They say that scientific "knowledge" is socially constructed—that it is driven by interests rather than feedback from the external world.

b. *The failure of naive inductivism as applied to thought experiment.*   There is a naive inductivism about thought experiment corresponding to the naive inductivism about regular experiments. Under this view, the intuitions generated by thought experiment are raw data against which theories are to be checked. If the theory conforms to these results, more consequences are inferred and tested by further thought experiment. If the theory conflicts with the results of a thought experiment, it must be rejected.

The refutation of this application of naive inductivism parallels the last. We first introduce the asymmetry of confirmation and disconfirmation to show that the results of thought experiment can only inductively support a theory that predicts them. We then loosen the control refutational thought experiments can have over theories by deploying Duhem's thesis. People can be persuaded of the existence of these background assumptions by a few unmaskings, such as Judith Thomson's:

> Suppose we know absolutely nothing else about blogs than just that all present and past blogs have been purple. Would it not be reasonable to expect that the next blog to come into existence will also be purple? If it seems so to you, this is surely because you are supposing that 'blog' is an English name of a natural kind which you just have not run into as yet, and that, though you do not know of them, there are background facts in virtue of which being a blog has something to do with having this or that colour. Contrast this with: 'I am now going to introduce the word "blog" into English in such a way that as a matter of fact all present and past blogs are purple. Is it reasonable to expect that the next blog to come into existence will also be purple?' Here you would be mad to

bet on it. In fact, the context being what it is, you would do well to bet that the next blog will not be purple.[26]

Theories enjoy the same immunity to strict refutation from thought experiments as they do from experiment. We can always shift the blame to the falsehood of an ancilliary theory, our limited imagination, bias, the trickiness of counterfactuals, or the like.

Imagination is as theory-laden as perception. Even at its crazyist, the mind still gets its shape from the culture that contains it. Thus, the systematic variation in the hallucinations of the American Indian, the medieval European, and the ancient Chinese.

Just as there is a two-way flow of influence from executed experiments to theory, there is a two-way flow from thought experiment to theory. The operation of reflective equilibrium is evident in the hedges used to describe judgments generated by thought experiment. A philosopher will grant that his adversary's thought experiment has some *initial* plausibility or *looks* like a counterexample or has *intuitive* pull. For example, most utilitarians grant that many of the thought experiments marshaled against utilitarianism have some force. That is why they try to "explain away" the hostile intuitions. Other utilitarians capitulate. But their conversion is not simply a matter of dropping theoretical prejudices; the change of mind is instead the result of a complicated assessment of a wide range of evidence.

As in the case of experiment, this complexity leads to anomaly toleration. Unfavorable thought experiments can be conceded as "difficulties" without being rejected. We can also detect some division of labor with thought experiments. An epistemologist out to solve the Gettier problem may rely on another epistemologist's lower-level generalizations formed about the Gettier thought experiments (such as those found in Robert Shope's summary of the Gettier literature).[27] A metaphysician may appeal to Unger's principle that existential beliefs are stronger than property beliefs, while ethicists work with summaries of factors influencing our judgment in thought experiments about self-defense. These well-established low-level generalizations have the same status that "experimental laws" have in physics. Unlike theoretical laws, experimental ones are theory-neutral in the sense that they can be understood or accepted without much reliance on theory.

Normally, the search for theoretical laws drives the development of experimental laws by setting the standards for what counts as interesting phenomena. However, there can be progress at the level of case description without the guidance of an overarching theory. In 1922 prevailing scepticism about Einstein's light quantum hypothesis led Robert Millikan to remark that

> we are in the position of having built a very perfect structure and then knocked out entirely the underpinning without causing the building to fall. It stands complete and apparently well tested, but without any visible means of support. These supports must obviously exist, and the most fascinating problem of modern physics is to find them. Experiment has outrun theory, or, better, guided by erroneous theory, it has discovered relationships which seem to be of

the greatest interest and importance, but the reasons for them are as yet not at all understood.[28]

Thought experiments are also conducted in this "atheoretical," Baconian fashion. Consider the experimental islands that continue to grow after the refutation of their founding principles (sufficient reason, plentitude, behaviorism).

Nonetheless, the larger-scale failure of naive inductivism about thought experiments spawns sceptical elaborations paralleling those common in the literature on experiment. Thought experiments are alleged to articulate the intuitions of an unrepresentative group such as Oxford dons, upper-class white males, or those who fund theorists. The theory-ladenness of imagination is used to smirk away thought experiments as epiphenomenal, circular, and self-deceptive. The constancy of their verdicts is explained as a kind of collective wish fulfillment, as components of a social fantasy system that is driven by desire rather than truth.

2. *Tinkering.*   Experiments are often inspired by serendipitous observations. Count Rumford's famous experiments supporting the kinetic theory of heat followed his observation of heat's being generated by the boring of cannon in a military workshop in Munich. According to the caloric theory, heat was a material substance added to matter. Heating an object consisted of adding some of this substance—called "caloric." The boring of canon seemed to disprove this theory. For the amount of heat did not dissipate as the boring continued. But one apparent counterexample is rarely sufficient to refute a theory. Proponents of the theory typically have several responses. One reply to Rumford's case was that the metal shavings do not hold caloric as well as massive metal. Cutting metal may therefore release heat. Rumford's rejoinder was a test featuring a blunt drill that produced fewer metal shavings. If the reply were correct, the blunt boring should produce less heat. But instead even greater heat was produced.

Rumford's experiment illustrates a general pattern. First, an apparent counterexample to a controversial theory is observed. Second, replies are anticipated. Third, an artificial variation of the natural counterexample is produced as a rejoinder. Thought experiments mature in the same dialectical pattern. For example, some ethicists noted that utilitarianism seems disproven by situations in which it is immoral to kill but permissible to let die. Utilitarians define 'right' as the maximization of goodness (or the minimization of badness), so that if the consequences are the same, there is no morally relevant difference between the two acts. Since death is the result in both killing and letting die, the distinction should be irrelevant. But there are many situations in which the distinction looks quite germane. Utilitarians reply that the distinction *seems* relevant because it covaries with factors that are relevant: intent, motive, certainty of result, ease of avoidance, number of omitters. The antiutilitarian rejoinder is a thought experiment that eliminates these interfering factors. For example, the "survival lottery" features a sophisticated organ donation scheme in which people consent to be part of a lottery.[29] "Winners" of the lottery are painlessly killed so that their organs can be redistributed among other lottery participants who need the organs. Although each participant has a tiny chance of being killed by the lottery, they are much more likely to be saved by it (since many crucial organs can be transplanted from one dead body). Opponents of utilitarianism say that most people would disapprove of the lottery even though it increases the average lifetimes of the participants. Since the scenario eliminates all the factors except the killing/letting die distinction, they claim that the survival lottery refutes utilitarianism. The same spiral of sophistication marks the history of Quinton's two-space myth.

One of the perils with excessive tinkering is that the situation will become too complicated to grasp firmly and so will fail to stimulate a strong judgment. This is often the point behind the complaint that a thought experiment is "too artificial." Effective tinkering does not take the form of accumulated amendments to the original case. Usually, it proceeds by a novel twist that elegantly eliminates the extraneous variables in one swoop.

A related observation to experiment sequence begins with the discovery of an anomaly. Benjamin Franklin's kite-flying experiment was inspired by a tradesman's report of knives becoming magnetized after they were struck by lightning. The twitching of Galviani's dead frog and the deflection of Oersted's magnetic needle began as unanticipated effects. Later, the effect was deliberately caused as part of question-raising experiments and then incorporated into more elaborate experiments. Anomalies also inspire thought experiments. Darwin eased his burden of proof by imaginatively blending and magnifying natural anomalies. For example, he softens the problem of furnishing transitional species by extending the strangeness of flying fish and land-adapted sea life. Suppose that the current species of flying fish had long ago evolved from mere gliders into perfectly winged animals and came to dominate the skies: "Who would ever imagined that in an early transitional state they had been inhabitants of the open ocean, and had used their incipient organs of flight exclusively, as far as we know, to escape being devoured by other fish."[30]

Thought experiments can themselves present anomalies and so spawn further thought experiments. Sometimes the *conjunction* of two individually nonanomalous thought experiments constitutes the anomaly. For example, Judith Jarvis Thomson motivates a string of hypotheticals by asking how we can reconcile our intuitive reactions to a certain pair of thought experiments that arose independently.[31] The first, due to Philippa Foot, features the driver of a trolley whose brakes have just failed. Ahead are five people on the track. The bank is too steep for them to escape. The good news is that the driver can switch the trolley to a sidetrack. The bad news is that one person is servicing this track and would be equally doomed by the steep bank. Is it permissible for the trolley driver to switch tracks? Now consider the transplant case. Five patients need five different organs that must come from a donor with a rare blood type. The good news is that their surgeon has found such an individual. The bad news is that the individual is healthy and will outlive all of the five dying patients if nature takes it course. Is it permissible for the surgeon to kill the one person in order to save the five others? Most who answered *yes* to the trolley question answer *no* to the transplant question. But what is the relevant difference between the cases?

3. *Standard formats.* Textbooks list generic designs: longitudinal studies, randomized clinical trials, Latin square designs. These "canned designs" serve many functions. They provide immunity from historically troublesome objections, ease the digestibility of the results by providing a familiar setting for the reader, and facilitate comparison with other experiments in the same format. One's choice of design also conveys "experimental implicatures," because the decision suggests which secondary variables are most likely to interfere with the effect, how previous experiments may have been flawed, how much evidence is needed for the acceptance of the thesis, and so on.

The most concrete level of classification centers on "paradigms," in which one experiment serves as an exemplar for a whole tradition of subsequent experiments. Aristotle's embryological experiment (described in chapter 8) and the four-card selection task (chapter 3) are capital instances. Gettier's counterexample and Darwin's hypothetical history of the eye are good examples of thought experiment paradigms.

The experiments following a paradigm stick to the same topic. Other experiments only serve as exemplars in that they neatly solve a technical problem. For example, a Catholic who wishes to apply the doctrine of double effect must discover whether the effect of an action is intended or merely foreseen. The standard solution is to ask whether you would still perform the action even if the effect could be magically prevented. Journalists try to figure out whether an event is really news by asking how people would have reacted to a prediction of that event. Alternatively, they appeal to the reaction of a Rip van Winkle. Theorists test whether a property is subjective by trying to picture the property in an uninhabited world—hence G. E. Moore's Two World experiment.

Classifications of regular experimental formats also work for thought experiments. For instance, there are experiments corresponding to all five of Mill's methods. The method of difference, for example, establishes a causal relationship by checking whether removal of the suspected cause eliminates the effect; we test whether a dead battery is the cause of a dead flashlight by seeing whether a new battery revives it. The method is also used when we only have a vague idea of what the effect might be. This exploratory use is illustrated by early physiological experiments that studied the functions of organs by removing them from test animals and waiting for something to go wrong. The ongoing mystery about the function of sleep is studied by observing how sleep-deprived animals and people behave. Loyalty to the method persists when ordinary experimentation is infeasible. Note how Ernst Mayr approaches the question why there are species:

> The fact that the organic world is organized into species seems so fundamental that one usually forgets to ask why there are species, what their meaning is in the scheme of things. There is no better way of answering these questions than

to try to conceive of a world without species. Let us think, for instance, of a world in which there are only individuals, all belonging to a single inter-breeding community. . . . Now let us assume that one of [the genetic] recom-binations is particularly well adapted for one of the available niches. It is prosperous in this niche, but when the time for mating comes, this superior genotype will inevitably be broken up. There is no mechanism that would prevent such a destruction of superior gene combinations, and there is, there-fore no possibility of the gradual improvement of gene combinations.[32]

Thus, supposing that species do not exist leads us to appreciate their role as preservatives: limiting genetic recombination allows adaptive diversity to pro-ceed without the destruction of the basic gene complex.

A second example involves some armchair moral psychology that can be found in Plato's *Republic* (Book II, 360C). To show that just action is only motivated by fear of punishment, Glaucon first alludes to the tale of the ring of Gyges, which concerns a ring with the power to make its bearer invisible.

Now suppose there were two such magic rings, and one were given to the just man, the other to the unjust. No one, it is commonly believed, would have such iron strength of mind as to stand fast in doing right or keep his hands off other men's goods, when he could go to the marketplace and fearlessly help himself to anything he wanted, enter houses and sleep with any woman he chose, set prisoners free and kill men with the powers of a god. He would behave no better than the other; both would take the same course. Surely this would be strong proof that men do right only under compulsion.

In a similar spirit we find ethicists wondering what would happen if no one kept their promises, linguists wondering how a language free of vagueness would work, and political theorists concentrating on the "state of nature" to arrive at the function of government.

Sometimes the point of proposing a thought experiment *format*, rather than particular thought experiments is that one wishes to rule out a whole class of claims. For example, some say that a statement is meaningful if and only if it can be at least partially confirmed. Ed Erwin objects that this is too broad, since all statements can be partially confirmed.[33] His reason is that for any statement you can imagine an authoritative computer supporting it and thereby confirming it. Here, the appeal is to the feasibility of thought experiments, not to any particular thought experiment. Erwin is giving a format, not an instance. A related technique is the internal visual display. This is used to show how any given proposition can acquire evidence. For example, some philosophers doubt that we could ever have evidence that the external world does not exist. The rejoinder has you suppose that little messages begin to appear in your subjec-tive visual field. Over time, you discovered that these messages were always correct even when they contain surprising predictions. One day 'The external world does not exist' is displayed. This is evidence for the highly philosophical thesis because the display has an excellent record.

Scale-effect thought experiments make the independent variable take on a peculiar value (0, 1/2, and ∞ are favorites) in order to show that an apparent qualitative difference is actually quantitative. Another pattern is to run two values in opposite directions in order to maximize the effect. Mach uses this technique to show that the animal/plant distinction rests largely on our subjective time scale. Use your imagination to further retard the slow movements of a chameleon and to hasten the grasping motions of lianas: "the observer will find the difference becoming blurred."[34] Other cases feature progressions of hypotheticals. Galileo explained how particles float in less dense media such as air and water by having us imagine the division of a cube. He first divides the cube into eight smaller cubes with three cuts, leaving the weight the same but doubling the cross section and thereby the resistance. Galileo then repeats the process over and over so that the resistance-to-weight ratio becomes enormous. Newton explained satellites by imagining a ball fired from a mountain top (Figure 9.2). The faster it is projected, the further it travels before landing. With enough velocity, it travels around the earth. With even greater velocity, it goes round and round indefinitely.

4. *Unanticipated uses.*   An experiment intended to address one question may be adapted to answer another question. Louis Pasteur's famous sealed flask experiment was not originally intended to answer the question whether life could originate by spontaneous generation. The initial point of the experiment was to show that putrefaction was a biological, rather than a chemical, process—in particular, that microorganisms are essential to the recycling of materials.

Thought experiments are also subject to diversion. Galileo's cube-cutting thought experiment is used to explain why crushed ice cools beverages quicker than big ice cubes. Ice cools liquids by absorbing their heat. The rate at which they absorb rises with the ratio of surface area to volume. So dividing the cube makes the ice cool the beverage faster. Also recall how Russell's five-minute hypothesis has been used and reused by different philosophers to make distinct epistemological, semantic, and ethical points.

Sometimes experiments will answer their question but not in the way they were designed to. The aim of the Hawthorne study starting in 1927 was to learn of factors that increase worker productivity.[35] The subjects were six women whose task was to assemble telephone relays. The researchers varied their illumination, temperature, rest periods, hours of work, wage rates, and so on. They were amazed to find that all changes were followed by improved performance. The researchers concluded that productivity was being increased by social factors (the attention lavished on the women by their observers, the development of esprit de corps, etc.).

Gilbert Harman's social knowledge case is an example of a thought experiment displaying this sort of twist. Harman wanted to learn about the conditions under which the justified true belief analysis of knowledge fails. He explained the Gettier counterexamples with principle P: "Reasoning that essentially involves false conclusions, intermediate or final, cannot give one knowledge."[36] Since some epistemologists had advanced Gettier cases that did not seem to involve any inference, Harman defended P by ferreting out *hidden*

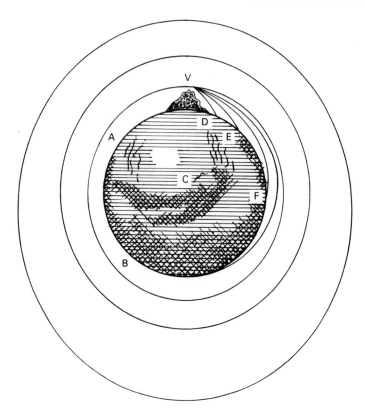

Figure 9.2

false steps. After showing how loyalty to principle P seemed to work out for known cases, Harman went on to invent even tougher cases. In one, government officials go on TV to falsely deny that their leader has been assassinated. However, since a reporter managed to phone in the truth prior to the announcement, some newspapers carry the real story. Jill has not been watching television but has read one of the newspapers. Hence, she has justified true belief that the leader was assassinated. But does Jill know? It seems that she is not saved by ignorance; Jill fails to know because of evidence she does not possess. Where, then, is the hidden false inference? Harman gets it by invoking another principle, Q: "One may infer a conclusion only if one also infers there is no undermining evidence one does not possess."[37] Thus, Jill's false step was the self-referential one of assuming there was no undermining evidence in the community. Nearly all epistemologists rejected this idea. They maintained that Harman had stumbled upon a new kind of counterexample to JTB, one that did not turn on the thinker's making a false step. They took Harman's case to show that proximity to misleading evidence can undermine knowledge. This opened up a new social dimension to knowledge.

C. Points of Difference

The key point of difference between ordinary experiments and thought experiments is that most ordinary experiments are executed and thereby provide fresh information. Since most of the differences between ordinary experiments and thought experiments emanate from this key difference, their delineation may constitute double-counting.

1. *Some ordinary experiments require more than one experimenter.* Every experiment has an experimenter. Of course, there might be more than one experimenter. Physics journals feature articles with twenty authors. As the individual's role in the experiment becomes more specialized, it becomes increasingly tempting to view the experiment as a collective act. Indeed, Peter Galison argues that the multimillion-dollar, accelerator-based experiment in high-energy physics is the enterprise of a complex community, rather than an individual.[38]

More striking, for our purposes, is that some experiments require more than one investigator to control bias:

> Two investigators may have to be involved when therapy cannot be provided blind or when a treatment affects a measurement in a constant manner. One investigator provides the treatment or takes the measurement and the second assesses the end points of the trial. An example is provided by a trial of dietary advice where one investigator gives therapeutic advice and a second assesses the outcome in terms of weight or blood pressure without knowing the treatment allocation.[39]

Some thought experiments are also conducted collectively to minimize expectancy bias. Since many subjects try to please or frustrate the experimenter, particularly cautious philosophers arrange to have their thought experiments run secondhand, that is, they ask someone else to pose the thought experiment to third parties.

Are there essentially collective thought experiments? The discussion of Quinton's Two-Space myth in chapter 6 illustrates how a thought experiment can be *developed* by more than one thinker. However, following the sequence of suppositions is a private matter. Even if we talk our way through the thought experiment, we could get diverging results. This possibility of divergence shows that there are two instances of the same kind of thought experiment. In contrast, nearly all ordinary experiments *can* be jointly executed; and some of them must be.

Since some thought experiments are complexes of smaller thought experiments, it would be possible to have a joint thought experiment in the sense that one person performs one module while other thought experimenters do the remaining parts. But the fact that the basic thought experiments cannot be jointly executed still constitutes an asymmetry.

2. *Thought experiments do not precipitate role conflict.* The social division between thinker and laborer created a role conflict for experimenters; for

insofar as scientists involved themselves with the toil and muck of empirical verification, they fell from their station as gentlemen. The conflict has been resolved in three ways. The historical favorite is just to refrain from executing experiments. Thus, the ancient Greeks and the medievals are notorious for endless debate on issues that could have been resolved by elementary experiments. In the seventeenth century, Robert Boyle developed a second solution: the thinker designs the experiment while his underlings carry out his instructions.[40] Thus, were born the scientific technicians who work quietly in the background without qualifying as true scientists. The third resolution of the thinker/laborer role conflict is to replicate the social distinction within the scientific community. Thus, theoreticians and experimentalists are both recognized as scientists, but the theoreticians have greater status.[41]

Happily, thought experiment never created a clash between the role of thinker and laborer. It has always been considered a cerebral affair. Thus, thought experiment has proceeded unhobbled from the earliest stages of the history of thought. There has never been the temptation to delegate part of it to underlings as "dirty work." Nor is it the locus of a class division within professional science.

3. *Luck has no substantial role in thought experiment.* Experimenters have lots of luck, most of it bad. Sometimes the experiment fails because attempts to change the independent variable fail. Sometimes the independent variable is manipulated, but an extraneous factor distorts its effect on the dependent variable. On yet other occasions, the relationship eludes tracking. Michael Faraday's early work on the relationship between magnetism and electricity was difficult because his equipment was not sensitive to the small effects he was seeking. Faraday was also misled by his underestimation of the speed of effect. Others had the same problem. In 1825 Jean Daniel Colladon performed an experiment that probably achieved his goal of producing electricity by magnetism—but he did not notice. To ensure that his powerful magnet would not affect his galvanometer, he placed the galvanometer in another room. This necessitated a time-consuming walk between rooms to check for an effect. Since the effect was instantaneous and fleeting, Colladon was always too late to see it.

Since thought experiments lack an execution element, nothing can go wrong with that element. The only hazards are at the level of contemplation. I might ruin a thought experiment through distraction, a chance biasing factor, or some sort of internal breakdown; but the scope for bad luck is quite narrow. Ordinary experiments have large scope for luck factors.

4. *Only ordinary experiments involve apparatus.* Another consequence of the "All talk, no action" nature of thought experiments is that they do not need instruments. Most ordinary experiments, in contrast, extort knowledge from nature "with levers and screws." Thought experimenters may use heuristic aids such as toy cars and computers, but they do not use any apparatus that plays a role in causally linking the independent and dependent variables.

This asymmetry leads to others. Some ordinary experiments are designed to enhance experimental apparatus. These "technically good" experiments improve established apparatus or introduce new devices. There is no exact counterpart to these experiments. The closest analogues are thought experiments that produce greater accuracy or precision by introducing novel fictitious devices such as the autocerebreoscope, black boxes, and teletransporters. They resemble experimental apparatus insofar as they are stock solutions to recurrent problems. Desert islands give the thought experimenter the isolation needed to bracket out concerns about social effects; hypothetical hypnotists provide a quick means for implanting psychological states; and imaginary drugs, potions, and spells are the lazy man's sources of amazing effects for his thought experiments.

New devices often cause a wave of new thought experiments that exploit the innovation. For instance, the Turing machine became the workhorse of automata studies. Hilary Putnam's twin earth scheme recently enabled philosophers to separate the contribution the mind makes to meaning from the contribution made by one's environment.[42] In Putnam's original thought experiment, we imagine that there is a distant planet, Twin Earth, that is identical to our planet except it contains XYZ instead of H_2O. XYZ only differs from H_2O at a subtle chemical level: it is a colorless, drinkable liquid that supports aquatic life and extinquishes fire. Given this indistinquishability, Twin Earthers have the same beliefs and desires as we have when we use the word *water*. Yet, in their usage, 'water' designates XYZ. Thus, the meaning of a word is not completely determined by what is in the head. Many philosophers have been intrigued by the extent to which semantics fails to reduce to psychology and so have surged forth with a title wave of Twin Earth variants that expose environmental contributions.

Thought experiment devices are as much a part of a philosopher's education as laboratory instrumentation is a part of the physicist's. Unlike the physicist's apparatus, the philosopher's is fictional. This fictionality makes the denizens of thought experiments resemble literary devices—such as freak accidents (to eliminate characters), evil twins (to explain away damning evidence), and dream sequences (to undo previous narration). Indeed, just as beginners in literature need to read a lot before they become they become competent literary critics, beginners need to be immersed in thought experiments before they can fluently appraise them. The same goes for neophyte experimentalists.

5. *Thought experiments are morally trivial.* Executed experiments involve action and so can greatly help or harm people and animals. Thought experiments proceed by reflection on an experimental plan and so are only open to, at most, *minor* moral praise or blame. Squeamish visualizers are sickened by the blood-and-guts scenarios favored by contemporary ethicists. But this is no more momentous than the revulsion experienced by readers of gory novels.

Nonetheless, thought experiments can have long-term effects because they do alter opinions and thereby influence action. Anscombe argues that thought experiments that are designed to poke holes in absolute prohibitions, such as "Never knowingly punish the innocent," are corrosive:

> The point of considering hypothetical situations, perhaps very improbable ones, *seems* to be to elicit from yourself or someone else a hypothetical decision to do something of a bad kind. I don't doubt this has the effect of predisposing people—who will never get into the situations for which they have made hypothetical choices—to consent to similar bad actions, or to praise and flatter those who do them, so long as their crowd does so too, when the desperate circumstances imagined don't hold at all.[43]

Those who regard absolute prohibitions as merely useful fictions also fear corruption. Michael Levin, an emotivist, argues that moral behavior is the outcome of training rather than reflection, so that children must be habituated into good behavior with the help of an easily followed ethical code.[44] Thought experiments that reveal conflicts and counterexamples suggest that conventional morality is not rationally binding and so provide yet another excuse for laziness and cowardice. Moral educators are more upbeat. They emphasize the positive influence hypothetical good decisions exert on real decisions.

Experimental evidence only bears obliquely on the topic. There are studies suggesting that children imitate the violence they see on television, that viewing pornography increases tolerance of rape, and that theoretical discussion of suicide increases the chance of suicide. But these studies only provide faint support for their own theses and so are even fainter evidence about the impact of thought experiments. Although one can imagine evidence that would clinch the issue, there is no conspicuous line of causation running from thought experiment to social behavior. The best explanation of why thought experiments rarely draw moral praise or blame is that as a matter of empirical fact, their consequences are much harder to predict than personal deeds or institutional actions.

This epistemological problem is prefigured in the debate over free speech. How can the costs of censorship be justified by unpredictable benefits of suppression? One strategy is to overwhelm the issue of probability by emphasizing the large stakes. Leo Szilard managed to suppress early research on nuclear fission with the warning that it could help the Nazis acquire the first atomic weapon. Certainly, the thought experiments that later took place in Los Alamos qualified as war secrets!

The second strategy for constraining speech softens the problem of prediction by concentrating on short-term effects. You do not need to be a wizard of social prognostication to be reasonably sure that embarassment, offense, and even panic will follow from certain utterances. This predictability carries over to thought experiment. One cannot take refuge in the hypothetical nature of the thought experiment's scenario, for conditionals and fiction can defame by implication and innuendo. Moreover, thought experiments go beyond their imaginary scenarios to support unconditional facts. That's what chapter 6 was all about.

Once short-term links are established, censors can lengthen the causal chain. Thus, some nonslanderous thought experiments have been condemned for the slander they may cause. This is the basis for David Lewis's complaint about the mischief caused by thought experiments concerning conditionally intended nuclear obliteration. He grants that philosophers writing on the issue have carefully separated their idealized scenarios from the real-life decision problems faced by our leaders. And Lewis grants that philosophers can contribute to our understanding of other issues raised by nuclear deterrence.

> But I wish we could leave the paradox of deterrence out of it. I am afraid that because paradoxical deterrence is philosophically fascinating, it will be much discussed; and because it is much discussed, it will be mistaken for reality. We don't need a bad reason to be discontented with our predicament and with our country's policies. After all, we have plenty of good reasons. And we don't need a picture of nuclear deterrence that implicitly slanders many decent patriots in the American armed forces and the White House.[45]

Since slander can have immediate bad consequences, the problem of long-range social forecasting is avoided. However, the scale of the immediate evil is normally too small to constitute a point of resemblance with regular experiments. In addition, the harmfulness is easily minimized by minor editing and audience targeting. In contrast, much of the harm associated with executed experiments is grave and unavoidable. The two strategies of focusing on great harms and short-term effects do yield a handful of thought experiments that merit censorship. But the existence of marginal cases does not overturn the major trend. Once in a while, we can predict that tapping a man on the shoulder will harm him as much as a punch in the nose; but that fails to overturn the general moral difference between punches and taps.

The epistemological problem is completely bypassed by nonconsequentialist condemnations of thought experiments. Perhaps a bad thought can constitute an intrinsically immoral act. A Kantian who disapproves of sexual fantasies (on the grounds that the fantasizer uses the woman as a means) may also disapprove of thought experimenters who cast unconsenting people into demeaning roles.

Another basis for condemnation is that the thought experiment *reveals* a vice such as bigotry. An illustration can be drawn from the early debate over evolution. A key difficulty of Darwin's theory was the absence of an adequate theory of inheritance. Although biologists knew that blending theories were inadequate (brown-eyed parents produce some blue-eyed babies), they had nothing better than qualified versions of this commonsense theory of inheritance. This set the background for an influential thought experiment by Fleeming Jenkin. He had the reader suppose that a white man became shipwrecked on a Negro island: "He would kill a great many blacks in the struggle for existence; he would have a great many wives and children. . . . But can anyone believe that the whole island will gradually acquire a white, or even a yellow population?"[46] To appreciate Jenkin's point, we have to go along with his assumption that a white man is superior to a Negro. We can then see that Jenkin was raising a legitimate objection that Darwin's theory: rare but useful traits will not spread through a large population because they are diluted into insignificance. But many people condemn the way that Jenkin makes his point. They think that the ease with which he assumes the superiority of the white man *reveals* racism. Maybe so. But we should bear in mind that the *sign* of a bad thing need not be itself bad.

Many people are reluctant to allow that thought experiments could be immoral because the possibility summons up images of "thought crime" as depicted by Orwell in *1984*. They feel that the realm of private thought is a morally free zone. Morality only applies when these thoughts are put into action. *Disseminating* the thought experiment could be blameworthy because others might be offended or there might be bad effects. But as long as you prevent the thought experiment from leaking out into the external world, you are beyond the reach of morality: "Suppositions don't kill people: people kill people."

This vision of mental space as being a sanctuary from morality could attract a metaphysical defense. During his Platonist phase, Russell stressed that the realm of universals is ethically inert; everything of moral significance is confined to the realm of particulars. However, the isolationist intuition is more plausibly traced to contingent facts of our psychology and to the type of media actually used to convey thought experiments. I have already stressed the psychological aspect: bad thoughts are difficult for outsiders to detect, difficult for the thinker to control, and lack clear causal connections to the production of censurable amounts of evil. If we lived in a psychokinetic world in which thought experiments depicting injuries to neighbors caused those neighbors to be injured, then the ethics of thought experiment would be a live issue. The point about media can be appreciated from an ethical problem raised in filming *The Andromeda Strain*. According to the novel, flashings of a red warning light induce an epileptic fit in the heroine. If the director used the correct frequency, he would have triggered fits amongst epileptic moviegoers. Hence, he sacrificed scientific accuracy for medical safety by using the wrong frequency. Notice that the ethical problem only arose when the fiction was switched from the medium of a novel to that of a movie. Since computer graphics specialists and producers of documentaries have already converted many thought experiments to new media, there is a chance that some future thought experiments will acquire a significant moral dimension. But given their present mode of presentation, thought experiments are morally trivial—of no more moral interest than poems, puzzles, and mathematical brain teasers.

7. *All thought experiments are fictionally incomplete.*   What was Tom Sawyer's blood type? Had Mark Twain depicted him as type O, he would have been O. But as the story now stands, the matter is indeterminate. This indeterminacy is deeper than that of Nero's blood type. We know that Nero had a specific blood type—we just cannot learn what it was. But our feeling is that Tom Sawyer lacks a specific blood type—that there is nothing to know. Unlike real things, fictional characters are incomplete. Since thought experiments are a species of fiction, it will always be possible to raise questions that cannot be

answered because of the story's incompleteness. Hence, we have an asymmetry: all executed experiments are metaphysically determinate; all thought experiments have gaps.

This contrast would be lethal if thought experiments *inevitably* left out relevant facts. Happily, underdescribed thought experiments are easy to repair. When the audience asks for more data, the designer of the thought experiment need only stipulate extra facts. Although there will always be some holes in a thought experiment, no particular hole is unfillable. The thought experimenter can plug in answers as fast as the critic can ask them.

Nevertheless, the inevitability of leftover holes does disturb some philosophers. Marilyn Friedman has alleged that this incompleteness makes moral thought experiments unstable.[47] The moral verdict that is supported by one stage of plugging can always be undermined by a subsequent stage of specification that favors the opposite verdict. This pattern of endless reversal looms for any thought experiment that fails to strictly entail the verdict. The only remedy suggested by Friedman is to exchange the thought experiment for a full-blown work of fiction. But this is like advising a cook to throw more water on a grease fire after learning that a little will not do. No matter how detailed the story gets, there will always be an opportunity for a surprising continuation. Indeed, there is a whole genre of mystery stories, such as the movie version of *Death Trap*, that specializes in long sequences of dramatic reversals.

Notice that the same sceptical considerations can be raised for inductive reasoning in general. If I conclude that Saddam Hussein is right-handed on the statistical grounds that 90% of men are right-handed, you can undermine my argument by pointing out that his wife reports that he is left-handed. Another person can then top this with evidence that Saddam's wife is a pathological liar. This addition can itself be topped. These potential reversals do not show that inductive reasoning is *viciously* unstable. Since the probability of the conclusion is relative to the premise set, enlarging our assumptions leads to new probabilities without any inconsistency. The mere possibility that new evidence will reverse our current judgment does not stop us from reasoning inductively—we have to consider the probability of such augmentation. In the case of thought experiment, we have advantage of being able to *stipulate* that "all other things are equal." Indeed, all thought experiments carry a tacit ceteris paribus clause because it would be a conversational misdeed to leave out a relevant fact, specifically, a violation of Grice's maxim, "Be complete."

An "all other things equal" clause requires a standard of relevance; we are to assume that the remaining facts do not alter the property in question—in moral cases, the rightness or wrongness of the act. But this presupposes that the target property has been achieved, that there is something which we are not to disturb in further elaborations of the case. Jonathan Dancy rejects this assumption on the grounds that you cannot get determinate properties from indeterminate ones.[48] Specifically, the rightness of an action depends on a complex group of nonmoral properties. Hence, if the thought experimenter leaves these properties in limbo, there is no rightness or wrongness to preserve in our possible continuations of the story because there is no right or wrong to begin with. In contrast with Friedman, Dancy is not worried about whether a continuation of the thought experiment will *overturn* our verdict; he denies that any verdict can be reached at all! Friedman says that fictional incompleteness leads to a judgment glut; Dancy says that it leads to a judgment gap.

Dancy only targets ethical thought experiments, but his argument works against nearly all thought experiments if it works against any of them. The indiscriminacy of the attack is a sign that a fallacy is afoot. And, indeed, Dancy has mixed up internal and external indeterminacy. From within Mark Twain's story, Tom Sawyer is a flesh-and-blood boy who is as determinate as you or me. After all, Twain was not writing about a group of exotic entities who inhabit an ontological twilight zone. He was writing about ordinary folks in pre–Civil War Missouri who see a big difference between themselves and indeterminate entities. Tom Sawyer is only indeterminate from the outside looking in. The point can be made without metaphor. Let 'Sp' stand for 'p is true according to the story' and '$^\wedge$p' mean 'p is indeterminate'. Dancy's argument relies on the following invalid schema:

It is indeterminate whether p is true in story.	$^\wedge$S p
Hence, in the story it is true that p is indeterminate.	$S^\wedge$p

By transposing the story and indeterminacy operators, Dancy invalidly transforms external indeterminacy (indeterminacy on the outside of the story operator) into internal indeterminacy. This bootleg indeterminacy creates the pseudo-problem of explaining how determinate properties can arise from indeterminate ones. Once we keep the operators straight, there is nothing to explain because there is no indeterminacy *within* the story.

So although fictional incompleteness is an impressive difference between thought experiments and ordinary experiments, it is not an important difference. This triviality is less surprising when we dwell on the fact that the metaphysical completeness of actual situations fails to imply their epistemological completeness. None of us can know exactly which possible world is the actual world. Since epistemic indeterminacy is the kind relevant to our thinking, its pervasiveness washes out the significance of metaphysical indeterminacy.

D. Bogus Points of Difference

Although there are legitimate differences between thought experiments and other kinds of experiment, there are also illusory contrasts.

1. *Lack of replication.* One apparent asymmetry is that although scientists report replications of ordinary experiments, they do not report replications of thought experiments. A striking sign of this asymmetry, some may say, is the impossibility of a fraudulent thought experiment.

This contrast rests on two errors. First, scientists do report agreements and disagreements with thought experiments. They just do not dedicate whole articles to these reports because there is not enough information to justify the journal space. The second weakness of the contrast is that it overestimates the amount of experimental repetition. Experimentalists rarely publish exact replications of an experiment. Usually, they try to improve on the old experiment by reducing the number of extraneous variables, increasing the simplicity of the procedure, or amplifying the effect. Once we count "new and improved" versions as replications, such as "Wigner's friend," we can see that thought experimenters have a lively interest in replication. Or, consider the twenty-five-year progression of Gettier counterexamples, the fifteen years of ever more elaborate trolley car problems, or the last ten years of Twin Earth cases.

Since thought experiments do not even purport to be executed, there is never any question of fraud with respect to the execution element. But since they can be faked or fouled in other ways, there is still a point to replication.

2. *Thought experimenters do not manipulate variables.* Since the thought experimenter does not causally influence the independent and dependent variables, there appears to be a glaring difference between thought experiment and ordinary experiments. This difference is that thought experimenters do not manipulate the variables in question.

However, some experimentation does not involve empirical inquiry. Chess-players experiment with openings in two ways. When "playing the man," they experiment with a view to learning a particular opponent's strengths and weaknesses. When "playing the board," they seek objectively good positions. This quest for value does not proceed statistically through the accumulation of a favorable win–loss–draw ratio. It is a priori assessment of the value of the opening. The formal nature of chess is obscured by the fact that players move the chess pieces about on a board. But the board and pieces are just thinking aids: experts playing mind chess do without them.

Consider expert puzzle-solvers who systematically manipulate colored blocks, dials, and cartons to "crack" the puzzles (the Fifteen Puzzle, Instant Insanity, Rubik's Cube). They achieve these general solutions with the aid of manipulations, but the movements are not essential. Some of them solve the puzzles with only mathematical notation. The exploratory behavior of the puzzle-solvers who do not use notation is experimentation; for they are trying to answer a question by manipulating variables. Yet the experiments are a priori in that they are not based on new empirical information. The process is no more empirical than using abacus beads to perform addition. The beads on an abacus are commensurate with written notation.

3. *Thought experiments are not quantitative.* Qualitative experiments test hypotheses. Others establish a quantity such as the gravitational constant or the atomic weight of gold. Nineteenth-century physics put great emphasis on this kind of experiment. Much of this research strikes twentieth-century scien-

tists as trivial. But some measurements have great scientific and philosophical significance. It was only in 1850 that the speed of nerve impulses was measured by Hermann von Helmholtz. Before that, their apparently infinite speed suggested the existence of a mysterious mind-stuff. The Michelson and Morely experiments provided empirical support for the constancy of the speed of light, thereby paving the way for the acceptance of relativity theory. These results relied on measurement devices. Since thought experiments lack apparatus, they seem unable to measure.

Nonetheless, there are quantitative thought experiments. Decision theorists measure subjective probabilities by having subjects place hypothetical bets. R. A. Fisher explained the prevalence of the one-to-one sex ratio with a thought experiment.[49] Suppose that more males than females are produced. A parent who produces more males in this population runs the risk of its son being one of the excess males. Daughters are more assured of reproductive success. So a parent who can produce more daughters than sons would be able to exploit the scarcity of females. If the population has more females than males, the opposite strategy works. The only unexploitable ratio is the one-to-one ratio.

Some quantitative experiments must be a priori because measurement devices are only applicable to finite quantities. Thus, Archytas tries to demonstrate that space is infinite by imagining a man traveling out to any supposed boundary and throwing a spear. In the first century B.C., Hero of Alexandria demonstrated that there is no minimum force necessary to move an object along a smooth horizontal plane.[50] Imagine a block on a frictionless surface. No matter how slightly you tilt the surface, the block slides downward.

IV. A Lopsided Tally

Although there are some points of difference between thought experiments and ordinary experiments, they are not major differences that defeat the attempt to classify thought experiments as experiments. In any case, the points of resemblance are substantial enough to overwhelm the differences. For the most part, comparing thought experiments to other experiments leads to insight, rather than error. 'Thought experiment' fails to be a systematically misleading expression for the simple reason that thought experiments are experiments.

Therefore my conclusion is that either thought experiments are experiments or 'experiment' ought to be precisified to include thought experiments. The short answer to this chapter's question is *yes*. I believe that both ends of the naive contrast between ordinary experiments and thought experiments have been exaggerated by an overheated debate between severe empiricists and profligate rationalists. When an empiricist looks at ordinary experiments, he is struck by the fact that they convey fresh information about the world. His belief that all knowledge of the world is derived from experience leads him to view experiment as an act emblematic of his philosophical position. So the empiricist pumps up the fresh information feature so that it expands over the other rationally persuasive features of experiment, concealing them from view yet drawing invisible support from them. Rather than challenging this distorted image of experiment, we find rationalists embracing it, presupposing its truth with the empiricists. The rationalist's theme becomes the narrowness of empiricism, that the empiricist overlooks alternatives to experiment.

The proper attitude toward experiment is to see that it rationally persuades in two ways. Its most striking mode of persuasion is the injection of fresh information about the world. But ordinary experiments also persuade by the methods associated with armchair inquiry: by reminder, transformation, delegation, rearrangement, and cleansing. Once we see that these factors were there all along, we can stop viewing thought experiment as introducing a *new* mode of persuasion. Thought experiment only reveals the hidden persuaders involved in ordinary experiments by taking away the most obvious persuader.

10

Fallacies and Antifallacies

If I could drop dead right now, I'd be the happiest man alive!

Samuel Goldwyn

In the previous chapters, I addressed the questions of what thought experiments are, how they work, their origin, and their relationship to ordinary experiments. There remains the question of their efficacy. What are the criteria for a *good* thought experiment? What are common flaws of bad ones?

I. The Biological Baseline

It is sound policy to separate 'What is an F?' from 'What is a good F?'. It is especially tempting to mix normative and descriptive considerations when F has an honorific use like that of 'art' and 'love'. Since 'thought experiment' is sometimes used in a lofty way, we had best remind ourselves that a bad thought experiment is still a thought experiment.

Nevertheless, normative considerations bear on the characterization of things that are designed to serve certain ends. Functional definitions such as "A clock is an instrument for telling time" characterize artifacts in terms of what they are designed to do. This gives us a broad clue as to which ones will be good and which bad. The good clocks will be those that efficiently execute their function of telling time. The bad fail to tell time or only do so in a cumbersome fashion.

Since thought experiments also have a function, we can say that a good thought experiment efficiently raises or answers its question. Thought experiments are intellectual instruments like arguments, demonstrations, explanations, and research programs. They are supposed to make you an authority by having you merely reflect on the experimental plan. Hence, thought experiments will be prized to the extent that they enlighten in this minimalist fashion.

The evolutionary account of thought experiments inaugurated by Mach provides clues about the conditions under which they work best. Thought experiments rely crucially on our sense of absurdity. Since this feel for what could or could not happen developed as a decision aid, it is geared toward situations demanding a practical choice. Thus, we are more at home with concrete, manipulable scenarios than with abstract, impersonal, uncontrollable scenes. This decision-making orientation explains why thought experiments abound in ethics, law, and economics. The explanation is not that these fields involve value judgments; for thought experiments are not nearly as common in aesthetics. The decision-making orientation is also evident from the interviewing tactics behind thought experiments in descriptive fields. For example, the hypotheticals in physics are often given a practical spin by presenting them as problems of escape, competition, or optimization. Metaphysicians couch questions of personal identity as choices (Which person do you want rewarded and which tortured?) and legal dilemmas (Which person is entitled to Smith's car?). Purely contemplative beings would have never evolved the capacity for thought experiment.

Biology gives us further clues when we concentrate on the kinds of problems our modal intuitions were designed to solve. The hunter–gatherer needed to track objects in space and time; to distinguish friend from foe; and to recognize predator and prey, shelter and foodstuffs. Moreover, he needed to manipulate objects and predict behavior. Thus, we have a rich stock of intuitions about space and time and biological and social relationships and a great aptitude for belief–desire psychology and causal reasoning. Humans are also endowed with excellent visual judgment and mastery of language. Hence, thought experiments that exploit these intuitions and abilities, whether directly or through resourceful indirection, are apt to fare well.

The outlook for other thought experiments is dicey. There is no selective pressure against errors about conditions that our ancestors never encountered. We have little intuitive sense of what would happen at abnormal temperatures, pressures, and speeds. People are notoriously fallacy-prone when dealing with extreme quantities. Nor has there been selective pressure to understand phenomena that transcend human sense organs. Electricity, magneticism, and radiation are all just a blur. Mother Nature also saw no reason to burden us with knowledge that lacked practical application. What use was knowledge of our internal organs or the composition of distant stars?

Fortunately, natural selection does not always condemn us to know only enough to squeak by. As I stressed in chapter 4, there are many ways in which "useless" traits can arise. Selection *of* a trait does not imply selection *for* that trait. Apparent biological limits can also be overridden by ill-understood cultural mechanisms. We have an elaborate social structure that strongly encourages the discovery and dissemination of knowledge. Thus, we now overcome traditional obstacles such as complexity and storage costs. Since knowledge breeds more knowledge, we enjoy a progression of breakthroughs. Thus, epistemological limits can only be drawn with nervous glances toward human cunning.

Nevertheless, biology sets a baseline of expected performance. Cultural factors may force us to revise our opinion upward. But even here, biology helps by warning of the need to distinguish naive intuitions from tutored ones. Physiologists have slowly surmounted eons of ignorance about our internal organs. Their experience and education puts thought experiment about the topic back in business. But a remnant of the biology barrier remains; we know that weight should only be assigned to the judgment of those who are elbow deep in anatomical practices. Thus, biology tells us when to be elitists and when to wax egalitarian.

Biological expectations can also be thwarted in the opposite direction. One would think that an animal as bright as man would be an excellent logician. But psychologists have exposed embarassing weaknesses at deduction and statistical reasoning. Evidently, Mother Nature has been content to have us schlep through with a hodge-podge of heuristics and a heavy reliance on memory. For all their wonderous efficiency, organisms do have appalling design flaws. Think of all the chokings that have been caused by the intersection between the mammalian food and wind pipes! There are also trade-offs. Becoming bipedal freed man's hands and let him grow a big head—but at the cost of back problems and badly drained sinuses. Given the pervasiveness of design flaws and trade-offs, we should expect the mind to have its share of structural faults.

Happily, biopsychologists learn from mistakes and can re-figure the biological baseline. If there were a graph of human intellectual performance, large areas would be blank because many questions cannot be asked by us. Most of what was charted would indicate a modest level of reliability, gently rolling up and down with the variables of need and familiarity. Here and there would be jumps into the realm of supernormal performance and drops into howling incompetence.

Since this chapter's topic is fallacies, I focus on the low points. Chapter 6, on the logical structure of thought experiments, contains an implicit demonstration of how fallacies can be systematically classified; for each solution to the paradox typically contains two parts. The first tries to show why a member of the quintet is false. The second part of the solution tries to explain why the member *seems* true. Since the analysis of illusion has a strong psychological element, the logical account of fallacies needs to be supplemented with soft generalizations about human reasoning. This loose talk will be balanced by greater-than-usual care at the metalevel.

II. Myths and Abuses

This higher level of care begins with a distinction between fallacies and two phenomena that are often confused with fallacies.

A myth is a common, false *belief*. Since all fallacies are *inference rules*, no myth is a fallacy. However, a myth can serve as a premise of an argument and so cause unsound reasoning. For example, biologists before Darwin agreed that organisms were created by God. This was evident from their intricate design: each creature's parts meshed together like components of a watch.

Given the premise of divine design, biology should be approached as a problem in reverse engineering. In normal engineering, you design a new machine. In reverse engineering, you infer the design of a preexisting machine (normally in the hope of copying it). Since God was the original architect, the biologists could assume that the blueprint was maximally efficient; and in general, they were not disappointed by God's ingenuity. Detailed inspections supported the hypothesis that each part of an organism had a function and that organisms as a whole had functions within the larger community of life. Reverse engineering encourages thought experiments because you are trying to figure out intentions by asking 'How would I have designed that?'. This tendency is amplified by the divine status of the original engineer because you can assume maximal knowledge, power, and goodness. Down from the mountain of idealization stream thought experiments attempting to establish the ideal liver or the ideal mollusk. But the flow of thought experiments dried up when Darwin gutted God out of biology. If there is no designer—if organisms do not owe their features to foresight—then they were never engineered to begin with! Therefore reverse engineering is just the wrong approach to biology. Thought experiments in search of biological blueprints falsely presuppose the existence of unique plans for each species. But according to Darwin's vision, living things are elements of a smooth spectrum of life that only looks like it contains qualitative leaps because hunks of that spectrum have fallen to extinction.

The divine design biologists were no dopes. Their myth was justified, given their pre-Darwinian evidence. Nevertheless, the thought experiments resting on these old myths are destined to look quaint, as out of date as chamber pots. The thought experiments of heretics who prematurely rejected the old orthodoxy have brighter careers. Giordono Bruno's ship-and-bridge hypothetical, for instance, looks ahead of its time. It was aimed against Aristotle's doctrine that each object seeks its own place in the universe. Stones seek the center of Earth and so naturally move downward in a straight line. Thus, two stones starting from the same point of origin must follow the same path. What the stones were doing before being dropped is irrelevant because stones have no memory. Bruno attacked this consequence of Aristotelian physics by imagining two men, one on a bridge, the other on the mast of a passing ship. Each man holds a stone and releases it when their hands meet. Although the stone released from the bridge falls straight into the water, the stone dropped from the ship lands at the bottom of the mast. The curved path of ship's stone is due to the fact that it retains the movement of the ship. Although this thought experiment was eventually accepted as a cogent refutation, it made no impact on the Aristotelians or even on the leading anti-Aristolelians (Tycho Brahe, Kepler, and Copernicus). Only Bruno, through a lucky leap of faith, had embraced the idea of an open infinite universe in which space is viewed as a homogeneous receptacle. So only he had a metaphysical system that completely undermined the distinction between natural and violent motion.

Regular experiments are equally vulnerable to myth. Prout's hypothesis states that all naturally occuring elements are composed of hydrogen atoms. This suggests that molecular weights are whole numbers (relative to hydrogen).

Thus, nineteenth-century chemists were guided by Prout's hypothesis in their precise measurements of the elements. The existence of recalcitrant elements that seemed to have fractional weights finally led to the discovery of isotopes. Isotopes have the same chemically properties but slightly different weights. Thus, their pervasiveness made the measurements of naturally occuring elements as pointless "as the determination of the average weight of a collection of bottles, some of them full and some them more or less empty" (Frederick Soddy).

Like myths, methodological abuses bear a misleading resemblance to fallacies. However, a method can be both constant and faithful even when abused. Consider the mental crutch. Multiplying Roman numerals is cumbersome. However, the Romans were able to fall back on the abacus for multiplication. Consequently they were never forced to develop a more efficient notation (like ours) which would have left them better off in the long run. Just as some philosophers use metaphor as a mental crutch, others use thought experiment as a mental crutch. Instead of taking the extra time to find a robust actual case, they fall back on anemic hypotheticals. Thought experiment is an easy way of getting short-term results. Philosophers already know how to do it, it's cheap, it's quick, it's accepted.

A second abuse is methodological chauvinism. Just as craftsmen overuse the tools of their apprenticeship, thinkers spurn more effective methods in favor of familiar ones. Loyalty toward thought experiment leads some thinkers to ignore data from other techniques, such as linguistic polls, comparative and historical studies, and experimental psychology. The illegitimacy of this resistance to looking at all of the evidence is often cloaked by appeals to the demarcation of disciplines. Philosophers using a varied data base are accused of extending their opinion beyond their ambit of competence. Thus knee-jerk reliance on thought experiment leads to a narrow-minded rejection of less philosophical looking techniques. There should be more behind the choice of thought experiment than force of habit.

Third, add the abuse of resource hogging. Some methods work only by crowding out the application or results of other methods. For instance, critics of job interviewing say that this method overloads the employer's memory. Hard data is displaced by the more vivid but less reliable impressions afforded by the interview. Thus, a scientist's preoccupation with a mildly illuminating thought experiment can make him neglect stronger evidence already in his possession.

III. Fallacious Thought Experiments

Any fallacy can be committed in a thought experiment—or ordinary experiment for that matter. Hence, there are no fallacies *unique* to thought experiments just as there are no diseases unique to miners. Physicians interested in the health of miners will nevertheless concentrate on the diseases that are especially common among miners. Since mining is an occupation that exposes the worker to a characteristic profile of risk factors, it makes sense to narrow attention to the diseases associated with those factors. Thought experimenters also face a special set of hazards. Studying the fallacies associated with these dangers to their reasoning yields benefits analogous to those obtained by the physician specializing in the diseases of miners.

A. Missupposition

A device such as a microscope can be used at various settings. The selections of novices tend to be either too high or low. Even after trial and error confers a delicate touch, the expert can still err by overgeneralizing. Mistaken expectations about what the instrument should reveal can warp the training of a whole generation of technicians by preventing them from recognizing an error or, more rarely, by causing them to reject an accurate reading. These caveats extend to thought experiment.

1. *Oversupposing.*   The godlike power of stipulation tempts one to annihilate obstacles through a sheer act of will. However, if you assume too much, you may inadvertently trivialize the very problem the thought experiment was intended to solve.

Often, the trivialization takes the form of circularity: the thought experimenter begs the question by presupposing what he aimed to prove. The fallacy is widely believed to have been committed by John Locke in his prince-and-cobbler tale. Locke wanted to prove that the psychological, rather than the physical, features of a person are crucial to his identity. Since the psychological and physical features are so tightly correlated in the actual world, Locke makes them come apart by means of an imaginary example featuring a prince and a cobbler. One morning the person with the body of the prince awakes with the cobbler's memories and the person with the body of the cobbler awakes with the prince's memories. Each remembers the past associated with the opposite bodies. Most people infer that the prince and cobbler have switched bodies. This conclusion supports Locke's contention that personal identity goes by psychological continuity. However, critics observe that you can only remember what you really did. Hence, by describing the cobbler as having the prince's memories we *presuppose* that he is the prince. So the thought experiment begs the question in favor of the psychological criterion of personal identity.

Trivialization can arise from the audience, as well as from the designer of the thought experiment. For they will be tempted to add amendments that ease the burden of problem solving. Recall the joke about the economist who was asked how he would survive on an island containing nothing but canned provisions: "Posit a can opener" was his reply. Greedy supposition is especially tempting in response to thought experiments that aim to expose an incompleteness. This is evident in Simplicius' attempt to rescue the principle of sufficient reason (which tells us that events cannot occur arbitrarily). Suppose a hair is composed of identical parts and is evenly pulled. Since the tension everywhere equal, there is no more reason for it to break at one point rather than any other. Hence, the principle of sufficient reason implies that the hair cannot break! Simplicius concedes that the hair breaks but replies, "Even hypothesizing a fictitious thing with parts thus identical, plainly an identical tension at the ends and in the middle is impossible."[1] Of course, it is legitimate forthrightly to reject the hypothetical situation as impossible. The fallacy only occurs when one pretends to be finding a hidden aspect of the situation that saves the threatened principle.

Since ethical verdicts can be reversed with additional information, moralists try to forestall embarassing consequences by surreptitiously introducing extra information. Thus, Anscombe met the hypothetical forced choice between punishing an innocent man and enduring nuclear war by saying, "Why not fake the punishment?" This maneuver is ineffectual because the designer of the original thought experiment is free to banish these interfering additions. It's *his* thought experiment, after all.

Oversupposition is also stimulated by emotional engagement with the characters in the fiction. Some people want a happy ending to the thought experiment and so make additions that ease the depicted difficulty. This is especially annoying to the thought experimenter trying to establish the possibility of rational suicide. He heaps misery upon hypothetical misery. But the mounting misfortunes only build hope for a miracle.

Force of habit even leads seasoned thought experimenters to overstipulate. For example, one standard technique for controlling nuisance variables is to idealize the entities in question. Thus, social contract theorists dispose of worries about stupidity and ignorance by assuming that the signers of the contract are perfectly rational and well informed. But as Thomas Jefferson observed, these saints would have no need for government! A similar dilemma arises for decision-theoretic acceptance rules. These tell you to believe a hypothesis when it has a sophisticated theoretical property (such as a high probability) that can only be ascertained by an ideal thinker. But why would an ideal thinker need acceptance rules? To apply a rule that takes you from rich inputs to coarse-grained outputs is to throw away information. Better to stick with the original information! Hence, the dilemma is that the rules cannot be followed by those who have limited resources and should not be followed by those who have unlimited resources.

Another rule of thumb is to start with simple cases and only later complicate the situation. However, the tactic occasionally exacerbates the problem. Rainbows are caused by light rays coursing through rain drops and being split into elements of the visible spectrum. Thus it is initially tempting to work out the details by assuming that each drop is perfectly round. But the angles of incidence and refraction would then form a square, ensuring that the ray never exits the drop! The real solution relies on the ugly fact that air drag distorts the shape of the raindrop so that the light can leave.

2. *Undersupposing.*   When the designer of the thought experiment fails to be specific enough, the usual flaw is indecisiveness. Either the audience recognizes the shortfall and complains about insufficient data, or they unwittingly read in extraneous details. If their creativity leads them to supply diverging details,

they become embroiled in a dispute or seduced into a consensus that is merely verbal. Even if they supply the same extraneous details, the audience will still have a mistaken view of what is implied by the conditions of the thought experiment.

Indeed, they may wind up confusing incompleteness with inconsistency. James Thomson's lamp paradox illustrates the hazard. His thought experiment was designed to prove the logical impossibility of performing an infinite number of tasks.[2] The lamp is one with a single button that turns the light on if it was off and off if it was on. Since the lamp starts in the off position, it will be on if the button is pressed an odd number of times and off if pressed an even number of times. Now suppose that Thomson manages to press the button an infinite number of times by making one jab in one minute, a second jab in the next half minute, a third in the next quarter minute, and so on. At the end of the two minutes of jabbing is the lamp on or off? It cannot be on because Thomson never turned it on without also turning it off. Nor can it be off: for after first turning it on, he never turned it off without also turning it on.

However, the appearance of contradiction is an illusion generated by the incompleteness of the supposition. Thomson's instructions only specify what happens at $2 - 1/2^{n-1}$ minutes, not the second minute itself. Paul Bennacerraf makes the point with a logical analogy. A man tells us that every number less than 1 is either fair or foul. In the sequence $1/2$, $1/4$, $1/8$, and so on, the first member is foul, the second fair, alternating so that $1/2^n$ is foul if n is odd and fair if n is even. Now, is the limit of the sequence fair or foul? It cannot be foul because there is a fair after every foul. But neither can it be fair because there is a foul after every fair. Contradiction? Nope. Since the instructions only cover the sequence, nothing is implied about a number outside the sequence.

Sometimes the blame lies with the audience rather than the designer of the thought experiment. Threatened or inattentive listeners may fail to comply with *all* instructions. A potent source of uncooperativeness is the confusion between the speech acts of supposing and reporting. A *report* of a dilemma is often best handled by checking its accuracy rather than by immediate deliberation. Maybe things are not as bad as was assumed and no hard choice is necessary. This wise policy tends to be overextended to thought experiments. Thus, Carol Gilligan's females respond to a hypothetical forced choice by searching for loopholes: maybe Heinz can avoid choosing between letting his wife die and stealing the expensive drug by getting a government loan or finding another treatment—indeed, his wife may not be as sick as she seems! These replies are misconceived because *stipulations* can be neither accurate nor inaccurate. Gilligan is free to admire her subjects's evasiveness as a sign of resourcefulness, sensitivity, or some other virtue. However, to assume that all virtues rest on an ethical *insight* is to overintellectualize morality.

B. Perspectival Illusions

Unfamiliar perspectives disorient. The historian's great distance from the events he reports in broad strokes makes them seem more abrupt. Since each of the intermediary steps in a gradual change is too small to be noticeable from the distant temporal vantage point, the (actually cumulative) contrast between the beginning events and end events creates the appearance of a rapid qualitative change. For instance, Larry Laudan argues that this illusion exaggerates the role of revolution in science and creates an appearance of irrationality; instead of seeing a series of individually rational changes at the micro level, one only sees a mighty, macro transformation of the Zeitgeist.[3] Hindsight makes past events appear inevitable because knowledge of how things turned out prevents other possibilities from coming to mind. As a corollary, past agents look stupid or self-deceived because they could not foresee what now strikes us as an obvious consequence. Historians are also vulnerable to "precursoritis." Once the bugs are worked out of an invention or idea, historians begin to find "the same thing" deeper and deeper in the past. Each precursor paves the way for the next precursor by slightly lowering standards for what counts as "the same thing." Hence, at the end of the slippery slope one concludes that "there is nothing new under the sun." The same mechanisms recently brought in a bumper crop of "multiple discoveries" by historians of science.[4] Science seems to be packed with amazing parallelisms which suggest that any idea would have been thought up by someone else at about the same time.

The thought experimenter's freedom to suppose what he wishes also gives him access to unfamiliar perspectives. By imagining how things would look if he (or the specimen) were much larger or smaller, he obtains cognitive advantages analogous to those bestowed by the telescope and microscope. By supposing that events run by more rapidly or slowly, his pattern recognition is aided in a way well known from time-lapse photography. But with these opportunities come risks of disorientation and miscalibration. Consider Leibniz's objection to *mechanism*—the view that all phenomena can be reduced to mechanical interactions:

> Moreover, it must be avowed that *perception* and what depends upon it *cannot possibly be explained by mechanical reasons*, that is, by figure and movement. Suppose that there be a machine, the structure of which produces thinking, feeling, and perceiving; imagine this machine enlarged but preserving the same proportions, so that you might enter it as if it were a mill. This being supposed, you might enter its inside but what would you observe there? Nothing but parts which push and move each other, and never anything that could explain perception.[5]

Leibniz's thought experiment involves enlargement of the specimen rather than shrinkage of the observer, but the effect on perspective is the same. His reasoning is that any thinking machine would have to be made up of something responsible for its mentality. Since we would find no thoughts, Leibniz concludes that thinking machines are impossible. Indeed, he goes on to infer that minds must be simple, immaterial things. The counterfactual fueling this thought experiment is 'If there were thinking machines and they were enlarged for inspection, then the inspectors would observe mental things responsible for the machine's thoughts'. However, this counterfactual illicitly assumes that properties of the whole must be possessed by the parts. David Cole develops this criticism by means of a counterexperiment:

> Imagine a drop of water expanded in size until each molecule is the size of a grindstone in a mill. If you walked through such a now mill-sized drop of water, you might see wondrous things but you would see nothing *wet*. But this hardly shows that water does not consist *solely* of H_2O molecules. Rather, it shows that a fallacy of composition is at work here and in Leibniz's thought experiment. Further, I think, it shows that whenever one takes the *perspective* of the subsystem or constituent, one is likely to find it hard to believe that the whole and all its properties can be accounted for by the properties of the constituents, given their arrangement.[6]

Cole's criticism consists of a refutation by logical analogy (which tells us that something or other is wrong) and a diagnosis of the error (which tells us what is wrong).

C. Framing Effects

The rising popularity of polls has made us aware of the importance of wording. A pollster who finds people answer *yes* to 'Is the cost of living increasing?' may find them answering *no* to 'Is the value of your money decreasing?'. Although the questions are synonymous, they lead us to dwell on different aspects of the situation. Hence, the framing effect is just the effect produced by rephrasings of the same question. We can see the framing effect at work in sales offers, diplomacy, and therapy. Alcoholics Anonymous advises against resolving to never drink again. Instead, start each day with the resolution to not drink today.

Experimental evidence buttresses this commonsense recognition of the framing effect. One study concerned preferences for radiation over surgery as a treatment for lung cancer.[7] The first group based their preference on survival statistics, while the second were given equivalent mortality statistics. Of those in the survival frame 18% preferred radiation over surgery, while 44% of the mortality frame group preferred radiation. The subjects cared more about reducing the probability of death than increasing the probability of survival! Interestingly, this frame effect was also obtained from experienced physicians.

D. Biases of Thought Experiment

Our desires color the reception of many thought experiments. Dennett explains part of the persuasiveness of Searle's Chinese Room as due to a yearning for a different status from machines. Antideterministic thought experiments get an illicit boost from our craving for freedom. To understand the warmth with which Maxwell's Demon was welcomed, we must understand what Victorians feared. In 1852 William Thomson (Lord Kelvin) had published a paper on the universal dissipation of mechanical energy.[8] Since mechanical energy continually dissipates into heat by friction and heat continually dissipates by conduction, the second law fates the universe to heat-death. Against the backdrop of this gloomy prospect, Maxwell's Demon seemed a savior. J. Loschmidt, the creator of a less anthropomorphic demon, rejoiced: "Thereby the terrifying *nimbus* of the second law, by which it was made to appear as a principle annihilating the total life of the universe, would also be destroyed; and mankind could take comfort in the disclosure that humanity was not solely dependent upon coal or the Sun in order to transform heat into work, but would have an inexhaustible supply of transformable heat at hand in all ages."[9] No doubt, all sources of clouded judgment have had some influence on some thought experiment or other. Hence, we ought to devote our attention to biases that have a larger-than-usual impact. But before turning to specific kinds of bias, a few general remarks about the concept are in order.

First, a bias is an *illegitimate* influence. Some deviants use the term more broadly to encompass any kind of partisanship and so are moved to paradoxical confessions such as "My bias is belief in the scientific method." This misguided humility gives some people the impression that all neutrals and only neutrals are free of irrationality. Thus, people will use the mere fact that you hold an opinion as reason to conclude that you are not a good judge of the issue. Worse, they think that a person who refrains from judgment is ipso facto unbiased. This overlooks the possibility that the neutrality is due to an irrelevant factor that is blocking the verdict indicated by the balance of evidence. Think of the mental constipation of cowards and worrywarts.

The context-relativity of 'bias' sets a second trap. Just as what counts as a bump, dirt, or curve varies with *purpose* and *custom*, what counts as a bias varies with purpose and custom. These standard-setters carry a presumption against counting very small perturbances as influences. Bias-mongers abuse this presumption by first lowering standards so that any illegitimate influence, regardless of size, counts as a bias and then acting as if the presumption is still in force by treating these biases as significant. The same pattern of abuse can be seen in scepticism about whether anything is really flat or certain. It is the business of the bias-monger to parlay the slightest eccentricity of judgment into grounds for discounting the opinion.

Hence, one of the ground rules for discussing biases is to distinguish between influences and illegitimate influences and to swear off pickiness. There is plenty of scientific precedent for working with biased data. For example, the fossil record is so heavily biased that there is a whole branch of paleontology— *taphonomy*—devoted to how burial processes and subsequent events prejudicially reflect the past. Most biases are less severe and are easily swamped by other factors. People overestimate the power of bias in the formation of judgment because it is one of the few cognitive weaknesses we readily understand. Since hidden and difficult factors rarely enter into our explanations of past mistakes, accessible foibles get overblamed. We should bear in mind that people are not cognitive cripples. Although they are susceptible to bias, they are also adept at correcting for bias. If there were no homeostatic controls on wandering judgment, people would be overwhelmed by the problems all animals must solve.

1. *Pet theory bias.* Theoreticians take pride in their work and so want their theories to succeed. This introduces the danger of wishful thinking; my desire to have the data confirm my theory tends to make me believe that the data confirms my theory. Furthermore, my theory will subtly (even if justifiably) alter my beliefs about the data. Therefore, if I use my changed beliefs as a test of my theory, I will be reasoning in a quiet circle. So far, only a general problem has been depicted. All theoreticians must guard against question-begging and wishful interpretations of their data. What makes the problem especially acute for thought experiment is the greater pliability of the data. When a theoretician predicts the prevalence of a certain property in a population, there will be some interpretative slack; but the amount will be kept in check by feedback from the population itself. Other people can examine the data and round up more evidence. In the case of thought experiment, however, there is less external constraint on interpretation. We only have an intangible imaginary scenario. Interpretation can wander, because it is no longer tethered to a public data base.

Since theorists suspect others of pet theory bias, they naturally wonder whether they are also victims of the distortion. So theorists have developed various ways of monitoring and controlling this menace. First, others are asked for their reaction to the thought experiment. It is especially informative to learn of the reactions of adversaries. Pet theory bias is probably at work if intuitions follow partisan lines. If theoretical allegiance is a poor indicator of reactions to the thought experiment, pet theory bias can be ruled out. Since second opinions can be biased by a desire to please or frustrate the questioner, the thought experiment needs to be posed in a way that conceals the questioner's opinion. Occasionally, this type of questioning can be developed into a full-blown survey. When the thought experiment is complicated or esoteric, you cannot use the opinion of ordinary people, because they are too prone to confusion and lack the background knowledge needed to understand the thought experiment. A second method of avoiding pet theory bias is to use variations of the thought experiment. Since reflective equilibrium creeps through a belief system slowly, one may be able to reach an intuition as yet unsullied by your theory. Third, one can recycle old thought experiments. If these were discussed in ignorance of your theory, the old intuitions can be put to a new purpose. Fourth, one can cancel out a favorable bias by introducing a comparable bias against the hypothesis. Lastly, one can gather wild specimens of the intuition. For example, a thought experiment mounted in support of the possibility of unfelt pain can be corroborated with medical texts describing operations that "block pain" and a line from *Captain Horatio Hornblower*, in which C. S. Forester writes, "Hornblower found the keen wind so delicious that he was unconscious of the pain the hailstones caused him." Such passages show that one need not be in the grip of theory to describe matters as the thought experimenter did.

Lingering suspicion of pet theory bias may be responsible for the greater weight attached to *new* confirmation. A theory that fits familiar data may owe its success to subliminal, ad hoc fudging by its proponents. Compare this post hoc rectitude with the achievements of a rich boy. How much of the success is due to behind-the-scenes help and how much was due to individual merit? Hence, a theory gets more credence from an experiment that confirms its bold prediction than from the past experiments with which it is shown to conform. For instance, relativity theory is made more persuasive by the twin paradox than by its conformity with Newton's thought experiment about launching satellites.

2. *Literary biases.* Since both stories and thought experiments can be sequences of suppositions, similarities should be expected. These resemblances create interference. Specifically, thought experiments tend to be dragged into the orbit of storytelling conventions.

a. *Bias in favor of continuity.* The audience of *Star Trek* stories believes that the characters travel by teletransportation. Captain Kirk enters the chamber, is deconstituted, and is then reconstituted at some distant place. Most philosophers working on personal identity think this is overly optimistic. They say Kirk is destroyed when deconstituted and then a replica is created at the "destination" point. The philosophers deny that any travel takes place because no one survives such a disruptive process.

Peter Unger has suggested that the intuition that we survive teletransportation is an artifice of fictional discourse.[10] When we listen to a story, we make assumptions that enhance literary qualities of the tale. One such assumption is character continuity. Rather than assume that there are a number of highly similar characters who succeed each other after each teletransportation, we assume Kirk survives. Since it is a story, we do not worry much about *how* it happens. We assume that the author could just stipulate things to be whichever way makes for the most engaging plot. Compare the freedom we grant science fiction writers over the criteria for personal identity with the liberties we grant them over natural law. Instead of demanding details, we let the authors surmount the speed of light with shallow literary devices such as hyperdrives and space-warps. Unger's point about the artifices of fiction can be extended to cover other suspicious thought experiments, such as those supporting the possibility of reincarnation, disembodied existence, and psychokinesis.

b. *Bias in favor of aesthetic values.* In addition to making the descriptive assumptions needed for a good story, we make the evaluative ones as well. Although we do not become outright aesthetes, we drift toward ethical values that harmonize with aesthetic ones. The thought experiment is construed as supporting certain metaphysical theses when the metaphysical theses really function as supporting backdrop for a drama in the foreground. Consider our tendency to extract fatalistic consequences from deterministic thought experiments. Fatalism is the view that people have no control over the future, that events will happen regardless of your actions. Determinism only locks in the future after your choices are figured in; it ascribes causal power to your actions. However, fatalism is an attractive literary theme that has sustained stories from *Oedepis Rex* to *Slaughterhouse Five*. Hence, deterministic thought experiments tend to be construed fatalistically to enhance their tragic qualities.

Our aesthetic affinity for adventure can be used to explain our reaction to the experience machine. Antihedonists envisage a machine that can produce a lifetime of pleasure by directly stimulating the brain. Although the stimulatee is actually lying in a hospital bed with electrodes planted in his head, he is having experiences that are qualitatively indistinguishable from those enjoyed by a jolly mountaineer. Does the experience machine produce a good life? Would you choose to live attached to the experience machine or would you rather have a normal life? The fact that most people prefer a normal life suggests that hedonism is mistaken. The experience machine seems to demonstrate that pleasure is not the only good: we value experiences that correspond to, and originate from, the real world. Hedonists contend that the thought experiments are tainted by a literary demand for adventure and plot development. The fellow attached to the experience machine just lies there like the sack of potatos in your cellar. Dull, dull, dull. The hedonist counters this prejudice by reminding us that what looks dull on the outside can be exciting on the inside. Take chess. The naive spectator only sees two men sitting and staring and intermittently moving wooden figures. But the competing grandmasters are having intense, absorbing experiences. The fact that something is a bore to watch does not show that it is boring or narrow or unsatisfying to those being watched.

For another ethical example, consider the question of whether the utilitarian should maximize absolute utility or average utility. The proponent of average utility has us compare two worlds. The first has a hundred billion people, each of whom is just barely better off alive than dead. For the sake of concreteness, assign each individual one unit of value. The second world has a billion people who are much better off—they live about as well as we do. Say, each enjoys ninety-nine units. Now according the absolute utilitarian, the more populated world is better than the less populated world (one hundred billion units of value vs. ninety-nine billion). But intuitively, the less populated world is better. We regard the larger world as overpopulated and think that members of the smaller world should try to prevent a population explosion that could lead their world to become like the big one. Average utilitarianism also counts the small world as better off than the big one because the small one has an average utility of ninety-nine while the big one has an average of only one. Thus, average utilitarians claim that the thought experiment refutes absolute utilitarianism. James Hudson finds the refutation unconvincing but speculates about how others come to be persuaded by it:

> One reason may be that the whole setting of the population problem encourages us to take a sort of God's-eye view of alternative possible worlds, and to ask ourselves which world God would or should have made actual. In this context it is inevitable that we attribute a transcendent purpose to human life, which after all is being created to fulfill God's purpose. It is natural to conceive of this

purpose in esthetic terms: God is to be thought of as a playwright, and human history is a drama of His creation. But however we think of the Divine purpose—whatever we think would be our purpose if we created the world—it is hard to see how mere numbers of people could contribute to it. Certainly if the purpose is esthetic, too large a cast of characters would spoil the drama, which requires rather a relatively small number of interesting characters interacting in interesting ways.[11]

Anticipate further prejudices. When irony or absurdity requires a surprising symmetry between cases, we nurse the analogy by suppressing differences and exaggerating similarities. A thought experiment that pleases aesthetic instincts will thus have an easier time "proving" arbitrariness. We are also wise to heed the aestheticians' warnings about confusing truth and beauty. Since falsehoods can be as brilliantly expressed as truths, we must guard against laundering aesthetic appreciation into assent. Thus, we worry whether the lovely thought experiments of Nietzsche, Einstein, and Judith Thomson *charm* us into conviction.

c. *Bias against complexity.*   Thought experiments have many of the limitations suffered by short stories. You cannot engage in leisurely character development. You are expected to get the point across quickly. This code of savage concision is reminiscent of the dogfight pilot's maxim: "See him, kill him, leave him." Processes that turn on high degrees of complexity and detail get left out. Compare thought experiment to television news: the medium leads broadcasters to overreport news that can be conveyed in spectacular imagery and short "sound-bites" and to underreport news that requires patient development and a visually bland diet of "talking heads."

Dennett alleges that this superficiality perverts thought experiments about free will. He commences his condemnation with a distinction between thought experiments that are supported by a background of precise, literal reasoning and vague thought experiments that are long on evocative imagery and short on argument:

> Here I want to point to a dangerous philosophical practice that will receive considerable scrutiny in this book: the deliberate oversimplification of tasks to be performed by the philosopher's imagination. A popular strategy in philosophy is to construct a certain sort of thought experiment I call an *intuition pump*. . . . Such thought experiments (unlike Galileo's or Einstein's, for instance) are *not* supposed to clothe strict arguments that prove conclusions from premises. Rather, their point is to entrain a family of imaginative reflections in the reader that ultimately yields not a formal conclusion but a dictate of "intuition."[12]

Dennett takes a thought experiment invented by Douglas Hofstadter as his flagship example of an intuition pump. Hofstadter's case is an extrapolation from the actual behavior of the digger wasp *Sphex ichneumoneus*. At first sight, the wasp seems a wise mother. After digging a burrow, she returns with a paralyzed cricket, so that when the wasp's egg hatches there will be well-preserved food. However, biologists have demonstrated that this apparently thoughtful behavior is really due to a rigid routine. Before depositing the cricket, the wasp leaves it at the threshold while checking whether the burrow has been disturbed. If all is well, she drags the cricket in. But if the cricket is moved a few inches away from the threshold, the wasp drags it back to the threshold and then rechecks the burrow before the final maneuver. If the cricket is again moved from the threshold, the wasp drags it to the threshold and checks the burrow a third time. Patient biologists have led the wasp through forty loops this way. Now the question is whether a superbiologist could unmask *our* apparent thoughtfulness as really a mechanical response to environmental impingements. Since we only differ in degree, it seems we could be "sphexish" and hence not really free agents. Dennett protests that our judgment is being formed

> from the very simplicity of the imagined case, rather than from the actual content of the example portrayed simply and clearly. Might it not be that what makes the wasp's fate so dreadful is not that her actions and "decisions" are *caused* but precisely that they are so *simply* caused? If so, then the acknowledged difference between the object of our intuition pump and ourselves—our complexity—may block our inheritance of the awfulness we see in the simple case.[13]

Dennett also puts some of the blame on our thirst for clarity. The clarity associated with simple models is well known to teachers: students have fewer false paths, fewer factors to track, less wait for the key point. So it is perhaps inevitable that

> when we think of causation, we tend to think of nicely isolated laboratory cases of causation, where a single, repeatable, salient effect is achieved under controlled circumstances. Or we think of particularly clear cases of everyday causation: Hume's billiard balls, sparks causing explosions, one big salient thing bumping into another big salient thing. We know that on closer examination we would find every corner of our world teeming with complicated, indecipherable, tangled webs of causation, but we tend to ignore that fact. Thus when we think of someone *caused* to believe this or that, we tend to imagine them being *shoved* willy-nilly into that state. The person thus caused to believe is analogized to the billiard ball caused to roll north, or the liquid caused to boil in the test tube.[14]

Notice the resemblance between Dennett's oversimplification objection and the oversimplification objection leveled at experiments in general: what holds in the laboratory setting need not hold in the real world. One of the reasons Aristotle did little experimentation was his distrust of contrived circumstances.[15] Ordinary folks are leery of results teased from the artificial realm of the laboratory. They hanker for real-life conditions and so prefer "field" observations over highly controlled experiments.

This concern for "ecological validity" animated the old criticism of the postmortem approach to medicine and anatomy. People reasoned that since there is a big difference between life and death, study of the living by the dissection of corpses is an unreliable route of investigation. Even today chiropracters dismiss the objections of anatomists on the grounds that these academics measure *dead* backbones. But since much of modern medicine developed from before-death and after-death comparisons of diseased people, we can see the chiropracters' reply is specious. Undoubtedly, there are some points of difference between live bodies and dead ones that have tripped physicians. Introduction of X-ray photography inaugurated a wave of unnecessary surgery to correct "ptosis"—a condition in which an organ has slipped down from its proper place. For instance, a man complaining of stomach pains might be X-rayed and then have the apparent slippage surgically corrected by hoisting it up a bit. Eventually, surgeons realized that organs of a standing person droop because of gravity. They had mistakenly inferred that the organs were misplaced because their anatomical studies had been dissections of supine cadavers.

3. *Bias in favor of familiar facts.*   When evaluating a counterfactual of the form 'If p were the case, then q would be the case', we are instructed to go to the nearest possible world in which p is true and then check whether q is true. If q is true, then so is the counterfactual. If q is false, then the counterfactual is false. Since the nearest possible world tends to be one that resembles the actual one, we have a legitimate expectation that familiar facts will be the same in both worlds. But in addition to this legitimate inertia of familiar facts, there are illegitimate factors. The overweighting of familiar facts is epitomized by the remark of a little girl in one of Lewis Carroll's tales: "I'm so glad I don't like asparagus. . . . Because, if I did, I should have to eat it—and I can't bear it!."

John Hick accuses atheists of falling into this trap when they argue that God does not exist because there is evil in the world.[16] Their reasoning begins with the premise that since God is defined as a perfect being, He has to be all-knowing, all-powerful, and all-good. But then God would know which possible world is best and have the power and the motive to bring it into existence. Yet since we can imagine a possible world without evil, it follows that we do not live in the best of all possible worlds. Therefore God does not exist. Hick objects that it only *seems* as if we can imagine a better world. The illusion is fabricated by our failure to thoroughly work through the consequences of the supposition of a world without evil. If there were no possibility of pain and injury, bullets could not pierce our skin, broken promises would not sow distrust, and reckless drivers would never cripple pedestrians. The world would be a disorderly playpen where God always breaks in to save the day. There would be no firm laws of nature for science to study. Worse, there would be no call for compassion, no need for courage and fortitude, in short, no irritating grain of sand about which the pearl of virtue may form. The world would be a delightful but aimless fairyland. The actual world is better than this dream reality because it contains human virtue. We tend to overlook the need for evil because we take human character traits for granted. Hence, Hick concludes that the atheist's thought experiments fail to show that there is unnecessary evil.

A second instance of this fallacy arises in the debate over preferential treatment. People sometimes demand compensation for a harm that occurred before they were conceived. For example, some contemporary blacks demand compensation for the harm done to them by slavery. The fellow making this compensation claim believes 'If slavery had not existed, I would have had a better life'. But this counterfactual is false because if slavery had not occurred, he would have never been conceived. (His parents would have never met or at least would have never produced the particular sperm–egg combination necessary for his existence.) As courts of law have upheld in "wrongful life" cases, you cannot be harmed by an event that was a precondition of your existence. People tend to assume that they would be around if history had been different because their own presence scores especially high on the criterion of familiarity.

E. Jumping the If/Ought Gap

One of Hume's best-known principles is that you cannot derive an 'ought' from an 'is'. More specifically, he denied the validity of arguments that infer an evaluative statement from purely descriptive premises. Just as Hume bewailed ethical treatises that slip quietly from how things are to how things ought to be, other philosophers deplore the transition from how things would be to how they should be.

Consider invocations of the Golden Rule. This principle tells me to evaluate actions by asking myself whether I would like to be acted upon in the same way. The Golden Rule readily forbids many injustices and has great popular appeal. Many critics of the rule concede that this exercise in imagination might be a useful heuristic but deny that its results are binding. True, I would not like to be murdered or tortured. But neither would I like to be fairly beaten in competition or audited. The Golden Rule fails to distinguish between forbidden and permissible harms. Imagining yourself in the other guy's shoes is a useful countermeasure to the ignorance bred by our natural egocentricism. (We are not social insects, after all.) But this rule is not a means of deriving ethical principles.

Sophisticated cousins of the Golden Rule are open to the same objection. Kant's first formulation of the categorical imperative enjoins you to act on only those maxims that you are willing to accept as universal laws. Hence, you are to suppose yourself legislating for an imaginary kingdom where all laws pass unviolated. "Ideal utilitarianism" works much the same way: laws that *would* produce the best consequences if universally obeyed are claimed to be the ones that we actually ought to adopt. The rules of utopia are deemed binding on the actual world. Ideal observer theories switch us from hypothetical situations to hypothetical agents. These agents differ from us in that they are perfectly rational and well informed. Knowing how they would choose tells us how we ought to choose. Social contract theory also belongs to this genre of "ethics by thought experiment" except that it looks specifically to an imaginary *past* stage of our actual society; for here we suppose that people draw up an agreement by which they will live together. Social contract theorists say that the fine print of this hypothetical agreement spells out the details of morality.

John R. Danley alleges that the appeal to a social contract confuses an invalid hypothetical process argument with the valid historical process argument.[17] If you win $20 from me in a fair game of poker, then the historical process of the game makes me owe you $20. But now suppose that it is only true that you *would* have won $20 from me if I had played poker with you. This hypothetical process does not obligate me to pay you $20. Again, it might be true that if you had offered me $100 for the painting in my attic before I learned that it was a masterpiece, then I would have sold it to you. But this does not obligate me to sell it for $100 after I learn that it is a masterpiece.

We are also tempted to jump the if/ought gap to justify paternalism. You act *paternalistically* toward a person when you justify your infringement of his rights on the grounds that you are a better judge of what will be in his best interests. Paternalism by parents on behalf of their immature children is readily accepted because there is a presumption that adults *are* the best judges of their own self-interest. This presumption can be overridden by the person's mental incapacitation or ignorance. The problem is to explain the pattern of overrides in a way that systematically preserves the presumption. The favored way of doing this is by appeal to hypothetical consent. When faced with a patient rendered unconscious by an injury, a physician can often justify his treatment by predicting that the patient will retroactively "consent" after regaining consciousness. But what about cases where the patient may die or become mentally incapacitated? What if he is already irreversibly incapacitated? Here, physicians resort to hypotheticals about what the patient would have agreed to or what the patient would have consented to had he been rational and well informed.

There is also a gap between 'if' and 'is'. Donald Davidson got snagged here when trying to amplify his argument that translators must regard their informants as having mostly true beliefs.[18] The reasoning for this step was that the translation is possible only if the informants are deemed rational and that this forces the translator to attribute mostly true beliefs to them. But since the translator and informants could be in complete but erroneous agreement, the argument does not preclude massive error. To rule out this possibility, Davidson thinks it sufficient to suppose that the translator is all-knowing (about nonlinguistic facts). Since an omniscient being has entirely true beliefs and must translate us as having mostly true beliefs, it follows that we cannot be in massive error. Critics agree that *if* there is an omniscient being and *if* he can translate our talk, then our beliefs are mostly true. But so what? Our *actual* beliefs are not guaranteed to be mostly true by the fact that in another possible world our beliefs are mostly true—namely, the nearest possible world in which we coexist with the omniscient translator.

Although the jump from 'if' to 'is' seems even more outrageous than the jump from 'if' to 'ought', there are valid arguments deriving a contingent descriptive statement from purely hypothetical premises: $(p \supset q) \supset r, (p \supset q)$, $\therefore$ r. If we count hypothetical 'oughts', as 'if' statements, we can use the same argument form to infer an 'ought' from an 'if':

1. If it is true that if pricked, I bleed, then you ought not to prick me.

2. It is true that if pricked, I bleed.

3. You ought not to prick me.

This example shows that we may not construe gap theses as providing sufficient conditions for the invalidity of arguments (as Hume seems to have done). The gap theses should instead be interpreted as statistical generalizations about the high failure rate of such arguments.[19]

Reification can create a double jump, from 'if' to 'is', and then from 'is' to 'ought'. Nineteenth-century idealists conceived of freedom as a matter of doing what you *really* want to do. Hence, your freedom is not infringed when others prevent you from acting out of ignorance of your true interests or the consequences of your action. Indeed, others can make you free by coercing you into satisfying the desires you would have if you were fully rational and well informed. The idealists spoke of this hypothetical agent as your "real self." So, they construed paternalistic frustration of your actual desires as liberating. Ironically, the idealists' defence of liberty becomes a rationale for totalitarianism. If we take the moral metaphysics of the "real self" seriously, then the inference hippety hops from 'if' to 'is' to 'ought'.

F. Overweighting Negative Thought Experiments

A negative experiment is one that fails to reveal an effect. For example, Lamarck's theory that acquired traits are inherited was tested by breeding many generations of "de-tailed" mice. The failure of the amputations to produce a strain of congenitally tailless mice refuted Lamarck's theory. The same gingerly executed appeal to ignorance underlies negative thought experiments such as Rachel's nephew removers.

A negative experiment is only cogent when the putative cause is given a fair chance to manifest itself. Hence, countervailing factors must be repressed and the opportunity for detection maximized. Sadly, these needs are surprisingly hard to satisfy. Many negative experiments have misleading outcomes because the effect is too small to be detected by the equipment of the day. For example, Robert Hooke's balances were not sensitive enough to show that the weight of an object varies inversely with its distance from earth. Often, too, the needed sensitivity can only be procured by tuning one's equipment into the phenomenon. But since the existence of the phenomenon has yet to be established, one must tightly navigate between fairness and circularity.

These problems with detection and hidden interference from countervailing causes lead scientists to attach less weight to negative experiments. This double standard in favor of experiments that claim to obtain an effect is also appropriate for thought experiments. True, thought experiments do not involve equipment; but our judgment still needs be tuned into subtle effects.

This is evident from the fact that thought experimenters rarely notice an unexpected effect. After Gettier published his refutation of the "justified true belief" definition of knowledge, philosophers found lots of "Gettier counterexamples" in earlier literature. They note that Bertrand Russell demonstrated that true belief is not sufficient for knowledge with the case of a man who infers the correct time from a stopped clock.[20] Russell did not notice he had a counterexample to JTB because he was not looking for one. Nor did Russell detect half the lessons coiled in his five-minute hypothesis.

Once we stop picturing the thought experimenter as an inwardly omniscient introspector, we can be more sanguine about the spectacle of a philosophy instructor coaching his students through hypothetical scenarios. Beginners are as unskilled at thought experiment as they are at ordinary experiment. They do not know what to look for, they don't know how to get it, and they don't know how to report what they do get. Compare philosophy with chemistry. A chemist cannot just let his pupils loose in the laboratory. He has to look over their shoulders and tell them what they are seeing.

G. The Additive Fallacy

One major use of thought experiments is to test the relevance of a factor. Aestheticians, for example, try to tell whether authenticity is relevant by imagining a perfect fake of the *Mona Lisa*. If authenticity matters, then the genuine painting and the forgery should differ in aesthetic value despite their observational equivalence. A similar use of contrast cases is used to determine the moral relevance of the distinction between killing and letting die, the epistemological relevance of a belief's origin, and the physical relevance of the left/right distinction. Shelly Kagan alleges that this method is fatally dependent on the ubiquity thesis: if a factor is relevant anywhere, then it is relevant everywhere;[21] for the purveyors of contrast cases think that the distinction is refuted by a single instance in which the factor has no effect. The ubiquity thesis seems plausible because we picture reasons as numbers in an addition equation. If pros and cons are literally summed, as $S = x + y + z$, then changing one of the factors must indeed affect the overall value. But why can't the factors be governed by a function that does not entail the ubiquity thesis? In $S = x \cdot y + z$, increases in y fail to have any impact when $x = 0$. Kagan suggests that self-defense may work like multiplication by zero. Normally, it is worse to kill someone than let him die. But there is no moral difference between killing your murderous assailant and letting him die.

According to Kagan, the alternative functions expose the arbitrariness of our powerful, unconscious preference for the additive model. Once we abandon this model, there is no longer reason to believe that a relevant factor always makes a difference in a particular case. In addition to undermining the method of contrast cases, the failure of the ubiquity thesis undermines transport arguments. Transport arguments are used to deal with complicated situations involving a number of factors of uncertain magnitude. The procedure is to measure the weight of the factors individually in simple situations, and then "transport" these quantities back to the original situation so that they can be added together. Whole-to-part transport arguments, such as the one based on Glover's bandits-and-beans scenario, are just as vulnerable.

There is a genuine fallacy here, but Kagan overestimates its extent and severity. Awareness of complementary goods and evils is widespread. Everyone knows that it is better to have a can and canopener together than to have each separately. Everyone knows the reverse holds for lockjaw and sea sickness. The reason why we tend to operate with the additive principle is that it is simple and effective. Hidden, robust complementarity is rare. That is just a contingent fact about the interaction between our values and our world. All experimental methodology is arbitrary in the sense that it would systematically fail in another possible world. The fallacy only begins when we make these inductive principles deductive.

How much should we worry about the additive principle's failures? Kagan says that the problem is grave enough to warrant abandoning transport arguments and the method of contrast cases. The proper degree of concern can be ascertained by letting ordinary experimentation be our guide. Experimentalists have their share of cautionary tales about interacting variables. But these are the exceptional cases, ones that do not stop them from robust allegiance to Mill's method of difference. If interaction effects are suspected, follow-up experiments settle the matter. Thought experimenters should follow the same policy of mild vigilance. But that is all.

H. The Blindspot Fallacy

There is a difference between what the characters within a thought experiment can know and what the audience can know. Generally, people on the outside of the story have the advantage of being able to know everything on the story-teller's word. The characters within the hypothetical cannot rely on this stipulative source. Indeed, sometimes the depicted situation will be epistemically inaccessible to the character. Consider Arthur Collins's thought experiment against the thesis that beliefs may be representations in the brain. If beliefs were representations in the brain, there could be a reliable belief-reading machine. Now suppose that as you gaze at the falling rain, the machine says you do not believe it is raining. You would then know 'It is raining, but I don't believe it.' But that is absurd. Hence, Collins concludes that that beliefs cannot be representations in the brain. However, this assumes that the fellow hooked up to the belief-reading machine could know that the machine is infallible, that he is operating it correctly, and so forth. It is tempting to assume that he can know these things because we know them from the outside by the thought experimenter's stipulation.

The Death case also illustrates the danger of shuttling back and forth between the privileged perspective of an outsider to that of the doomed traveler within the thought experiment. The relativity of unknowability introduces a further wrinkle into the epistemology of hypotheticals. According to David Lewis, the hypothetical variants of the deterrence paradox involve situations in which knowledge is logically but not technologically possible. Hence, there is the added danger that we will slide from what is knowable in a distant, possible world to what is knowable in a near, possible world.

IV. Antifallacies

A fallacy is a bad inference rule that looks like a good one. An antifallacy is a good inference rule that looks like a bad one. All antifallacies are pseudofallacies, but not vice versa. The fallacy of animism seems to be committed by the bowler who gestures at his straying ball as if directing midcourse corrections. But the bowler is only *expressing* his hopes and fears. Since he is not making any inference about the effect of his "body English" on the ball, there is no fallacy—and no antifallacy.

A. General Characterization of Antifallacies

The concept of an antifallacy is a useful counterweight to the negative thinking caused by the fallacy chapters of logic textbooks. By concentrating exclusively on bad inferences that look like good ones, students bathe in a biased sample. Restricting their assignments to "search and destroy" missions makes them trigger-happy.

Judging inferences is an exercise in quality control. Here is the contingency table:

	Good Inference Rule	Bad Inference Rule
Looks good	True positives	False positives
Looks bad	False negatives	True negatives

There is no testing flaw if all items are in the top left or bottom right cells. However, members in the other two cells are mistakes and so comprise the price of accepting a bad item or rejecting a good one. If this cost is significant, it may pay to improve accuracy. When we apply this table to argument evaluation, the study of fallacy becomes the attempt to reduce the number of false positives, that is, our acceptance of bad arguments. If we only cared about minimizing this type of error, we could simply spurn all arguments. But we also want the benefits of inference and so must risk some error to get truth. This motivates study of the lower left-hand cell: the domain of antifallacies.

Logic teachers illustrate the fallibility of casual validity judgments with pseudoinvalidities such as "Everybody loves a lover. George does not love himself. Therefore, George does not love Martha." Lots of cogent statistical reasoning looks fallacious. Many U.S. citizens distrust polls when they learn that an inference about the preferences of fifty million voters has been based on a sample of two thousand. They think that the sample size must be some fixed percentage of the population (say, 5%) and so conclude that the pollsters have committed the fallacy of using too small a sample.

Antifallacies are not confined to cases of hidden expertise. For example, anthropomorphization of animals may be an antifallacy. This tendency to put a human face on animal behavior has been criticized as an overextension of an analogy but evolutionary biology has shown that the analogy is surprisingly robust: most differences between man and beast have been explained away as differences of degree rather than kind. Of course, evolutionary biology will not vindicate the extreme anthropomorphization committed by pet lovers. But it may well reveal that our gut inferences about animals are reliable enough. Our conservativeness serves us well when it dovetails with a hidden necessity.

As a prelude to the discussion of the antifallacies surrounding thought experiments, consider the popular practice of dismissing 'What if?' questions on the grounds that they are *hypothetical*. This is a puzzling objection. Its mystery survives the follow-up 'So what?'—for hypophobes only have circular continuations such as 'You are making an assumption' or 'It did not actually happen' or 'It probably will not happen'. This conversational dead end resembles the one reached when you ask for clarification of their triumphant "That's the exception that proves the rule." Your inquiry as to how a counterexample to a generalization can be evidence *for* the generalization is fruitless. Many people are not reasoning at all when they utter these one-liners. Either they are just faking an objection or they are engaged in a verbal ritual such as the standard exchange of insults between quarreling cab drivers. Another emotive explanation takes "That's hypothetical" as akin to the complaint that a story is merely make-believe. Many people have an aesthetic preference for true stories and so are disappointed by the tale's fictionality. (The literati occasionally echo this desire in their calls for "realism.")

However, not every utterance of 'But that's *hypothetical!*' can be analyzed noncognitively. There remain cases that do involve genuine inferences. Here, the principle of charity forces us to elucidate interpretations that minimize irrationality. Hence, we naturally turn to cases where the exclamation is a good (though maybe sloppily formulated) objection. The first point to note is that the apparent triviality of the observation is relative. Highly educated people have had heavy traffic with hypotheticals. Less-educated people have more trouble recognizing and handling hypotheticals. Maybe they deploy 'That's hypothetical' as a cautionary marker that is helpful to *them* in the way 'That's allegorical' and 'That's self-referential' is helpful to intellectuals. Ernst Cassirer claims that the problem is acute in primitive thought:

> Human knowledge is by its very nature symbolic knowledge. It is this feature which characterizes both its strength and its limitations. And for symbolic thought it is indispensable to make a sharp distinction between real and possible, between actual and ideal things. A symbol has no actual existence as a part of the physical world; it has a "meaning." In primitive thought it is still very difficult to differentiate between the two spheres of being and meaning. They are constantly being confused: a symbol is looked upon as if it were endowed

with magical or physical powers. But in the further progress of human culture the difference between things and symbols becomes clearly felt, which means that the distinction between actuality and possibility also becomes more and more pronounced.[22]

If what holds for Cassirer's primitives holds to some degree for our less-educated contemporaries, we can view 'That's hypothetical' as a warning that is helpful in ways that the well schooled are apt to overlook.

Cassirer illustrates the interdependence between the concrete and abstract spheres with empirical research on aphasia. Some patients lose more than vocabulary or syntactic abilities; they lose the ability to entertain mere possibilities. Cassirer mentions the plight of a hemiplegiac with a paralyzed right hand, who could not utter the words 'I can write with my right hand' even though he could say 'I can write with my left hand'. Other aphasiacs are unable to imitate or copy anything that is outside their immediate, concrete experience. Although they do well with concrete tasks, they become confused by abstract ones.

Cassirer's discussion of the aphasiacs may tempt some to say 'That's hypothetical' betokens mental deficiency or linguistic perversion. One of the marks of language is "displaced speech," in which the communicator refers to things outside of his immediate surroundings. *Outside* could be understood spatially or temporally—or modally. To confine yourself to the actual is to be a linguistic shut-in, an alethic agoraphobic. Perhaps the more aggressive cases might be labeled modal xenophobes; they stomp on counterfactuals with the same satisfaction boys display when they splatter exotic insects.

My discussion of the if/ought and if/is gaps contain some examples in which the charge 'That's hypothetical' may be appropriate. People sometimes worry about what is, in effect, the hypothetical process fallacy. But a more promising context is contingency planning. The question 'What should we do if x happens?' can be properly dismissed when we know that x will not happen. For example, 'What will become of the washer if the basement floods?' is dismissed if we know that the house is atop a tall hill. When the stakes are low, we refuse to deliberate about highly improbable contingencies. For instance, 'How will we get home if the car gets *two* flat tires?' is apt to be derided as "too hypothetical." A second promising context is law. Lawyers have a penchant for paradoxical-sounding objections, such as the complaint that an argument is "too simple" or "proves too much." These are stamps for a variety of defects. The lawyers' complaint that a question is "hypothetical" must be understood in this spirit, because they obviously rely on hypothetical questions to establish causation and state of mind. What they are usually objecting to is that the hypothetical question is being used to support an irrelevant proposition or that it calls for unreliable speculation. Sometimes the lawyer is merely *reminding* the jury that the event in question is hypothetical rather than actual. In any case, the hypothetical nature of the question is not the ultimate object of complaint. The real irritant is actually a cluster of features that are *correlated* with hypotheticality.

Since many politicians are ex-lawyers, legal parlance perculates up into political vocabulary and then foams over into mass consciousness. Public officials are fond of declining questions on the grounds that the queries are hypothetical: 'Would you run if nominated?', 'What will you do if they strike?', 'Will you resign if found guilty?'. Often 'That's a hypothetical question' is prudent evasion—a less curt version of 'No comment'. When the politician's response is legitimate, its grounds reduce to either the point about planning or to the legal hodge-podge. My purpose in mentioning the political usage of the "That's hypothetical" objection is to supply a mechanism by which the objection is widely disseminated. The news services ensure that nearly everyone is exposed to political interviews and so nearly everyone has witnessed its deployment. From there, it's "Monkey see, monkey do." People use the incantation if they only have a vague notion of their target—or if they just have nothing better to say. They will use 'That's *hypothetical*' as they would a shotgun: you do not have to know how it works to defend yourself with it. Unfortunately, ignorance breeds indiscriminate usage.

Having done my bit for charity, I conclude that there is some basis for the 'It's only hypothetical' objection—but only enough to explain its popularity as an overexpansion from its tiny beachhead of legitimacy. Many arguments with highly hypothetical premises are antifallacies; they look fallacious but are not. Happily, the hypothetical antifallacy is almost entirely restricted to nonintellectuals. The same cannot be said for the next underestimate.

B. The Far Out Antifallacy

One of most popular objections to a thought experiment is that it is "*too* hypothetical," "unrealistic," or "bizarre." The fact that so many people make this objection suggests that they have some *common* complaint against far out thought experiments. But when you press for details, you find a hodge-podge of reservations—some warranted, most not. Although distinct from the crude, hypothetical antifallacy, this master antifallacy has similarities that make it the rich man's version. The goal of this section is to unravel the lines of reasoning that have been balled up under this umbrella objection.

1. *The trivial interpretation.*   "Bizarre suppositions are flawed" is an *untruism*.[23] An untruism is a statement that is ambiguous between a true but trivial reading and a substantive but controversial reading. My favorites are "Men are men and women are women" and "What will be will be." Defenders of untruisms tend to retreat to the trivial reading when challenged and then drift back to the substantive reading when the pressure is off. Others seem to take the blur between the readings as their object of belief. Therefore, let us carefully distinguish between 'His act of supposing p was bizarre' and 'He supposed p, which is a bizarre proposition'. One can bizarrely suppose a proposition that is not bizarre. A mathematician who begins his number theory proof with the supposition that even numbers envy odd numbers makes a bizarre (because irrelevant) supposition. One can also nonbizarrely suppose a bizarre proposi-

tion. Consider how John Passmore demonstrates the falsifiability of 'Every-thing doubled in size last night'.[24] Passmore has us imagine that the relative sizes of objects have changed at random: pins are as long as poles, rings are bigger than tires, thimbles wider than cups. Although he would not know whether some of the objects doubled or some of them halved their size, Passmore would know the falsity of 'Everything doubled in size last night' because he knows that universal doubling preserves relative sizes. Although Passmore is imagining a bizarre possible world, his supposition is a paradigm of rational behavior.

When 'bizarre' modifies the *act* of supposing p (rather than the object of the supposition p), we get a reading under which the thought experimenter is breaking one of the conventions governing supposition. Since the failure to satisfy these conventions constitutes a flawed performance, bizarre supposi-tions are flawed under this reading. But under the dominant reading, in which the content of the thought experiment is being described as bizarre, there need be no flaw. This point of caution also holds with the other debunking adjec-tives: *crazy, fantastic, kooky, loopy, nutty, outlandish, queer, unrealistic, wacky, weird*. If bizarre thought experiments are to be debunked, they should be debunked cleanly, without equivocation, without the illicit boost caused by the scope confusion.

The distinction is also therapeutic for fact-fetishism. Practical chaps gain a sense of security from the mere fact that their claims concern down-to-earth, realistic situations. One source of the feeling is the fallacy of reverse verbalism. The fallacy of verbalism is committed when properties of words are ascribed to their referents (as when clouds are described as "vague"). Reverse verbalism occurs when properties of the referents are ascribed to the words. In this case, the fallacy is transferring 'down to earth' and 'realistic' from the situation under discussion to the discussion itself. A realistic topic does not ensure realistic comments. Hence, some other justification is needed for preferring realistic thought experiments over fantastic ones or for preferring actual cases over imaginary ones.

2. *The impossibility interpretation.* The strongest reservation behind 'That thought experiment is too farfetched' is that the hypothetical situation is impossible. Although the impossibility response is one of the legitimate resolu-tions of a thought experiment discussed in chapter 6, people tend to equivocate by latching on to the wrong kind of impossibility. 'Impossible' has to be relativized to the proper background constraints. It is a practical impossibility for all the oxygen molecules to segregate to one corner of the room, thereby suffocating me. But it is physically possible. An attack on a thought experi-ment that shows the supposition to be logically impossible is sure to be successful. But the choice of a weaker impossibility courts the danger of too weak a response.

R. M. Hare is exasperated into this mistake by the swarm of hypotheticals that harry every utilitarian:

It shows the lack of contact with reality of a system based on moral intuitions without critical thought, that it can go on churning out the same defences of liberty and democracy *whatever* assumptions are made about the state of the world or the preferences of its inhabitants. This should be remembered whenever some critic of utilitarianism, or of my own views, produces some bizarre example in which the doctrine he is attacking could condone slavery or condemn democracy. What we should be trying to find are moral principles which are acceptable for general use in the world as it actually is.[25]

Paul Taylor commits a less-severe version of the error when he defends utilitarianism from thought experiments that pit utility against justice: "Given a clearheaded view of the world as it is and a realistic understanding of man's nature, it becomes more and more evident that injustice will never have, in the long run, greater utility than justice. Even if the two principles of justice and utility can logically be separated in the abstract and even if they can be shown to yield contradictory results in hypothetical cases, it does not follow that the fundamental idea of utilitarianism must be given up."[26] Some may be tempted to say that Taylor has fallen for the simple hypothetical antifallacy because he seems to be discounting the counterexamples to utilitarianism merely on the grounds that they are hypothetical. However, Taylor continues his defense by drawing attention to a mechanism by which injustice lowers overall utility: injustice causes resentment, resentment breeds opposition to social rules, which in turn creates social strife and thereby misery. Given this regularity, there will be no conflict between justice and utility in the possible worlds psychologically and sociologically similar to our own. Hence, Taylor's real thesis is that utilitarianism can only be required to get the right results for this local group of possible worlds, not the more distant ones. However, utilitarians must satisfy a stricter standard because they are claiming to *define* 'morally right'. Definitions must hold for all possible worlds. Lower standards are appropriate when stating *laws* about the term in question. The utilitarians could switch to these lower standards if they admit they only wished to generalize about the nature of moral rightness for human beings who are pretty much like us. But then their enterprise looks more like applied ethics. Philosophers doing business ethics and medical ethics need only form generalizations that hold for business contexts and medical contexts, because their restricted interests allow them to use factual assumptions that narrow their domain of discourse. Applied ethics is a legitimate enterprise and can be conducted without commitment to a general theory of right and wrong. However, we are still curious about the general theory, and the utilitarians have presented themselves as providing the big answer. If they are not, then they should be condemned for misleading advertising and for applying the wrong techniques—even if we agreed that utilitarianism works for the special case of folks like us. Philosophers would still wonder *why* "utilitarianism" works in our special case and would seek the more general theory of rightness.

Mark the danger of evasive oscillation. When challenged, slippery characters retreat to the thesis that their theory is only intended for a special case.

Once the pressure is off, they drift to the more ambitious stand that the theory works in general.

3. *The inaccessibility interpretation.* The biological baseline creates a presumption against certain kinds of exotic thought experiments. The baseline tolerates improbability, incongruity, and the strangeness that accrues from simplification. But it frowns on thought experiments that demand transcendental powers of cognition: "Biopsychology good—parapsychology bad."

Bizarreness is also an appropriate target when it clouds up the modal status of the imagined situation. One kind of obscurity is remediable. Recall Alvin Goldman's contention that bizarre suppositions risk a clash between our understanding of the rules of language and the strategies that help us abide by these rules. Future psycholinguistic research may vindicate Goldman's hypothesis of interference. On the other hand, it may reveal ways in which bizarreness *helps* linguistic judgment. Professional mnemonists (as well as some researchers) claim that bizarreness strengthens the method of loci.[27] They say that it is easier to remember a dog and a car by forming a mental image of the dog driving rather than of the dog chasing the car. If so, bizarre suppositions might have the advantage of being better memory cues. (Notice that this causal claim goes beyond Amartya Sen's argument for *tolerating* bizarreness as a side effect of measures taken to damp down extraneous variables.)

The other sort of obscurity is irremediable. The vagueness of a predicate's entailments is responsible for borderline cases of 'impossible' such as 'Some vats are tiny'. As before, the obscurity will lead many people to reserve judgment. Others will be bolder and precisify the vagueness away. Theoreticians are free to alter familiar concepts if the modifications produce intellectual benefits such as increased concision, testability, and intertheoretic coherence. Greater and greater benefits are needed to offset increasingly radical reform of the familiar concept. Most theorists think that changing clear positive cases to clear negatives (or vice versa) is more radical than changing borderline cases to clear cases.[28] Hence, they take greater license with borderline cases. Thus, if a thought experiment involves a situation that is borderline between being possible and impossible, a theoretician is more apt to stipulate it as impossible on the grounds that the change secures theoretical benefits. By complaining that the thought experiment is too far out, the critic could be cautioning against an overestimate of its destructive power. The point is that it is easier to stipulate away a difficulty posed by a borderline case than one posed by a clear case. The thought experiment is not impotent; it is just less virile than it appears.

Happily, the issue of bizarreness can often be sidestepped. For the strangeness of a scenario is often confined its nonessential aspects. We are then free to revise the supposition into an unobjectionable format.

4. *The rhetorical version.* The queerness of a thought experiment is sometimes cited as a purely *tactical* drawback, rather than a justificatory matter. Thought experiments aim at persuading people and so can be criticized on the

grounds that they are not apt to achieve the intended psychological effect. Although go-go hypotheticals are nirvana to control freaks, conservatives only cosy up to counterfactuals that concern familiar situations. Farfetched thought experiments do not engage them. The point of this objection is purely psychological. Some of my students are unpersuaded by highly impersonal, dry hypotheticals. They are bored by universes containing just a few elementary particles varying only in their direction of spin. So if I want to persuade them, I'd better juice up my thought experiments. Likewise, if I want to persuade the homebodies who only warm to familiar surroundings, need to tone down the exotic bits and find a Norman Rockwell setting. Hence, under this rhetorical interpretation, the objector is only warning against inflexible salesmanship. His point is that even if fantastical thought experiments constitute proof, they are not *persuasive* to people of a certain temperament.

Fair enough. But note that few intellectuals have a hang-up about far-out thought experiments. Capitulation is more common. For example, Thomas Nagel tried to explain why we fear death but serenely accept prenatal nonexistence by claiming that only death deprives us of more life. Robert Nozick objected. Imagine people who develop from spores that normally lie dormant for thousands of years. Upon hatching, these people have a lifespan of a hundred years. A speed-up technique is then discovered that gives the early hatchers thousands of years of active life. Now suppose that you are a spore person who learns that you hatched naturally and so missed thousands of years of earlier active life. Although you would lament this deprivation, you would not feel the same way toward it as your death. Rather than dismissing Nozick's hypothetical as too farfetched, Nagel agrees that "something about the future *prospect* of permanent nothingness is not captured by the analysis in terms of denied possibilities."[29]

5. *The digression objection.*   What brings "the lava of the imagination" to the surface? Plato categorized the force as madness: "For the poet is a winged and holy thing, and there is no invention in him until he has been inspired and is out of his senses, and the mind is no longer in him: when he has not attained to this state, he is powerless and is unable to utter his oracles" (*Jon* 534). Plato's thesis about artistic creation continues to be influential. Hence, the analogy between literature and hypothetical scenarios breathes life into the idea that there is a "touch of the Muses' madness" in the soul of the thought experimenter. This impression grows with the bizarreness of the scenario because of the resemblance to delusions of the insane. Thus, a "crazy" thought experiment tends to be aestheticized into irrelevance.

This dismissive attitude is reinforced by the ambiguities of 'ideal'.[30] An ideal F is an F that perfectly satisfies a set of criteria. Thus, an apple is an ideal dessert because it is nutritious, sweet, and cheap. Since criteria are normally laid down to guide the pursuit of our goals, 'ideal' usually covers pleasing things. But this is not a matter of entailment, because goals need not be shared. For example, ideal centrifuges and test tubes only please laboratory staff. Since 'perfect' is an absolute term, fewer objects qualify as ideal as we raise standards. Apples do cause tooth decay, centrifuges wobble and whir. Thus, intensive discussion about ideal objects *tends* to become increasingly hypothetical. When the initial criteria are geared to a noninstrumental goal, the result can be utopian escapism. For instance, Emily Bronte wrote poems and stories about the imaginary kingdom of Gondal because this ideal world was a refuge from her bleak home life. However, most idealizations are kept in focus by the specific functions they serve. Galileo's stipulations are intended to remove particular distractions and complications. For example, he stipulates away friction and air resistance to simplify the behavior of moving bodies. Nonetheless, the irrealism of the instrumental and recreational uses of 'ideal' lead equivocators to picture the idealizer as a dreamy quitter. Thought experimenters and poets come to be regarded as kindred spirits who react to ugly realities by withdrawing into fantasies.

6. *Semantic holism.*   This reading of 'Bizarre thought experiments do not move me' unites several of the preceding themes by reorganizing the statement into a claim about language use.

Consider the Wittgensteinian insistence that words be understood as they function within a language game. This encourages us to view concepts as having a certain niche. Concepts wither when transplanted into alien soil. Those who make claims about how these deanimated notions behave should be compared to worshippers who eventually "see" religious statues move after long periods of gazing.

P. H. Nowell-Smith took this line against those who appealed to desert island cases. The issue arose over his claim that it always makes sense to ask 'Why obey this rule?'. Nowell-Smith notes that one of the most common ways of objecting to this position is to invent a peculiar scenario in which a rule should be obeyed even though no advantage accrues. Suppose, for instance, that you have been marooned on an island with a man who is now dying. His solace has been a garden. The man's deathbed request is that you carry on his gardening. You promise. He dies. You let his flowers perish. Now, isn't it obvious that you have done something wrong even though you knew the lying promise was devoid of bad consequences? Utilitarians respond by raising doubts about whether you really know there would be no bad effects. Can you be quite sure that your character will not be undermined? How do you know that you will be able to keep the secret? You might be rescued and let the secret slip, thereby weakening the institution of promising. But the purveyor of desert island stories patiently stipulates these doubts away with amendments to the thought experiment.

Nowell-Smith suggests that we instead exclude the hypothetical scenario as irrelevant on the grounds that ordinary language does not cater to these improbable situations. The rules of English do not determine any answer. He invites us to compare the desert islander's question to "What would you say if half of the standard tests for deciding whether a piece of copper wire is electrified gave a positive answer and half a negative answer?" or "What would you say if you added a column of ten figures a hundred times and got one answer fifty times and another the other fifty times?"

> The answers to these questions could only be either "I must see a doctor at once" or "I simply do not know what I should say; for the logic of my language for talking about electricity or adding does not allow for this sort of thing. If it occurred, I should have to treat some sentence which normally expresses an analytic proposition as expressing a synthetic one; but I certainly cannot say which."
>
> In the same way I confess to being quite unable to decide *now* what I should say if a desert-island situation arose. Moral language is used against a background in which it is almost always true that a breach of trust will, either directly or in the more roundabout ways which utilitarians suggest, do more harm than good; and if this background is expressly removed my ordinary moral language breaks down.[31]

Nowell–Smith is denying that there is any truth or falsity to the counterfactual. If the situation were to arise, there would be changes in language that give the sentence a truth value. Nowell–Smith denies that we can predict whether the language would change in a way that made the sentence true or made it false. In any case, since this revised language would not be identical to our actual language, we would only be learning the truth value of a verbally similar counterfactual.

Linguists who study the evolution of languages have had little success so far, but they may one day be in a position to predict semantic shifts. For example, they may be able to predict that in English of the year 2200 'dog' will become a general term covering any pet. But that would not show that 'dog' in contemporary English covers cats and goldfish. The same holds for semantic shifts involving borderline cases. Evolutionary linguists might know that in the English of 2200 'food' will have been precisified to make coffee a clear case of food. But this foreknowledge leaves coffee a borderline case of food. Hence, Nowell–Smith's emphasis on our inability to predict linguistic changes is a red herring. Future precisifications of currently vague terms do not alter the current meanings of those vague terms.

Fodor's moderate scepticism about thought experiments is also guilty of topic switching. Recall that Fodor bases his scepticism on three premises. The first is meaning holism, the view that a change in central theory tends to cause a change in meaning of the vocabulary expressing that theory. The second premise is that farfetched thought experiments imply the falsehood of central theory. The third premise is that we cannot predict how we would revise our beliefs in response to a falsification of a central theory. As Fodor notes, these premises imply that we cannot predict what we would say after a dramatic falsification of central theory. But our future or hypothetical language is irrelevant. Recall Abraham Lincoln's riddle "If you called a tail a 'leg', how many legs would a dog have?". The answer is four: calling a tail a 'leg' does not make it one. The moral is that when we judge that hypothetical language changes, we hold the actual language constant. Otherwise, we would be able to "prove" that 'Two plus two might equal five' by merely supposing that we used 'five' to denote what we now denote with 'four'. Hence, even if we grant Fodor's conclusion that we cannot predict what we would say *in our revised language*, it does not follow that we are restricted to tame counterfactuals.

This charge of equivocation can be supplemented with a logical analogy. If Fodor is right to distrust the farfetched suppositions associated with the appeal to ordinary language, he should also distrust the equally weird scenarios envisioned by those who try to separate laws from accidental generalizations and basic rules from derived ones and those trying to gauge the depth of a necessity. All of these clarificatory activities engage in curve fitting. They try to reconcile theory and modal intuitions, so thought experiments play a role in this process of reflective equilibrium. Thus, they face choices about how to weight the data. To dismiss all of our intuitions about far out scenarios as "outliers" is to attribute an extensive illusion to human beings. Such attributions conflict with the principle of charitable interpretation.

As a corollary, we dispose of the worry that thought experiment is an intrusive measure of possibility. A charismatic speaker cannot change a borderline possibility into a clear one. Changing the habits of his speech community only alters which sentences express which propositions.

C. Strangeness In, Strangeness Out?

Often a thought experimenter will try to refute a theory by conjoining it with a strange supposition and then extracting a strange consequence from this conjunction. A common reply is to admit that the consequence follows, admit that the consequence is strange, but then blame the strangeness on the strange supposition, not the theory to which it has been haplessly conjoined. If you force a theory into a shotgun marriage with a weird scenario, you should expect peculiar progeny.

However, "Strangeness in, strangeness out" is false when construed deductively. Counterfactuals with bizarre antecedents need not entail bizarre consequents: 'If there were six sexes, then there would be more than one sex'. Indeed, since tautologies are entailed by any proposition, the strangest antecedent can have the tamest consequent. Since the consequent of any deductively valid conditional must be at least as probable as its antecedent, the improbability of the antecedent cannot be exceeded by that of the consequent. Indeed, in all but the trivial cases, the consequent must be more probable than the antecedent and so less strange. Hence, if the consequent is stranger than the antecedent, the strangeness cannot be drawn entirely from the antecedent.

We should also note that humdrum antecedents sometimes have bizarre consequents. For example, it is not odd to suppose that a man on crutches who is suffering from phantom limb pain locates his painful "foot" as resting on a book lying on the floor. Nor is it odd to suppose that a second sufferer standing beside him also locates his foot as resting on that book. But if this were the case, then we get the bizarre consequence of two distinct pains being located in the same spot.

Much of the appeal of the slogan can be traced to two heuristics used in causal reasoning. The unusualness heuristic nominates peculiar antecedents of the effect as the cause. Thus, a baseball player who eats raw fish prior to pitching a perfect game is apt to eat raw fish before the next game. The strangeness of a thought experiment scenario is a magnet for the unusualness heuristic; critics will say that the very peculiarity of the stipulated situation is the cause of the trouble. The representativeness heuristic nominates causes on the basis of their similarity to the effect. For example, the slow of speech were fed lizard tongues. People think that "great events ought to have great causes, complex events ought to have complex causes, and emotionally relevant events ought to have emotionally relevant causes."[32] The belief that strange causes should have strange effects harmonizes with this list. It is a short step from this causal context to the principle that a strange supposition will have strange consequences. Philosophers appeal to this principle when the application of their theory to a strange situation yields an apparently absurd consequence. They urge us to blame the situation for the absurdity, not the theory. The take-home message is that this excuse is overrated. "Strangeness in, strangeness out" only has inductive force. The slogan is useful as a reminder that the strangeness of the consequent *might* be *partly* due to the strangeness of the antecedent. To convict the strange scenario, one must first show that the consequent is not stranger than the antecedent and that the antecedent and consequent are strange in the same way.

D. The Voyeur Antifallacy

In chapter 8 I spoke of the immigration of the supposition operator from SBp to BSp. This transition is needed to free the thought experimenter from the constraints of observability. If I am supposing that I am bringing about a state of affairs (conducting an experiment, for example), then I must suppose the existence of the preconditions of this action. Hence, the BSp format forces the thought experimenter to assume that he exists, that his perceptual apparatus is operative, and that the general background conditions of life hold as usual. These constraints are lifted with the BSp format. The voyeur antifallacy occurs when a BSp thought experiment is criticized as if it were a SBp thought experiment. The objector alleges that the thought experimenter is failing to follow his own instructions because he is telling us to imagine that some precondition of perception fails. If the precondition really fails, says the objector, we cannot imagine the situation. It only seems that we are imagining the situation because we are sneaking a look in violation of our supposition.

The appearance of fallacy may be strengthened by the resemblance to the blindspot fallacy in which the audience's epistemic state is illictly transferred to the characters in the hypothetical. I have already discussed how the voyeur antifallacy is fueled by the mistaken view that all imagination is pretended perception. It can also be motivated by the belief that imagination is a mode of perception, just as vision, hearing, and smell are modes of perception. The error stands out in a criticism of G. E. Moore's Two World thought experiment.[33] Moore wished to refute Sidgewick's thesis that value resides only in conscious states. Hence, he has us imagine a beautiful but uninhabitated world and then an ugly but uninhabited world. Since it would be better for the beautiful world to exist, value must be possible without experience. Oliver Johnson objects that Moore's procedure cannot be executed:

> The world of his choice, when it comes into being later, either will or will not have the qualities that he imaginatively pictures in it. If it has the qualities, then it is a world that has been viewed by an observer, even though only by means of imagination. If it does not have these qualities, then we have no grounds on which to attribute any value to it at all, since we have no idea what it is like.
>
> ... The root difficulty with Moore's two world illustration is that he asks the person who is to choose between the worlds to perform a task he simply cannot accomplish—to imagine the unimaginable, to visualise two worlds with certain aesthetic qualities and at the same time to choose between these worlds without allowing that visualisation to affect his choice.[34]

Moore's thought experiment is innocent because imaginative perception is not perception. A necessary condition of perception is that the process gives the perceiver an opportunity to pick up new information. One also expects a corresponding sense organ and access to the perceived scene by other senses. Almost the only reason for believing that imaginative perception is perception is the surface grammar of expressions such as 'imaginative perception', 'before the mind's eye', and 'picturing the scene in one's mind'. These are systematically misleading expressions, grist for Ryle's mill.

E. The Kabuki Antifallacy

One peculiarity of Kabuki theater is that the stagehands handle props in full view of the audience. The regular Japanese audience is not distracted, but those unfamiliar with Kabuki have difficulty ignoring the stagehands. Either they mistake the stagehands as actors (and think the play stupid or unintelligible) or regard the stagehands as aesthetic intruders.

This accusation of shoddy showmanship is naive. The Kabuki theatergoer is supposed to know what to pay attention to and what to ignore. People attending a Broadway production of *West Side Story* know better than to attribute amazing musical and dancing abilities to the gang members. When the cast breaks out into song, they either let only a bit of the activity count as part of the plot or bracket out the event altogether. Likewise, the "Greek chorus" that meanders in and out of plays are not regular characters in the story even though they are part of the play. The interpretation of stories requires a surprising degree of selectivity.

If we approach thought experiments as tiny stories, we should then expect that they will also have some red herrings tossed at them. For example, if a physics student rejects the following thought experiment on the grounds that time travel is impossible, he will be accused of "missing the point":

> Due to a malfunction of a time machine, you find you have become a gunnery sergeant for a Roman Legion that is sieging a walled city. You are in charge of a catapult and must hurl a projectile to the *top* of a wall. The wall has a height of H, a width of W, and is located a distance L from your catapult.
>
> What is the minimum initial speed necessary to hit the *top* of the wall? If the initial speed is actually *twice* this, what are the minimum and maximum angles that will allow you to hit the top of the wall? Neglect air resistance throughout.[35]

The time machine is just a stylistic flourish and should be ignored like a mechanic ignores a car's hood ornament. Next, consider the picky pupil's reaction to the following scenario: "Suppose in a nightmare you find yourself locked in a light cage on rollers on the edge of a rapidly eroding cliff. Assuming that no external forces act on the system consisting of you and the cage, what could you do to move the cage away from the edge? What must you avoid doing? If you weigh 140 lb and the cage weighs 210 lb and is 10 ft long, how far can you move the cage?"[36] The pupil objects that since dream events do not obey the laws of physics, nothing follows from this thought experiment. Once again, we conclude that he has confused the means of representation with the object of representation.

Ordinary standards for interpreting thought experiments are generous. In addition to seeking interpretations that sidestep the inventor's minor mistakes, we rescue "the main point" from substantive error and ignorance. Plato's Ring of Gyges, for instance, has serious run-ins with contemporary opthamology. To see, the eye lens must bend light to form an image on the retina. But a transparent lens has the same index of refraction as the air (and so cannot bend light) and a transparent retina cannot absorb light. Hence, the invisible man could not see! But an ethicist who tried to rebut Plato by stressing the disadvantages of blindness would be laughed down.

There are two rationales for the amusement. The first and more secure rationale concedes that the Ring of Gyges case is flawed but stresses the triviality of the flaw. Plato's main point is easily salvaged by substituting another scenario that does not run afoul of contemporary science. We are so confident in the possibility of a backup that we do not bother to actually construct it. Instead, we stick with the original, scientifically defective version because the original is more easily understood and recognized.

The feisty rationale admits no error at all. Compare the opthamological criticism of Plato with a criticism Carl Gans levels at Conan Doyle: "In 'The Adventure of the Speckled Band' Sherlock Holmes solves a murder mystery by showing that the victim has been killed by a Russell's viper that has climbed up a bell-rope. What Holmes did not realize was that Russell's viper is not a constrictor. The snake is therefore incapable of concertina movement and could not have climbed the rope. Either the snake reached its victim some other way or the case remains open."[37] As David Lewis observes, Gans is right, given one analysis of what counts as the implicit truths of the story. The explicit truths of the story are the ones explicitly stated in the author's text. The implicit truths are the ones that can be inferred (Sherlock Holmes had toes, Watson did not own a television, etc.)—but inferred from the explicit truths plus what? Under one analysis, you add a suitably edited list of truths about the *actual world*; that is, you delete whatever actual truths conflict with the explicit statements of the story and then merge the two lists to form a base from which you can fill in the gaps. So what is true in "The Adventure of the Speckled Band" is pretty much what is true in the actual world except for whatever adjustment is needed to preserve the truth of text's explicit statements. Now since the story never explicitly states that Holmes was right about the snake climbing up the rope, we should conclude that he was wrong because in the actual world, Russell's viper cannot climb ropes.

Many people balk at this consequence. They prefer a second way of analyzing truth in fiction. This one is like the first except you relativize to "common knowledge" rather than the actual world. A proposition is common knowledge between an author and his audience if both believe it, both believe they both believe it, and so on. (Despite the name, "common knowledge" can be false.) This makes unknown and little-known facts irrelevant to the story. Hence, under this second analysis, Holmes was right.

Which analysis is right for thought experiments? The pugnacious defender of Plato favors the second analysis. He says that we should relativize to the overt beliefs of Plato's community, not to the actual world. If so, then the bearer of the ring can see even though he is invisible. I am not sure whether the pugnacious response is correct. Perhaps some thought experiments should be relativized to common knowledge and others to the actual world. Happily, the first rationale works even if the feisty response proves too adventuresome.

V. A Parting Comparison

Thought experiments are profitably compared to compasses. A compass is a simple but useful device for determining direction. Nevertheless, it systematically errs in the presence of magnets. The compass's scope is limited; it becomes unreliable near the North Pole, in mineshafts, when vibrated, in the presence of metal, and under more arcane conditions. The compass does not point precisely north—only close enough for nearly all navigational problems. Lastly, its mode of operation was long a mystery and the object of superstition. The first compasses were Chinese but were principally used for necromancy and fortune-telling. Even nowadays, we find that their operation can be explained by only a tiny percentage of the educated. Yet ignorance of how it works does not stop anyone from using the compass. Nearly everyone on earth can navigate with its guidance. The few who know how compasses work only have a small (but occasionally crucial) advantage over those who do not. Even these experts will wish to use the compass as one element in a wider portfolio of navigational techniques that provide ample opportunity for cross-checking.

Analogously, thought experiments are simple but useful devices for determining the alethic status of propositions. Sadly, they systematically err under certain conditions and so are best used with sensitivity to their foibles and limited scope. Thought experiments do not directly indicate the alethic status of propositions; they instead provide evidence about the conceivability of a proposition that only imperfectly corresponds to its possibility. This correspondence is close enough for most purposes. But under rare conditions, thought experiment leads us badly astray. Happily, knowledge of its imperfections usually prevents, softens, or remedies serious error. Even so, the circumspect thinker will use thought experiment as one technique among many, basing his final judgment on the *collective* behavior of his indicators. Like compasses, there is mystery as to how thought experiment works. Were it not for the inexorable utility of particular thought experiments, this mystery might provide a basis for general scepticism about thought experiment. But (as with compasses) we see that theoretical inquiry can ground the use of thought experiments with more than an appeal to practicality.

NOTES

Chapter 1

1. Stephen Toulmin, *Foresight and Understanding* (Bloomington: Indiana University Press, 1961), 36.

2. For instance, A. Koslow denies that the first law follows by arguing that the second law tacitly assumes that a *net* force is acting. See his "Law of Inertia: Some Remarks on Its Structure and Significance," in *Philosophy, Science, and Method*, ed. Sidney Morgenbesser et al. (New York: St. Martin's, 1969).

3. Leibniz makes this criticism in "Nouvelles de la république des lettres" (July 1687). It appears as "Letter of Mr. Leibniz on a General Principle Useful in Explaining the Laws of Nature through a Consideration of the Divine Wisdom; to Serve as a Reply to the Response of the Rev. Father Malebranche" in *Philosophical Papers and Letters* vol. 2, trans. and ed. Leroy E. Loemker (Dordrecht: D. Reidel, 1969), 351–54.

4. Friedrich Nietzsche, *Gay Science* 341. This translation appears in Ivan Soll's "Reflections on Recurrence," in *Nietzsche*, ed. Robert Solomon (New York: Anchor Books, 1973), 323.

5. *Nietzsches Werke* (Leipzig: Alfred Kroner Verlag, 1919), 12, 119.

6. Georg Simmel *Schopenhauer und Nietzsche: Ein Vorrogszyklus* (Leipzig: Verlag von Duncker & Humblot, 1907), 250–51. John Carroll has pointed out that the proof goes through without the 2*n* wheel: "Let *c* be the circumference of each disk and suppose for a reductio that they do line up along the string at some time in the future. Let *t* be the time it takes for them to realign, let *j* be the number of revolutions made by the second disk, and let *k* be the number of revolutions made by the first disk. Since the distance divided by the time is equal to the rate, $(k \cdot c)/t = n$ and $(j \cdot c)/t = n/\pi$. But this is impossible; *j* is a positive integer, but since *k* is also a positive integer k/π is not" (personal communication).

7. The most recent example of this tradition is provided by J. L. H. Thomas in "Against the Fantasts" *Philosophy* (1991), 349–67.

8. Alasdair MacIntyre, *After Virtue* (Notre Dame, IN: University of Notre Dame Press, 1981), 2. MacIntyre goes on to note that phenomenology and existentialism would also fail to detect a problem.

9. Jonathan Dancy, "The Role of Imaginary Cases in Ethics," *Pacific Philosophical Quarterly* 66 (1985): 146.

Chapter 2

1. David Hume, *A Treatise of Human Nature*, 1.4.2.190.

2. Richard E. Nisbett and Timothy DeCamp Wilson discuss the position effect in "Telling More Than We Can Know: Verbal Reports on Mental Processes," *Psychological Review* 84 (1977): 243–44 and the bystander effect on pp. 231–59.

3. Auguste Comte, *Cours de philosophie positive* 1.34–38.

4. James refers to this and other evidence for the indivisibility of attention in *The Principles of Psychology*, vol. 1 (New York: Henry Holt, 1890), chap. 11. Many apparent cases of divided attention were explained away as rapid alteration of attention. This strategy is found as late as Robert Woodworth's and Harold Schlosberg's *Experimental Psychology*, 3d. ed. (London: Methuen, 1955).

5. See John Stuart Mill, *Auguste Comte and Positivism*, 3d ed. (1882), 64. William James endorsed Mill's reply in *Principles of Psychology*, vol. 1, p. 189.

6. John Carroll persuaded me to substitute 'constant' for the psychologist's 'reliable', and 'faithful' for their 'valid'. The psychological terminology is misleading because it conflicts with ordinary usage, as well as technical philosophical usage.

7. Auguste Comte, *Cours de philosophie positive*, 1.34–38, quoted in William James's *Principles of Psychology*, vol. 1, p. 188.

8. T. Okabe, "An Experimental Study of Imagination," *American Journal of Psychology* 21 (1910): 563–96.

9. Ibid., p. 576.

10. Annette Baier, "Hume, the Women's Moral Theorist?," in *Women and Moral Theory*, ed. Eva Feder Kittay and Diana T. Meyers (Totowa, NJ: Rowman & Littlefield), 49.

11. James Robert Brown defends the Godelian position in *The Laboratory of the Mind* (London: Routledge, 1991). Also see Alexandre Koyre's *Metaphysics and Measurement* (London: Chapman & Hall, 1968).

12. John Wesley Powell, *Truth and Error* (Chicago: Open Court, 1898), 1–2.

13. Charles Schmitt, "Experimental Evidence for and Against a Void: The Sixteenth-Century Arguments" *Isis* 58 (1967): 352–66.

14. Steven Shapin and Simon Schaffer explain Hobbes's scepticism in *Leviathan and the Air-Pump* (Princeton: Princeton University Press, 1985), 116.

15. Putnam introduced the robot cats in "It Ain't Necessarily So," *Journal of Philosophy* 59 (1962): 658–71. His indecisiveness gives way to a positive verdict in "The Meaning of 'Meaning',," in *Language, Mind, and Knowing*, ed. Keith Gunderson (Minneapolis: University of Minnesota Press, 1975).

16. Bernard Williams, *Problems of the Self* (Cambridge: Cambridge University Press, 1973) 46–63.

17. Mark Johnston, "Human Beings," *Journal of Philosophy* 84 (1987): 67.

18. Ibid., 81.

19. Unger makes this and other psychological points about thought experiment in "Toward a Psychology of Common Sense," *American Philosophical Quarterly* 19 (1982): 117–29.

20. W. V. Quine, Review of *Identity and Individuation*, *Journal of Philosophy* 69 (1972): 489–90.

21. Derek Parfit, *Reasons and Persons* (Oxford: Clarendon Press, 1984), 200.

22. James Jeans, *Mysterious Universe* (Cambridge: Cambridge University Press,

1930), 4. Jeans attributes the statement to Huxley. The myth is debunked by Charles Kittel in his *Thermal Physics* (New York: John Wiley, 1969), 65–66.

23. The circumference of an object is $2\pi r$. Let r_1 be the radius of the earth and r_2 be the radius of the ribbon. The circumference of the ribbon equals the circumference of earth plus twelve inches: $2\pi r_2 = 2\pi r_1 + 12$. Since the right-hand side equals $2\pi (r_1 + 12/2 \pi)$, dividing both sides by 2π reveals $r_2 = r_1 + 12/2\pi$. Hence, the difference in radii equals $r_2 - r_1 = 12/2 \pi = 1.9$ inches.

24. Gilbert Harman, "Moral Explanations of Natural Facts—Can Moral Claims Be Tested Against Moral Reality?" *Southern Journal of Philosophy* 24 Suppl. (1986): 60.

25. Ibid., 61.

26. Lawrence Kohlberg, "Stage and Sequence," in *Handbook of Socialization Theory and Research*, ed. D. A. Goslin (Chicago: Rand McNally, 1969), 379.

27. Carol Gilligan, *In a Different Voice* (Cambridge: Harvard University Press, 1982).

28. G. E. M. Anscombe, "Modern Moral Philosophy," *Philosophy* 33 (1958): 17.

29. Bernard Williams, *Utilitarianism: For and Against* (Cambridge University Press, 1976), 97.

30. Lewis Thomas relates the incident in *The Youngest Science: Notes of a Medicine Watcher* (New York: Viking, 1983), 22.

31. Thomas Kuhn, "The Function of Dogma in Scientific Research," in *Scientific Change*, ed. Alistair C. Crombie (New York: Basic Books, 1963).

32. David Hume, *A Treatise of Human Nature*, ed. L. A. Selby-Bigge (New York: Oxford University Press, 1978), 32.

33. Peter van Inwagen, *An Essay on Free Will* (Oxford: Clarendon Press, 1983), 154.

34. Nicholas Rescher discusses ineffable questions in *The Limits of Science* (Berkeley: University of California Press, 1984), chap. 2.

35. Peter Strawson, *Individuals* (New York: Anchor Books, 1959), chap. 2.

36. Antoine Arnauld, "Objections IV, and Replies" in *The Essential Descartes*, ed. Margaret D. Wilson (New York: New American Library, 1983), 267.

37. I. Moar and G. H. Bower, "Inconsistency in Spatial Knowledge," *Memory and Cognition* 11 (1983): 107–13.

38. Albert Casullo, "Reid and Mill on Hume's Maxim of Conceivability," *Analysis* 39 (1979): 212–19.

39. W. D. Hart promotes the perceptual analogy in *The Engines of the Soul* (Cambridge University Press, 1988), 28.

40. Labov's scepticism is advanced in *Sociolinguistic Patterns* (Philadelphia: University of Pennsylvania Press, 1972) and *What Is a Linguistic Fact?* (Lisse: Peter de Ridder, 1975).

41. A contribution to this project is found in Guy Carden and Thomas G. Dieterich's "Introspection, Observation, and Experiment: An Example Where Experiment Pays Off" *PSA 1980* 2 (1980): 583–97.

42. Jerry Fodor, "On Knowing What We Would Say," in *Readings in the Philosophy of Language*, ed. Jay F. Rosenberg and Charles Travis (Englewood Cliffs, NJ: Prentice-Hall, 1971).

43. Alvin Goldman, "Psychology and Philosophical Analysis," *Proceedings of the Aristotelian Society* 39 (1989): 195–209.

44. J. L. Austin, *Philosophical Papers* (Oxford: Clarendon Press, 1961), 133n.

45. Ibid., 146n.

46. Hilary Putnam, *Reason, Truth, and History* (Cambridge: Cambridge University Press, 1981), 1.

47. Ludwig Wittgenstein, *Philosophical Investigations* 142.

48. Idem., *Zettel* 350.

49. James Broyles, "An Observation on Wittgenstein's Use of Fantasy" *Megaphilosophy* 5 (1974): 296.

50. Carl Hempel, "The Theoretician's Dilemma," in *Minnesota Studies in the Philosophy of Science II*, ed. Herbert Feigl et al. (Minneapolis: University of Minnesota Press, 1958).

51. Charles B. Schmitt quotes this passage from Toletus' commentary on Aristotle's *Physics* in "Experimental Evidence for and Against a Void: The Sixteenth-Century Arguments," *Isis* 58 (1967): 352–66.

52. Galileo to Kepler, quoted in E. A. Burtt, *Metaphysical Foundations of Modern Physical Science* (London: Kegan Paul, 1932), 66–67.

53. Carl Hempel, *Aspects of Scientific Explanation* (New York: Free Press, 1965), 165. Others are content to stress that actual cases are much superior to hypothetical ones, e.g., David L. Hull, *What the Philosophy of Biology Is* (Dordrecht: Kluwer Academic, 1989), 309–21.

54. Pierre Duhem, *The Structure of Physical Theory*, trans. Philip P. Wiener (New York: Atheneum, 1974), 200–205.

55. Ron Naylor, "Galileo's Experimental Discourse," in *The Uses of Experiment*, ed. David Gooding et al. (Cambridge: Cambridge University Press, 1989).

56. Hans Hahn, "Logic, Mathematics, and Knowledge of Nature," in *Twentieth-Century Philosophy*, ed. Morris Weitz (New York: Macmillan, 1966), 226.

Chapter 3

1. Ernst Mach, *Popular Scientific Lectures*, 5th ed., trans. Thomas J. McCormack (La Salle, IL: Open Court, 1943), 220.

2. Idem, *The Science of Mechanics*, 9th ed., trans. Thomas J. McCormack (London: Open Court, 1893), 36.

3. Idem, *Knowledge and Error*, trans. C. M. Williams (Dordrecht: D. Reidel, 1976), 140–41.

4. Ibid.

5. Ibid., 34.

6. Ibid., 35.

7. Idem, *Science of Mechanics*, 611–12.

8. Hume dismisses the issue in *A Treatise of Human Nature* . . . , ed. L. A. Selby-Bigge (New York: Oxford University Press, 1978).

9. See Ernst Mach, *Contributions to the Analysis of the Sensations*, trans. C. M. Williams (La Salle, IL: Open Court, 1984), 36–39n.

10. Unger introduced the duplicate in "Experience and Factual Knowledge," *Journal of Philosophy* 64 (1967): 152–73.

11. I elaborate this historical background in "Thought Experiments and the Epistemology of Laws," *Canadian Journal of Philosophy*, 22 (1992): 15–44.

12. Ludwig Boltzmann, "On Statistical Mechanics," in *Ludwig Boltzmann: Theoretical Physics and Philosophical Problems*, ed. B. McGuiness, trans. P. Foulkes (Boston: Reidel, 1974).

13. Charles Darwin, "Essays on Theology and Natural Selection," in *Metaphysics, Materialism, and the Evolution of Mind*, ed. Paul H. Barrett (Chicago: University of Chicago Press, 1974), 160.

14. Mach, *Science of Mechanics*, 35–36.

15. Ibid., 94.

16. Emphasized in Ernst Mach, *Principles of the Theory of Heat* (Dordrecht: D. Reidel, 1986), 374.

17. Sextus Empiricus says that Pythagoras and Empedocles held that "there is a certain community uniting us not only with each other and with the gods but even with the brute creation. There is in fact one breath pervading the whole cosmos like soul, and uniting us with them."

18. Henrik Steffens, *Alt und Neu* (Breslau, 1821), vol. 2, p. 102. The microcosm theme dates back as far as Anaximenes.

19. Mach, *Knowledge and Error*, 72.

20. For references, see Roger Shepherd, "Evolution of a Mesh: Mind and World," in *The Latest on the Best*, ed. John Dupre (Cambridge: MIT Press, 1987), 251–76. The spider analogy is Shepherd's.

21. John Joseph O'Neill, *Prodigal Genius: The Life of Nicola Tesla* (New York: Ives Washburn, 1944), 51.

22. M. A. Just and P. A. Carpenter, "Cognitive Coordinate Systems: Accounts of Mental Rotation and Individual Differences in Spatial Abilities," *Psychological Review* 92 (1985): 137–72; R. A. Finke and K. Slayton, "Explorations of Creative Visual Synthesis in Mental Imagery," *Memory and Cognition* 16 (1988): 252–57.

23. Brentano makes the objections in *Psychology from an Empirical Standpoint* (New York: Humanities Press, 1973), 12.2 29.

24. Mach, *Knowledge and Error*, 125.

25. Ibid., 136–37.

26. This and other miniature world experiments are described in William H. Ittelson and Frankling P. Kilpatrick, "Experiments in Perception," in *Readings in Perception*, ed. David C. Beardslee and Michael Werthimer (New York: D. Van Nostrand, 1958), 432–44.

27. Mach, *Knowledge and Error*, 142.

28. Peter C. Wason and Philip N. Johnson-Laird discuss the selection problem in their *Psychology of Reasoning* (Cambridge: Harvard University Press, 1972).

29. R. A. Griggs and J. R. Cox, "The Elusive Thematic-Materials Effect in Wason's Selection Task," *British Journal of Psychology* 73 (1982): 407–20.

30. The experiment is Roy D'Andrade's as described in D. E. Rumelhar, *Analogical Processes and Procedural Representations*, Center for Human Information Processing Technical Report No. 81 (San Diego: University of California, 1979).

31. Evelyn Golding, "The Effect of Past Experience on Problem Solving" (presented at the annual conference of the British Psychological Society, 1982).

32. Richard A. Griggs, "The Role of Problem Content in the Selection Task and THOG Problem," in *Thinking and Reasoning*, ed. Jonathan St. B. T. Evans (London: Routledge & Kegan Paul, 1983), 28.

33. A number of these studies are found in Dedre Gentner and Albert L. Stevens, eds., *Mental Models* (Hillsdale, NJ: Lawrence Erlbaum Associates, 1983).

34. See Kathleen Wilkes, *Physicalism* (London: Routledge and Kegan Paul, 1978), chap. 7.

35. See, e.g., E. A. Lunzer, "The Development of Consciousness," in *Aspects of Consciousness*, ed. Geoffrey Underwood and Robin Stevens (New York: Academic Press, 1979), 175–77; see also Herbert Ginsburg and Sylvia Opper, *Piaget's Theory of Intellectual Development* (Englewood Cliffs, NJ: Prentice-Hall, 1988).

36. J. Prytz Johansen says that there may be as few as two mental terms in the

Maori vocabulary; see his *The Maori and His Religion in Its Non-Ritualistic Aspects* (Copenhagen: I Kommission Hos Ejnar Munksgaard, 1954), chap. 10. Also see Jean Smith, "Self and Experience in Maori Culture," in *Indigenous Psychologies: The Anthropology of Self*, ed. Paul Heelas and Andrew Lock (New York: Academic Press, 1981).

37. George C. Williams floats this idea in *Adaptation and Natural Selection* (Princeton: Princeton University Press, 1966).

38. Leda Cosmides, "Deduction or Darwinian Algorithms? An Explanation of the "Elusive" Content Effect on the Wason Selection Task" (Ph.D. diss., Harvard University, 1985).

39. N. H. Horowitz, "On the Evolution of Biochemical Syntheses" *Proceedings of the National Academy USA* 31 (1945): 153–57.

40. Lewis Carroll, "What the Tortoise Said to Achilles," *Mind* 4 (1895): 278–80.

41. W. V. Quine, *Word and Object* (Cambridge: MIT Press, 1960), 27. His discussion of radical translation takes up the entire second chapter.

42. Idem, *Philosophy of Logic* (Englewood Cliffs, NJ: Prentice–Hall, 1970), 81.

43. Allan Franklin discusses the discovery in depth in order to demonstrate that there are crucial experiments; see his *Neglect of Experiment* (Cambridge: Cambridge University Press, 1986), chap. 1.

44. Sheldon Krimsky, "The Nature and Function of 'Gedankenexperimente' in Physics" (Ph.D. diss., University of Michigan, 1970), 229; see also p. 22.

45. Erwin Hiebert, "Mach's Conception of Thought Experiments in the Natural Sciences," in *The Interaction Between Science and Philosophy*, ed. Yehuda Elkana (Atlantic Highlands, NJ: Humanities Press, 1974), 339.

Chapter 4

1. Charles Darwin, *Origin of Species* (London: John Murray, 1859), 171.

2. Neither principle holds in full generality. David Lewis discusses both in *On the Plurality of Worlds* (New York: Basil Blackwell, 1986).

3. Stephen Hawking, *A Brief History of Time* (New York: Bantam, 1988), 93.

4. For lists of philosophers who deny that a contingent premise can entail a necessity, see R. Routley and V. Routley, "A Fallacy of Modality," *Nous* 3 (1969): 129–53 and Donald McQueen, "Evidence for Necessary Propositions," *Mind* 80 (1971): 59–69.

5. This example is drawn from Georg Polya's *Mathematics and Plausible Reasoning* (Princeton: Princeton University Press, 1954). The book teems with inductive arguments for necessary propositions.

6. G. H. von Wright, "Deontic Logic," *Mind* 60 (1951): 1–15.

7. Cherly S. Alexander and Henry Jay Becker, "The Use of Vignettes in Survey Research," *Public Opinion Quarterly* 42 (1978): 93.

8. Ibid., 94.

9. This sample is given by T. A. Nosanchuk in "The Vignette As an Experimental Approach to the Study of Social Status: An Exploratory Study," *Social Science Research* 1 (1972): 110.

10. Jean Piaget, *The Moral Judgment of the Child*, trans. Marjorie Gabain (New York: Free Press, 1965), 121–26.

11. George C. Williams, *Adaptation and Natural Selection* (Princeton: Princeton University Press, 1966), 15.

12. Elliot Sober elegantly dissolves the apparent incompatibility in "The Evolution of Rationality," *Synthese* 46 (1981): 95–120.

13. Richard M. Hare, *Moral Thinking: Its Levels, Method, and Point* (Oxford: Oxford University Press, 1981).

14. Malcolm gives a more detailed version of this definition in *Knowledge and Certainty* (Englewood Cliffs, NJ: Prentice-Hall, 1963), 236.

15. Isaac Maleh presents the disappearing sun example in *Mechanics, Heat, and Sound* (Columbus, OH: Charles E. Merrill, 1969), 78–79.

16. Amartya Sen, "Rights and Agency," *Philosophy and Public Affairs* 11 (1982): 14.

17. Cicero, *De oratore*, trans. H. Rackham (Cambridge: Harvard University Press, 1948), 357.

18. Roger Shepherd, "The Imagination of the Scientist," in *Imagination and Education*, ed. Kieran Egan and Dan Nadaner (New York: Teacher's College Press, 1988), 180.

19. Harry Frankfurt, "Alternate Possibilities and Moral Responsibility," *Journal of Philosophy* 66 (1969): 829–39.

20. Robin M. Hogarth, *Judgment and Choice* (New York: John Wiley, 1987), 100.

21. Frank Jackson, "What Mary Didn't Know," *Journal of Philosophy* 83 (1986): 291.

22. C. Mason Myers restricts his conventionalist account to philosophical thought experiments in "Analytical Thought Experiments," *Metaphilosophy* 17 (1986): 109–18.

23. Daniel C. Dennett, *Elbow Room* (Cambridge: MIT Press, 1984), 40.

24. Ibid.

25. J. S. Sachs, "Recognition Memory for Syntactic and Semantic Aspects of Connected Discourse," *Perception and Psychophysics* 2 (1967): 437–42.

26. James H. Jeans, *The Mathematical Theory of Electricity and Magneticism* (Cambridge: Cambridge University Press, 1925), 24.

27. Konrad B. Krauskopf and Arthur Beiser, *The Physical Universe* (New York: McGraw-Hill, 1986), 596.

28. Jonathan Glover, "It Makes No Difference Whether or Not I Do It," *Proceedings of the Aristotelian Society*, 49 Suppl. (1975): 174–75.

29. James Rachels, "Active and Passive Euthanasia," *New England Journal of Medicine* 292 (1975): 78–80.

30. For a general discussion of orientation effects, see Irvin Rock, "The Perception of Disoriented Figures," in *Image, Object, and Illusion*, ed. Richard Held (San Francisco: W. H. Freeman, 1974), 71–78.

31. The original version appears in John Wisdom, "Gods," *Proceedings of the Aristotelian Society* (1944–45): 45, 185–206.

32. W. K. Clifford, *Lectures and Essays*, ed. Leslie Stephen and Frederick Pollock, vol. 1 (London: MacMillan, 1901), 100.

33. The philosophical fertility of absolute concepts is demonstrated in Peter Unger's *Philosophical Relativity* (Minneapolis: University of Minnesota Press, 1984). The application to rationality is developed in my "Rationality As an Absolute Concept," *Philosophy*, 66 (1991): 473–86.

34. Bertrand Russell, *The Problems of Philosophy* (Oxford: Oxford University Press, 1912), 147.

35. Ibid., 148.

36. Alfred Wegener quoted in Ronald Gere, *Explaining Science* (Chicago: University of Chicago Press, 1988), 230 (Giere's emphasis). Giere introduces the concept of cognitive resources on p. 213.

37. Here I fan the evolutionary embers of Michael Stocker, "Emotional Thoughts," *American Philosophical Quarterly* 24 (1987): 59–69.

Chapter 5

1. Thomas Kuhn, *The Essential Tension* (Chicago: University of Chicago Press, 1977), 242.

2. Ibid., 252.

3. Ibid., 254.

4. Ibid., 253.

5. Ibid., 254–55.

6. Ibid., 255.

7. Hans Vaihinger, *The Philosophy of "As If,"* trans. C. K. Ogden (London: Routledge & Kegan Paul, 1924), 65.

8. Ibid., 66.

9. James Cargile, "Definitions and Counter-Examples," *Philosophy* 62 (1987): 179–93.

10. M. C. Escher, *The Graphic Work of M. C. Escher*, trans. John Brigham (New York: Ballantine Books, 1969).

11. Joseph Heller, *Catch-22* (New York: Dell, 1955), 47.

12. Chisholm's classification appears in his chapter "The Problem of the Criterion" in *Theory of Knowledge* (Englewood Cliffs, NJ: Prentice-Hall, 1966). Rescher makes extended use of this taxonomic scheme in *The Strife of Systems* (Pittsburgh: University of Pittsburgh Press, 1985).

Chapter 6

1. David Hume, *Treatise of Human Nature*. 1.1, ed. L. A. Selby-Bigge (New York: Oxford University Press, 1978), 6.

2. Ibid.

3. Edmund Gettier, "Is Justified True Belief Knowledge?," *Analysis* 23 (1963): 121–23.

4. John Searle, "Minds, Brains, and Programs," *The Behavioral and Brain Sciences* 3 (1980): 417–24.

5. This thought experiment is described in Milton Rothman, *The Laws of Physics* (New York: Basic Books, 1963), 187.

6. James Clerk Maxwell, *Theory of Heat* (London: Longmans, Green, 1871), 153–54.

7. James Maxwell to Peter Tait, 11 December 1867, quoted in C. G. Knott, *Life and Scientific Work of Peter Guthrie Tait* (Cambridge: Cambridge University Press, 1911), 215.

8. James Maxwell to John Strutt, 6 December 1870, quoted in R. J. Strutt, *John William Strutt* (London: E. Arnold, 1924), 47.

9. David Cole, "Thought and Thought Experiments," *Philosophical Studies* 45 (1984): 431–44.

10. Newton describes the bucket experiment in *Principia*. vol. 1, trans. Florian Cajori (Berkeley: University of California Press, 1966), 10.

11. Ibid., 6.

12. Ibid., 12.

13. Ernst Mach, *The Science of Mechanics*, 9th ed., trans. Thomas J. McCormack (London: Open Court, 1942), 284.

14. Ibid., 341.

15. Allen Franklin stresses this distinction in *The Neglect of Experiment* (Cambridge: Cambridge University Press, 1986), 109–10.

16. Allan Gibbard and William Harper, "Two Kinds of Expected Utility," in *Ifs*, ed. William Harper et al. (Dordrecht: D. Reidel, 1981), 152–90.

17. George Schlesinger, "The Unpredictability of Free Choices," *British Journal for the Philosophy of Science* 25 (1974): 209–22.

18. G. Seddon presses this example in "Logical Possibility," *Mind* 81 (1972): 481–94. Kathleen Wilkes cites Seddons in her cross-examination of thought experiments featuring linguistically sophisticated chimps in *Real People: Personal Identity Without Thought Experiments* (Oxford: Clarendon Press, 1988), 31.

19. I argue for this solution in *Blindspots* (Oxford: Clarendon Press, 1988), chap. 11. The wide applicability of this strategy is one of the book's themes.

20. David Lewis, "Devil's Bargains and the Real World," in *The Security Gamble*, ed. Douglas MacLean (Totowa, NJ: Rowman & Allanheld, 1984), 141–54.

21. I argue that an empiricist/rationalist divergence lies behind Einstein's divorce from Mach in my "Thought Experiments," *American Scientistt* 79 (1991): 250–63.

22. Delo E. Mook and Thomas Vargish try to soothe the reader with this thought in *Inside Relativity* (Princeton: Princeton University Press, 1987), 103.

23. Mark Johnston presents this objection in "Human Beings," *Journal of Philosophy* 84 (1987): 68.

24. Krimsky's analysis of Stevin's chain of balls appears in "The Nature and Function of 'Gendankenexperimente' in Physics" (Ph.D. diss., University of Michigan, 1970), 156–64.

25. Ernst Mach, *History and Root of the Principles of the Conservation of Energy*, trans. Philip E. B. Jourdain (Chicago: Open Court, 1911), 23.

26. David Lewis and Jane Richardson, "Scriven on Human Unpredictability," *Philosophical Studies* 17 (1966): 70–71.

27. The five-minute hypothesis was first mentioned in Bertrand Russell, *The Analysis of Mind* (London: Unwin, 1921), 159.

28. Peter Unger, *Ignorance* (Oxford: Clarendon Press, 1975), 72.

29. Peter Galison, *How Experiments End* (Chicago: University of Chicago Press, 1987).

30. Anthony Quinton, "Spaces and Times," *Philosophy* 37 (1962): 141.

31. Richard Swinburne, *Space and Time* (New York: St. Martin's, 1981), 30.

32. David Lewis, *Philosophical Papers*, vol. 2 (New York: Oxford University Press, 1986), 299–304.

Chapter 7

1. Irenaus Eibl-Eibesfeldt gives an overview in *Ethology*, trans. Erick Klinghammer (New York: Holt, Rinehart & Winston, 1975), chap. 10.

2. The history of the issue is presented in Virginia P. Dawson, *Nature's Enigma* (Philadelphia: American Philosophical Society, 1987).

3. J. L. Gorman, "A Problem in the Justification of Democracy," *Analysis* 38 (1978): 48.

4. J. O. Urmson, "On Grading," *Mind* 59 (1950): 145–69.

5. My source for the inheritance rules is W. C. Holdsworth, *A History of English Law* (Boston: Little, Brown, 1923), 177–83.

6. Marianne Wiser and Susan Carey give a Kuhnian account of the transition to this technical terminology in "When Heat and Temperature Were One," in *Mental Models*, ed. Dedre Gentner and Albert Stevens (Hillsdale, NJ: Lawrence Erlbaum, 1983), 267–97.

7. Seymour Benzer, "The Elementary Units of Heredity," in *The Chemical Basis of Heredity*, ed. W. D. McElroy and B. Glass (Baltimore: Johns Hopkins Press, 1957), 70–93.

8. Henri Poicare, *Science and Hypothesis* (New York: Dover, 1952), 65.

Chapter 8

1. Galileo's definition is endorsed by Alexandre Koyre (*Metaphysics and Measurement* [London: Chapman & Hall, 1968], 90) and Hans Reichenbach (*The Rise of Scientific Philosophy* [Berkeley: University of California Press, 1964], 97).

2. The authenticity of the story is defended in Antoni Sulek, "The Experiment of Psammetichus," *Journal of the History of Ideas* 50 (1989): 645–51.

3. Frank Jackson, *Conditionals* (New York: Basil Blackwell, 1987), 94–95.

4. Mario Bunge requires active interference: "By definition, experiment is the kind of scientific experience in which some change is deliberately provoked, and its outcome observed, recorded and interpreted with a cognitive aim" (*Scientific Research II: The Search for Truth* [New York: Springer Verlag, 1967], 251). So does Reichenbach (*Rise of Scientific*, 97).

5. These experiments are described in Gerald Oster, "Phosphenes," *Scientific American*, February 1970, 82. Contemporary researchers hope that phosphenes could be eventually used to cure certain kinds of blindness.

6. HYPO is discussed by E. L. Rissland and K. D. Ashley in "Hypotheticals As Heuristic Device," *Proceedings of the American Association for Artificial Intelligence*, (Philadelphia, 1986).

7. The surprising confirmation is widely reputed to have caused the capitulation of particle theorists. However, John Worrall argues that its impact has been overestimated. See his "Fresnel, Poisson, and the White Spot," in *The Uses of Experiment*, ed. David Gooding et al. (New York: Cambridge University Press, 1989), 135–57.

8. Mach mentions the incident in *Knowledge and Error*, trans. C. M. Williams (Dordrecht: D. Reidel, 1976), 210n.

9. Quoted by John Locke in *An Essay Concerning Human Understanding*, vol. 2 (New York: Dover, 1959), 186–87.

10. G. M. White, "Immediate and Deferred Effects of Model Observation and Guided and Unguided Rehearsal on Donating and Stealing," *Journal of Personality and Social Psychology* 14 (1978): 58–65.

11. Gerald Holton discusses Millikan's consolidation of lucky breaks in *The Scientific Imagination* (Cambridge: Cambridge University Press, 1978), chap. 2.

12. Noretta Koertge recounts the development of Galileo's experimental methodology in "Galileo and the Problem of Accidents," *Journal of the History of Ideas* 38 (1977): 389–408.

13. Ampère, quoted in Pierre Duhem, *The Aim and Structure of Physical Theory* (Princeton: Princeton University Press, 1954), 196.

14. Galileo, *Two New Sciences*, trans. Stillman Drake (Madison: University of Wisconsin Press, 1974), 228.

15. Kathleen Wilkes, *Real People: Personal Identity Without Thought Experiments* (Oxford: Clarendon Press, 1988), 2. The same view is expressed by Richard Schlegel in *Completeness in Science* (New York: Appleton–Century–Crofts, 1967), 163.

16. Max Black, *The Problems of Analysis* (London: Routledge & Kegan Paul, 1954), 83.

17. George Berkeley, *Philosophical Writings*, ed. T. E. Jessop (London, 1952), 212–13.

18. Gilbert Ryle, *The Concept of Mind* (London: Hutchinson, 1949), 273.

19. Ibid., 264.

20. David Lewis, *Philosophical Papers*, vol. 1 (New York: Oxford University Press, 1983), 266.

21. Kendall Walton, "Fearing Fictions," *Journal of Philosophy* 75 (1978): 5–27.

22. George Dickie, *Art and the Aesthetic* (Ithaca: Cornell University Press, 1974), 34.

23. Koyre, *Metaphysics and Measurement*, 150.

24. James Trefil, *A Scientist at the Seashore* (New York: Charles Scribner's Sons, 1984), 183.

25. Einstein, quoted in Ronald W. Clark, *Einstein: The Life and Times* (New York: Harry N. Abrams, 1984), 231.

26. Ralph W. Clark leads his readers through this mental exercise in "Fictional Entities: Talking About Them and Having Feelings About Them," *Philosophical Studies* 38 (1980): 347–48.

27. Wertheimer analyzes the interviews in the chapter "The Thinking That Led to the Theory of Relativity" in *Productive Thinking* (New York: Harper & Brothers, 1945). See also Gerald Holton, "On Trying to Understand Scientific Genius," in *Thematic Origins of Scientific Thought* (Cambridge: Harvard University Press, 1988).

28. Stephen M. Kosslyn, *Ghosts in the Mind's Machine* (New York: Norton, 1983).

29. Roland Hall, "Excluders," *Analysis* 20 (1959): 1–7. This makes thought experiments logically similar to single-blind, double-blind, and triple-blind experiments, where the force 'blind' is the negative one of excluding knowledge by the subject, treator, and statistician, respectively.

30. Judith Jarvis Thomson, "A Defense of Abortion," *Philosophy and Public Affairs* 1 (1971): 47–68.

31. Donald Campbell, "Evolutionary Epistemology," in *The Philosophy of Karl Popper*, ed. Paul Schilpp (La Salle, IL: Open Court, 1974), 413–63.

32. A sample is found in John Alcock, *Animal Behavior: An Evolutionary Approach* (Sunderland, MA: Sinauer Associates, 1984), chap. 13.

33. The hypothesis is advanced in Nicholas Humphrey, *Consciousness Regained* (New York: Oxford University Press, 1983), chap. 2. Paleontological interest is evident from Donald Johanson and James Shreeve, *Lucy's Child* (New York: Avon Books, 1989), 270–80.

34. Gerald Massey, "Are There Any Good Arguments That Bad Arguments Are Bad?," *Philosophy in Context* 4 (1975): 61–77.

Chapter 9

1. Gilbert Ryle, "Systematically Misleading Expressions," in *Twentieth Century Philosophy*, ed. Morris Weitz (New York: Free Press, 1966), 201.

2. Ludwig Wittgenstein, *Philosophical Investigations* 27.

3. Karl Popper uses the terms interchangeably in the appendix "On the Use and Misuse of Imaginary Experiments, Especially in Quantum Theory" in *The Logic of Scientific Discovery* (New York: Harper & Row, 1965). Sheldon Krimsky argues that all imaginary experiments are thought experiments (but not vice versa) in *The Nature and Function of "Gedankenexperimente" in Physics* (Ann Arbor: University of Michigan Microfilms, 1970), 19.

4. Konrad B. Krauskopf and Arthur Beiser, *The Physical Universe*, 5th ed. (New York: McGraw–Hill, 1986), 546.

5. Alan J. Parkin, *Memory and Amnesia* (Oxford: Basil Blackwell, 1987), 42.

6. Jerry Fodor, "Imagistic Representation," in *Imagery*, ed. Ned Block (Cambridge: MIT Press, 1981), 75.

7. Michael Lockwood, "Sins of Omission?," *Aristotelian Society* 57 (1983): 212.

8. Steven C. Frautschi, *The Mechanical Universe* (Cambridge: Cambridge University Press, 1986), 489.

9. The best-known textbook applying this pedagogical strategy is Fred Miller and Nicholas Smith, *Thought Probes: Philosophy Through Science Fiction Literature* (Englewood Cliffs, NJ: Prentice–Hall, 1981).

10. They appear in Janet A. Kourany's anthology *Scientific Knowledge* (Belmont, CA: Wadsworth, 1987), 65–76.

11. Oliver Lodge, *Pioneers of Science* (New York: Dover, 1960), 90.

12. Galileo, *Dialogues Concerning Two New Sciences* 145.

13. James Trefil, *A Scientist at the Seashore* (New York: Charles Scribner's Sons, 1984), 27–28.

14. Dworkin draws the analogy in "No Right Answer?," in *Law, Morality, and Society*, ed. Peter Hacker and Joseph Raz (Oxford University Press, 1977).

15. *Encyclopedia of Philosophy*, ed. Paul Edwards (New York: Macmillan, 1966), s.v. "Copernicus, Nicholas."

16. H. M. Smith, "Synchronous Flashing of Fireflies," *Science*, 16 August 1935, 151.

17. Robert P. Breckenridge, *Modern Camouflage* (New York: Farrar & Rinehart, 1942), 4–6.

18. Thomas Settle, "An Experiment in the History of Science," *Science*, 6 January 1961, 19–23.

19. Alexandre Koyre, *Metaphysics and Measurement* (London: Chapman & Hall, 1968), 84.

20. J. MacLachlan, "A Test of an 'Imaginary' Experiment of Galileo's," *Isis* 64 (1973): 374–79.

21. J. S. Bruner and Leo Postman, "On the Perception of Incongruity: A Paradigm," *Journal of Personality* 68 (1949): 206–23.

22. R. G. Swinburne discusses examples of anomaly toleration in "Falsifiability of Scientific Theories," *Mind* 72 (1964): 434–36.

23. John Worrall, "The Pressure of Light: The Strange Case of the Vacillating 'Crucial Experiment,'" *Studies in the History and Philosophy of Science* 13 (1982): 133–71.

24. Hobbes's criticisms and the institionalization of experiment are described in Steven Shapin and Simon Schaffer, *Leviathan and the Air-Pump* (Princeton: Princeton University Press, 1985).

25. Andrew Pickering, "The Hunting of Quarks," *Isis* 72 (1981): 236. The theme is also taken up in his *Constructing Quarks* (Chicago: University of Chicago Press, 1984), chap. 4.

26. Judith Thomson, *Philosophy in America* (Ithaca: Cornell University Press, 1964), 292.

27. Robert Shope, *The Analysis of Knowledge* (Princeton: Princeton University Press, 1983).

28. Robert Millikan, *The Electron*, 7th ed. (Chicago: University of Chicago Press, 1922), 230. Einstein's hypothesis gained acceptance after 1924.

29. John Harris develops the scheme in "The Survival Lottery," *Philosophy* 50 (1975): 81–87.

30. Charles Darwin, *Origin of Species* (New York: Appleton-Century, 1937), 218–19.

31. Judith Jarvis Thomson, "Killing, Letting Die, and the Trolley Problem," *Monist* 59 (1976): 204–17.

32. Ernst Mayr, "Species Concepts and Their Applications," in *Conceptual Issues in Evolutionary Biology*, ed. Elliot Sober (Cambridge: MIT Press, 1984), 539.

33. Ed Erwin, *The Concept of Meaninglessness* (Baltimore: Johns Hopkins University Press, 1970), 37.

34. Ernst Mach, *Knowledge and Error*, trans. C. M. Williams (Dordrecht: D. Reidel, 1976), 40.

35. The experiment is described in Fritz J. Roethlisberger and William J. Dickson, *Management and the Worker* (Cambridge: Cambridge University Press, 1947), 39.

36. Gilbert Harman, *Thought* (Princeton: Princeton University Press, 1973), 120. The social knowledge cases appear on pp. 142–45.

37. Ibid., 151.

38. Peter Galison, *How Experiments End* (Chicago: University of Chicago Press, 1987).

39. Christopher J. Bulpitt, *Randomised Controlled Clinical Trials* (The Hague: Martinus Nijhoff, 1983), 70.

40. See Steven Shapin, "The Invisible Technician," *American Scientist* 77 (1989): 554–63.

41. Ian Hacking discusses the scientific caste system in *Representing and Intervening* (New York: Cambridge University Press, 1983), chap. 9.

42. Hilary Putnam, "The Meaning of 'Meaning,'" in *Language, Mind, and Knowledge*, ed. K. Gunderson (Minneapolis: University of Minnesota Press, 1975).

43. Gertrude E. M. Anscombe, "Modern Moral Philosophy," *Philosophy* 33 (1958): 13.

44. Michael Levin, "Ethics Courses: Useless," *New York Times*, 25 November 1989, p. 23.

45. David Lewis, "Devil's Bargains and the Real World," in *The Security Gamble*, ed. Douglas MacLean (Totowa, NJ: Rowman & Allanheld, 1984), 148.

46. Fleeming Jenkin, "The Origin of Species," in *Darwin and His Critics* (Cambridge: Harvard University Press, 1973), 315–6.

47. Marilyn Friedman, "Care and Context in Moral Reasoning," in *Women and Moral Theory*, ed. Eva Kittay and Diana Myers (Totowa, NJ: Rowman & Littlefield, 1987), 190–204.

48. Jonathan Dancy, "The Role of Imaginary Cases in Ethics," *Pacific Philosophical Quarterly* 66 (1985): 141–53.

49. Ronald A. Fisher, *The Genetical Theory of Natural Selection* (Oxford: Oxford University Press, 1930).

50. See A. G. Drachmann, *The Mechanical Technology of Greek and Roman Antiquity* (London: Hafner, 1963), 46.

Chapter 10

1. *Commentaria in Aristotelem Graeca* (Royal Prussian Academy), vol. 7, *Simplicii in Aristotelis "De Caelo" Commentaria*, ed. I. L. Heiberg (Berlin: 1894), 533–34.

2. James Thomson, "Tasks and Super-Tasks," in *Zeno's Paradoxes*, ed. Wesley C. Salmon (New York: Bobbs–Merril, 1970); see also Paul Benacerraf's "Task, Super-Tasks, and the Modern Eleatics" (which I draw on below).

3. Larry Laudan, "Dissecting the Holist Picture of Scientific Change," in *Scientific Knowledge*, ed. Janet A. Kourany (Belmont, CA: Wadsworth, 1987), 276–95.

4. Dean Keith Simonton details some of the loose descriptions in the multiple discovery literature in the *Scientific Genius* (New York: Cambridge University Press, 1988), chap. 6.

5. Leibniz, *Monadology*, trans. Schrecker (New York: Bobbs–Merrill, 1965), sec. 17 (written ca. 1714).

6. David Cole, "Thought and Thought Experiments," *Philosophical Studies* 45 (1984): 432.

7. B. J. McNeil et al., "On the Elicitation of Preferences for Alternative Therapies," *New England Journal of Medicine* 306 (1982): 1259–62.

8. William Thomson, "On a Universal Tendency in Nature to the Dissipation of Mechanical Energy," *Philosophical Magazine* 4 (1852): 256–60.

9. J. Loschmidt, "Ueber den Zustnd des Warmegleichgewichtes eines System von Korpern," *Akademie der Wissenschaften, Wien, Mathematisch-Naturwissisnschaftliche Klasse, Sitxungsberichte* 59 (1876): 135.

10. Peter Unger, *Identity, Consciousness, and Value* (New York: Oxford University Press, 1990), 83–87.

11. James Hudson, "The Diminishing Value of Happy People," *Philosophical Studies* 51 (1987): 133.

12. Daniel Dennett, *Elbow Room* (Cambridge: MIT Press, 1984), 12.

13. Ibid.

14. Ibid., 35–36.

15. This distrust was elaborated in medieval and Renaissance thought. See Peter Dear, "Jesuit Mathematical Science and the Reconstitution of Experience in the Early Seventeenth Century," *Studies in the History and Philosophy of Science* 18 (1987), 133–75.

16. John Hick, *Faith and Knowledge* (Ithaca: Cornell University Press, 1957), chap. 7.

17. John R. Hanley, "An Examination of the Fundamental Assumption of Hypothetical Process Arguments," *Philosophical Studies* 34 (1978): 187–95.

18. Davidson deploys the thought experiment in "A Coherence Theory of Truth and Knowledge," in *Kant oder Hegel?*, ed. Dieter Heinrich (Stuttgart: Klett–Cotta Buchaudlang, 1981), 423–33. Richard Foley and Richard Fumerton criticize it in "Davidson's Theism?," *Philosophical Studies* 48 (1985): 83–89.

19. I defend the statistical concept of fallacy in *Blindspots* (Oxford: Clarendon Press, 1988), chap. 4.

20. Bertrand Russell, *Human Knowledge: Its Scope and Limits* (New York: Allen & Unwin, 1948), 154.

21. Shelly Kagan, "The Additive Fallacy," *Ethics* 99 (1988): 5–31. Others have argued in the same vein, but Kagan is the most forceful.

22. Ernst Cassirer, *An Essay on Man* (New Haven: Yale University Press, 1944), 57.

23. The term was introduced in Jonathan Barnes and Richard Robinson, "Untruisms," *Metaphilosophy* 3 (1972): 189–97.

24. John Passmore, "Everything Has Just Doubled in Size," *Mind* 74 (1965): 257.

25. R. M. Hare, *Moral Thinking* (Oxford: Clarendon, 1981), 167–68. Also see Russell Hardin's section entitled "Peculiar Examples and Ethical Argument" in his *Morality Within the Limits of Reason* (Chicago: University of Chicago Press, 1988). For nonutilitarian restrictionism, see Alan Donagan, *The Theory of Morality* (Chicago: University of Chicago Press, 1977), 36.

26. Paul Taylor, "A Problem for Utilitarianism," in *Philosophy: The Basic Issues*, 2d ed., ed. Edward D. Klemke et al. (New York: St. Martin's, 1986), 406.

27. E. J. O'Brien and C. R. Wolford, "Effects of Delay in Testing on Retention of Plausible Versus Bizarre Mental Images," *Journal of Experimental Psychology* 8 (1982): 148–52. Others deny the relevance of bizarreness, e.g., N. E. Kroll et al., "Bizarre Imagery: The Misremembered Mnemonic," *Journal of Experimental Psychology* 12 (1986): 42–53.

28. I, however, do not. I study the connection between borderline cases and thought experiments in "Vagueness and the Desiderata for Definition," in *Definitions and Definability*, ed. J. H. Fetzer, D. Shatz, and G. Schlesinger (Dordrecht: Kluwer Academic, 1991), 71–109.

29. Thomas Nagel, *Mortal Questions* (New York: Cambridge University Press, 1979), 8–9.

30. Henry Kyburg maintains that there is an evaluative sense (*ideal* = "nice") and a comparative sense in *Science and Reason* (New York: Oxford University Press, 1990), 168–69. My position is that *ideal* is relative, not polysemous.

31. P. H. Nowell-Smith, *Ethics* (Baltimore, MD: Penguin Books, 1954): 240–41.

32. Richard Nisbett and Lee Ross, *Human Inference* (Englewood Cliffs, NJ: Prentice-Hall, 1980), 115–16.

33. Moore presents the Two World thought experiment in *Principia ethica* (Cambridge: Cambridge University Press, 1903), 83–84.

34. Oliver Johnson, "Aesthetic Objectivity and the Analogy with Ethics," in *Philosophy and the Arts*, ed. Godfrey Vesey (New York: St. Martin's, 1973), 168.

35. A. Douslas Davis, *Classical Mechanics* (New York: Harcourt, 1986), 14.

36. Steven C. Frautchi, *The Mechanical Universe* (New York: Cambridge University Press, 1986), 289.

37. Carl Gans, "How Snakes Move," *Scientific American* 222 (1970): 93. My discussion of Gans's objection follows David Lewis, *Philosophical Papers*, vol. 1 (New York: Oxford University Press, 1983), 261–80.

SELECT BIBLIOGRAPHY

Anscombe, G. E. M. "Modern Moral Philosophy." *Philosophy* 33 (1958): 1–19.

Baier, Annette. "Hume, the Women's Moral Theorist?" In *Women and Moral Theory*, ed. Eva Kittay and Diana Myers. Totowa, NJ: Rowman & Littlefield, 1987.

Blackmore, John T. *Ernst Mach: His Life, Work, and Influence*. Berkeley: University of California Press, 1972.

Brown, James Robert. *The Laboratory of the Mind*. London: Routledge, 1991.

Broyles, James. "An Observation on Wittgenstein's Use of Fantasy." *Metaphilosophy* 5 (1974): 291–97.

Cargile, James. "Definitions and Counter-Examples." *Philosophy* 62 (1987): 179–93.

Casullo, Albert. "Reid and Mill on Hume's Maxim of Conceivability." *Analysis* 39 (1979): 212–19.

Chisholm, Roderick. *Theory of Knowledge*. Englewood Cliffs, NJ: Prentice-Hall, 1966.

Cole, David. "Thought and Thought Experiments." *Philosophical Studies* 45 (1984): 431–44.

Dancy, Jonathan. "The Role of Imaginary Cases in Ethics." *Pacific Philosophical Quarterly* 66 (1985): 141–53.

Danly, John R. "An Examination of the Fundamental Assumption of Hypothetical Process Arguments." *Philosophical Studies* 48 (1985): 83–89.

Dennett, Daniel. *Elbow Room*. Cambridge: MIT Press, 1984.

Evans, Jonathan St. B. T. *Thinking and Reasoning*. London: Routledge & Kegan Paul, 1983.

Fodor, Jerry. "On Knowing What We Would Say." In *Readings in the Philosophy of Language*, ed. Jay F. Rosenberg and Charles Travis. Englewood Cliffs, NJ: Prentice-Hall, 1971.

Franklin, Allan. *The Neglect of Experiment*. New York: Cambridge University Press, 1986.

Friedman, Marilyn. "Care and Context in Moral Reasoning." In *Women and Moral Theory*, ed. Eva Kittay and Diana Myers. Totowa, NJ: Rowman & Littlefield, 1987.

Gentner, Dedre, and Albert L. Stevens. *Mental Models*. Hillsdale, NJ: Lawrence Erlbaum, 1983.

Giere, Ronald. *Explaining Science*. Chicago: University of Chicago Press, 1988.

Goldman, Alvin. "Psychology and Philosophical Analysis." *Proceedings of the Aristotelian Society* 39 (1989): 195–209.

Gooding, David, et al. *The Uses of Experiment*. New York: Cambridge University Press, 1989.

Hacking, Ian. *Representing and Intervening*. New York: Cambridge University Press, 1975.

Hardin, Russell. *Morality Within the Limits of Reason*. Chicago: University of Chicago Press, 1988.

Hare, Richard M.. *Moral Thinking: Its Levels, Method, and Point*. Oxford: Oxford University Press, 1981.

Harman, Gilbert. "Moral Explanations of Natural Facts—Can Moral Claims Be Tested Against Moral Reality?" *Southern Journal of Philosophy* 24 suppl. (1986).

Hempel, Carl. *Aspects of Scientific Explanation*. New York: Free Press, 1965.

Hiebert, Erwin. "Mach's Conception of Thought Experiments in the Natural Sciences." In *The Interaction Between Science and Philosophy*, ed. Yehuda Elkana. Atlantic Highlands, NJ: Humanities Press, 1974.

Holton, Gerald. *Thematic Origins of Scientific Thought*. Cambridge: Harvard University Press, 1988.

Hudson, James. "The Diminishing Value of Happy People." *Philosophical Studies* 51 (1987): 123–37.

Hull, David L. "A Function for Actual Examples in Philosophy of Science." In his *What the Philosophy of Biology Is*. Dordrecht: Kluwer Academic, 1989.

Johnston, Mark. "Human Beings." *Journal of Philosophy* 84 (1987), 59–83.

Kagan, Shelly. "The Additive Fallacy." *Ethics* 99 (1988): 5–31.

Koertge, Noretta. "Galileo and the Problem of Accidents." *Journal of the History of Ideas* 38 (1977): 389–408.

Koyre, Alexandre. *Metaphysics and Measurement*. London: Chapman & Hall, 1968.

Krimsky, Sheldon. "The Nature and Function of 'Gedankenexperimente' in Physics." Ph.D. diss., University of Michigan, 1970.

Kuhn, Thomas. *The Essential Tension*. Chicago: University of Chicago, 1977.

Lewis, David. "Truth in Fiction." In his *Philosophical Papers* vol. 1. New York: Oxford University Press, 1983.

Lyons, William. *The Disappearance of Introspection*. Cambridge: MIT Press, 1986.

Mach, Ernst. *Contributions to the Analysis of the Sensations*. Trans. C. M. Williams. LaSalle, IL: Open Court, 1984.

——. *History and Root of the Principle of the Conservation of Energy*. Trans. Philip E. B. Jourdain. Chicago: Open Court, 1911.

——. *Knowledge and Error*. Trans. C. M. Williams. Dordrecht: D. Reidel, 1976.

——. *Popular Scientific Lectures*. 5th ed. Trans. Thomas J. McCormack. London: Open Court, 1948.

——. *Principles of the Theory of Heat*. Trans. Philip E. B. Jourdain. Dordrecht: D. Reidel, 1986.

——. *The Science of Mechanics*. 9th ed. Trans. Thomas J. McCormack. London: Open Court, 1942.

MacLachlan, J. "A Test of an 'Imaginary' Experiment of Galileo's." *Isis* 64 (1973): 374–79.

Myers, C. Mason. "Analytical Thought Experiments." *Metaphilosophy* 17 (1986):109–18.

Nisbett, Richard, and Lee Ross. *Human Inference*. Englewood Cliffs, NJ: Prentice-Hall, 1980.

Popper, Karl. "On the Use and Misuse of Imaginary Experiments, Especially in Quantum Theory." In his *The Logic of Scientific Discovery*. New York: Basic Books, 1959.

Ray, Christopher. *The Evolution of Relativity*. Philadelphia: Adam Hilger, 1987.

Rescher, Nicholas, *The Strife of Systems*. Pittsburgh: University of Pittsburgh Press, 1985.

Ryle, Gilbert. *The Concept of Mind*. London: Hutchinson, 1949.

Salmon, Wesley C. *Zeno's Paradoxes*. New York: Bobbs–Merril, 1970.

Schmitt, Charles. "Experimental Evidence for and Against a Void: The Sixteenth-Century Arguments." *Isis* 58 (1967): 352–66.

Sen, Amartya. "Rights and Agency." *Philosophy and Public Affairs* 11 (1982): 3–39.

Shapin, Stevin. "The Invisible Technician." *American Scientist* 77 (1989): 554–63.

Shapin, Steven, and Simon Schaffer. *Leviathan and the Air-Pump*. Princeton University Press, 1985.

Smart, John, and Bernard Williams. *Utilitarianism: For and against*. Cambridge: Cambridge University Press, 1976.

Sorensen, Roy. *Blindspots*. Oxford: Clarendon Press, 1988.

———. "Moral Dilemmas, Thought Experiments, and Conflict Vagueness." *Philosophical Studies* 63 (1991): 291–308.

———. "Thought Experiments." *American Scientist* 79 (1991): 250–63.

———. "Thought Experiments and the Epistemology of Laws." *Canadian Journal of Philosophy*. 22 (1992): 15–44.

———. "Vagueness and the Desiderata for Definition." In *Definitions and Definability*, ed. James Fetzer, David Shatz, and George Schlesinger. Dordrecht: Kluwer, 1991.

Unger, Peter. *Identity, Consciousness, and Value*. New York: Oxford University Press, 1990.

———. *Philosophical Relativity*. Minneapolis: University of Minnesota Press, 1984.

———. "Toward a Psychology of Common Sense." *American Philosophical Quarterly* 19 (1982): 117–129.

Vaihinger, Hans. *The Philosophy of "As if."* Trans. C. K. Ogden. London: Routledge & Kegan Paul, 1924.

Wilkes, Kathleen. *Real People: Personal Identity Without Thought Experiments*. Oxford: Clarendon Press, 1988.

Worrall, John. "The Pressure of Light." *Studies in the History and Philosophy of Science* 13 (1982): 133–71.

SUBJECT INDEX

NAME INDEX

Agrippa of Nettesheim, 120
Alberti, Leon Battista, 194
Ampere, André-Marie, 196
Anscombe, G. E. M., 33, 243–44, 258
Aquinas, Thomas, 138
Arago, François, 191
Archimedes, 56–57, 215
Archytas of Tarentum, 155–56, 161, 250
Aristotle, 67, 114, 224
  experiments of, 188–89, 237, 267
  on infinity, 155
  on logic, 125
  on motion, 9–10, 61, 109, 114–15, 126–27
  on terrestial/celestial distinction, 78, 179, 197–98, 255
  thought experiments of, 148–49
  on vacuums, 27–28, 46–47, 61, 148–49
Arnauld, Antoine, 39
Augustine, 67
Austin, J. L., 44–46, 93

Bacon, Frances, 230, 235
Baier, Annette, 26
Benacerraf, Paul, 259
Bentham, Jeremy, 91
Benzer, Seymour, 180
Berkeley, George, 202–5, 209, 219
Bernoullli, Daniel, 153
Biot, J. B., 191
Black, Max, 201–2
Boerhaave, Hermann, 171–72
Bohr, Neils, 143
Boltzmann, Ludwig, 56
Boole, George, 110
Bourget, Paul, 222
Boyle, Robert, 28, 242
Brahe, Tycho, 203, 255
Bronte, Emily, 282
Brown, James Robert, 292n11
Brentano, Franz, 61
Broyles, James, 46
Bruno, Giordono, 255

Bunge, Mario, 300n4
Buridan, Jean, 197
Burtt, Cyril, 206

Campbell, Donald, 212
Cargile, James, 121–22
Carroll, John, 291n6, 292n6
Carroll, Lewis, 71–72, 268
Cassirer, Ernst, 275–76
Casullo, Albert, 39–41
Chisholm, Roderick, 130–31
Cicero, 91–92
Clausius, 142
Clifford, W. K., 103
Cole, David, 143–44, 261
Colladon, Jean Daniel, 242
Collins, Arthur, 273
Comte, Auguste, 24–25, 59, 68
Condorcet, Marquis de, 152
Copernicus, 78, 90, 227, 255
Coriolis, Gaspard Gustave de, 172–73
Cosmides, Leda, 70

Dancy, Jonathan, 20, 247–48
Danley, John R., 270
Darwin, Charles, 76, 212, 232
  epistemology of, 56
  evolutionary theory of, 77–78, 232, 236, 254–55
  on heredity, 56, 245–46
  thought experiments of, 70–71, 100, 106, 236
Davidson, Donald, 214
Dear, Peter, 304n15
DeCamp, Timothy, 192n2
Dedekind, Richard, 158
Descartes, René, 10–11, 22–23, 39, 160
Dennett, Daniel, 18, 95, 99, 261, 266–68
Dickie, George, 206
Dostoyevski, Fyodor Mikhailovich, 223–24
Douglass, Frederick, 126
Downey, Mort, 218

315